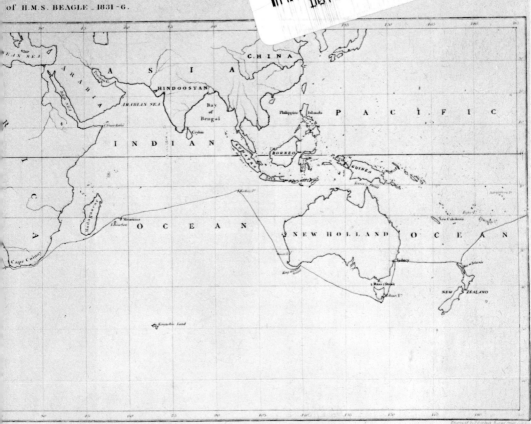

THE GARDEN CENTER OF
· GREATER CLEVELAND ·

In Memory of

Alice Jean Land
from
The Indoor Light Gardening Society

THE ELEANOR SQUIRE LIBRARY

DARWIN AND HIS FLOWERS

Other books by Mea Allan

Fiction

LONELY (*Harrap*)
CHANGE OF HEART (*Harrap*)
ROSE COTTAGE (*Ward Lock*)
BASE RUMOUR (*Ward Lock*)

Biography

THE TRADESCANTS (*Michael Joseph*)
THE HOOKERS OF KEW (*Michael Joseph*)
TOM'S WEEDS (*Faber & Faber*)
PALGRAVE OF ARABIA (*Macmillan*)
E.A. BOWLES AND HIS GARDEN (*Faber & Faber*)
PLANTS THAT CHANGED OUR GARDENS (*David & Charles*)

FISONS GUIDE TO GARDENS (*Leslie Frewin*)
GARDENS OF EAST ANGLIA (*East Anglia Tourist Board*)

DARWIN
AND HIS FLOWERS

THE KEY TO NATURAL SELECTION

Mea Allan

Taplinger Publishing Company
New York

First published in the United States in 1977 by
TAPLINGER PUBLISHING CO., INC.
New York, New York

Copyright © 1977 by Mea Allan
All rights reserved
Printed in Great Britain

Library of Congress Catalog Card Number: 77-77261
ISBN 0-8008-2113-0

Contents

Illustrations		9
Acknowledgements		13
Foreword by Dr Sydney Smith		15
Introduction		17
1.	The Origins of Charles Darwin	21
2.	First Discoveries	32
3.	The Man who Walked with Henslow	43
4.	The Plants of the *Beagle*	
	i. The Pampas and the Land of Fire	55
	ii. Rivers and Mountains	72
	iii. The Galapagos and the Long Road Home	90
5.	Home is the Sailor	106
6.	'These facts origin of all my views'	117
7.	'At last gleams of light have come'	133
8.	'Hurrah for my Species work!'	146
9.	The Quiet War	161
10.	The Great Migrations	176
11.	'A millionaire in odd and curious little facts'	191
12.	Hops, Hooks and Tendrils	207
13.	Man's Gigantic Experiment	221
14.	Mayhem and Murder in the Plant World	232
15.	Darwin's Hero, rule and method	249
16.	Legitimate and Illegitimate Marriages	263
17.	The Sleep of Plants and other Matters	277
18.	Memorial	291
	Illustration Acknowledgements	300
	Glossary of Botanical Terms	301
	Bibliography	305
	Index	309

Illustrations

COLOUR PHOTOGRAPHS   *page*

Catasetum tridentatum (*macrocarpum*) (George Hurn)   65
Calceolaria darwinii (Harold Langford)   65
Bell Mountain (David M. Moore)   80
Orchis morio (Clare Williams)   129
Orchis morio (Grace Woodbridge)   129
Ophrys apifera (E. A. Ellis)   129
Ophrys apifera (Grace Woodbridge)   129
Angraecum sesquipedale (Grace Woodbridge)   204
Cycnoches ventricosum (G. E. Nicholson)   204
Odontoglossum grande (Harold Langford)   204
Mormodes histrio (David M. Menzies)   204
Variations in Antirrhinum   241
3:1 ratio of normal and peloric forms (both photographs by
    John Innes Institute)   241
Dionaea muscipula, Venus's Fly-trap (Harold Langford)   244
Trigger-hairs of Venus's Fly-trap (Oxford Scientific Films)   244
The Bladderwort *Utricularia vulgaris* (Oxford Scientific Films)   244
Tentacles of the Sundew *Drosera rotundifolia* (Oxford Scientific
    Films)   245
The Sundew captures an ant (Oxford Scientific Films)   245
The Peruvian Pitcher Plant, *Nepenthes pervillei* (Mary Gillham)   245

MONOCHROME PHOTOGRAPHS AND LINE DRAWINGS   *page*

Linnaeus's description of *Pinus sylvestris* (Drawing by Brian
    Hughes)   21
Erasmus Darwin (From the portrait by Joseph Wright of
    Derby)   23
Robert Waring Darwin (From the miniature at Down House)   25
Charles Darwin aged six with his sister Catherine (From the
    chalk drawing by James Sharples)   27
The seaweed *Fucus lorius* and 'peppercorns' (Drawing by Keith
    Roberts)   32
John Stevens Henslow in 1849 (From the portrait by T. H.
    Maguire)   41

The pollen grains of *Orchis morio* and other pollen grains
(Drawing by Keith Roberts)   43

Captain Robert FitzRoy (From the drawing by Captain Philip
Parker King)   50

Josiah Wedgwood of Maer Hall (From the portrait by William
Owen)   52

A scene in the Brazilian forest (Drawing by Brian Hughes)   55

H.M.S. *Beagle* in Sydney Harbour (From the watercolour by
Owen Stanley)   57

Map of South America showing Darwin's inland expeditions
(From *Charles Darwin's Diary of the Voyage of H.M.S.
Beagle*, ed. Nora Barlow)   64

Four Darwin plants from Tierra del Fuego (Photographs by
David M. Moore)   69

Cruckshanksia glacialis (Drawing by Victoria A. Matthews)   72

Opuntia darwinii (Drawing by Keith Roberts)   78

Four Darwin plants drawn by W. H. Fitch (From J. D. Hooker's
Flora Antarctica)   82–3

The Cordillera of the Andes (Engraving after Conrad Martens, in
the *Narrative of the Surveying Voyages of H.M.S. Beagle and
Adventure between the years 1826 and 1836*)   84

The Galapagos Archipelago: its winds, sea currents and ocean
depths (Drawing by Brian Hughes)   90

Seed capsule of the Toothache Tree (From Darwin's original
drawing) (Crown copyright. Reproduced with the permission of the
Controller of Her Majesty's Stationery Office and
of the Director, Royal Botanic Gardens, Kew)   92

Four Darwin plants from the Galapagos Islands (Drawings
from the *Flora of the Galápagos Islands*. Photographs by
Uno Eliasson)   96–7

Darwin's herbarium sheet of *Scalesia darwinii*   106

Scalesia darwinii (Photograph by Uno Eliasson)   110

Excerpt from Darwin's first Transmutation Notebook   117

Darwin's notes on *Rhododendron azaloides*, 1841   117

Emma Darwin in 1840 (From the chalk drawing by George
Richmond)   126

Charles Darwin in the same year (From the chalk drawing by
George Richmond)   127

Part of a letter from J. D. Hooker to Charles Darwin, 1845   133

Down House (Photograph by Leonard Darwin)   139

The Sand Walk (Photograph by Leonard Darwin)   140

Charles Darwin with his eldest child, William (From a
daguerrotype)   143

George, Darwin's second son   143

The nectar-secreting pea flower (Drawing by Keith Roberts)   146

A Darwin note on sea-water germination   152

Charles Darwin, Charles Lyell and Joseph Hooker (From a
painting in the Darwin Museum, Moscow)   156

Darwin's contemporaries: Asa Gray; Thomas Henry Huxley;
    Alfred Russel Wallace   158–9
Darwin's evolutionary tree   161
The old study at Down House (Photograph by Leonard Darwin)   166
A Darwin experiment in progress (Photograph by Leonard
    Darwin)   167
Darwin's genealogical tree (Interpretation by Keith Roberts)   174
Plant migration during the Ice Age (Drawing by Brian Hughes)   176
Geographical distribution of seeds (Drawing by Brian Hughes)   181
Components of an archetypal flower (Diagram by Keith
    Roberts)   186
Similarity of seedlings (Drawing by Keith Roberts)   187
A bumble-bee pollinating *Orchis mascula* (Drawing by Brian
    Hughes)   191
The Darwin family at Down House   195
Pollination mechanisms of orchids (From Charles Darwin's
    The Fertilisation of Orchids)   197, 201, 203
'Twiners entwining twiners . . .' (Drawing by Brian Hughes)   207
The greenhouse at Down House   210
Contrivances by which plants climb (Photographs by Grace
    Woodbridge)   212–13
The evolution of leaves in *Tropaeolum tricolorum* (Drawing by
    Keith Roberts)   216
Tendril of the White Bryony, from Darwin's *Climbing Plants*
    (Drawing by Vicky Fisher)   219
The development of garden flowers (Drawing by Brian Hughes)   221
Variation in cultivated peas (From *The Variation of Animals
    and Plants under Domestication* by Charles Darwin)   225
Diagram showing the Mendelian 3:1 ratio (By Keith Roberts)   231
Drosera rotundifolia (By James Sowerby in J. E. Smith's *English
    Botany*)   232
Inflected tentacles on a leaf of *Drosera rotundifolia* (From
    Charles Darwin's *Insectivorous Plants*)   236
A scanning electron-micrograph of *Drosophyllum lusitanicum*
    (By members of The Botany School, The University of
    Oxford)   240
An inflected leaf of *Pinguicula vulgaris* (From *Insectivorous Plants*) 243
A scanning electron-micrograph of a Butterwort's leaf surface
    (By members of The Botany School, The University of
    Oxford)   246
The grids of *Genlisia ornata* (From *Insectivorous Plants*)   247
The Morning Glory and Darwin's 'Hero' (Drawing by Brian
    Hughes)   249
Yucca filamentosa (Photograph by Grace Woodbridge)   256
The pollination of the Cuckoo-pint and Foxglove (Drawing by
    Keith Roberts)   259
Two bee plants which flower earlier when crossed
    (Photographs by Grace Woodbridge)   261

The sex symbols of flowers (From Linnaeus's *Örtabok*, 1725)   263

Legitimate and illegitimate marriages of the dimorphic
Primrose (From Darwin's *Different Forms of Flowers*)   266

Darwin's true-species Oxlip with its cousins the Primrose and
Cowslip (Photographs by Grace Woodbridge)   267

The three forms of marriage of *Lythrum salicaria* (From
Different Forms of Flowers)   271

Plants employing different kinds of pollinators (Photographs
by Grace Woodbridge)   275

Acacia leaves awake and asleep (Drawing by Brian Hughes)   277

Darwin's great discovery of circumnutation (Drawing by Keith
Roberts)   279

Experiment by Darwin on the radicle of a bean (Photograph by
Leonard Darwin)   280

Darwin measures growth and sleep movements (Photograph by
Leonard Darwin)   285

The leaves of the Wood Sorrel awake and asleep (Drawing by
Keith Roberts)   286

Leaves of the Telegraph Plant awake and asleep (From
Darwin's *The Movements of Plants*)   287

Charles Darwin in 1868 (Photograph by Julia Margaret
Cameron)   293

The endpapers are reproduced from the Track Chart in Captain Fitzroy's *Narrative of the Surveying Voyages of His Majesty's Ships Adventure and Beagle*, 1839

Acknowledgements

It is to Cambridge that I am most deeply indebted for help given me during the writing of this book. First, my sincere thanks are to that great Darwinian Dr Sydney Smith who inspired the choice of Darwin's botany as a subject for exploration and was my guide along the way. I am grateful to Nora, Lady Barlow as another valued encourager and reader of the typescript, and as lender of ungettable Darwin books. To work with those inscribed in Darwin's own handwriting brought me close to him as I followed his footsteps. I extend thanks to E. B. Ceadel, Librarian of the University Library, and to his staff for their assistance, including Margaret Pamplin and particularly Peter J. Gautrey for his untiring help at all times, for giving me the benefit of his own knowledge of Charles Darwin and for steering me through the vast Darwin archives under his care. It was on this original material, made available by members of the Darwin family, that the substance of much of my book was founded. My thanks are also given to Dr S. M. Walters, Director of the University Botanic Garden, and to Dr Peter Yeo and C. J. King; at the Botany School to Peter D. Sell, Assistant Curator, for his help with the Darwin herbarium, and to Mrs Heap, Library Assistant. Finally and not least at Cambridge I am grateful to John S. L. Gilmour for the kindliest of help and ready advice on botanical questions and for his meticulous scrutiny of the typescript. For the loan of unpublished Darwin letters I thank Arthur V. Hooker and T. H. Rivers. Other unpublished correspondence, at the Library of the Royal Botanic Gardens, Kew, I was able to consult with the assistance of V. T. H. Parry, Librarian and Archivist, and Mrs K. E. Mortimer whose friendly help was never-failing. I thank also the librarians at the Central Library, Edinburgh; at Shrewsbury, Norwich, Lowestoft and Ipswich, and especially Mrs Elizabeth A. Atchison, Librarian, the John Innes Institute, whose most willing assistance over the years of research was invaluable. For special information I thank the Linnean Society, the Royal Geographical Society, the Royal Entomological Society, and the Zoological Society, and personally J. C. Thackray, the Institute of Geological Sciences; Dr John D. Bradley, the Commonwealth Institute of Entomology; J. K. Burras, Superintendent of the Botanic Garden, Oxford; and Dr David M. Moore, Plant Science Laboratories, Department of Botany, the University of Reading, for information on South American plants. Others I wish to thank are E. David Kohn, Peter Robson, Ralph Gould of Hurst, Gunson, Cooper, Taber; David Stanbury for information on the crew of the *Beagle*, H. E. Chipperfield for entomological help for this and other books, and Thomas Hoog of Haarlem for information about the Darwin

Tulip and a letter to Francis Darwin. For permission to visit Down House I thank the Royal College of Surgeons of England and, while there, Sir Hedley Atkins for being my guide, and Sydney Robinson and Philip Titheradge for their friendly help. All through my researches it was my privilege to visit the John Innes Institute where, thanks to Professor R. Markham, Director, I was able to see demonstrations in plant genetics relating to Darwin's experiments on the antirrhinum being carried on by Brian J. Harrison of the Genetics Department who as a Darwin disciple was, besides, a source of valuable help and advice. I thank, too, Dr Graham Hussey of the Department of Applied Genetics for the interest he took in my project. Finally I most gratefully thank the Trustees of the Leverhulme Research Awards for the Fellowship they awarded me in 1973 which was of considerable financial help in the early stages of research, and once more my friend Grace Woodbridge to whom, as a tireless co-researcher and collaborator, this book and I owe so much.

Foreword

It was Charles Darwin's peculiar genius to search for and discover unifying principles that not only explained natural phenomena known and visible but which generated insights into phenomena as yet to be encountered. His work is continued nowadays at a level accessible only to those trained in the arcane and costly techniques of cell biology. The whole organisms, the plants and animals of our common daily experience, tend therefore to be neglected in this age accustomed to the mysteries of the universe at the one extreme and the submicroscopic constructs of the gene and its time-dependent products of differing proteins at the other. So it is not the currently fashionable nostalgia for the Victorians (though Darwin was twenty-eight and already famous as a circumnavigator when Victoria became Queen) which makes the appearance of Mea Allan's book so appropriate and felicitous in its timing, but an impelling need for us to see life whole and continuous throughout all levels of scale and all ascertainable periods of time.

My acquaintance with Charles Darwin arose from the study during the past sixteen years of the great assemblage of manuscripts, annotated books, letters and unnumbered scraps of paper which accumulated throughout his working life. The greater proportion is in the Cambridge University Library to which it was given in 1942 by members of the Darwin family and by the Pilgrim Trust. The papers more relevant to Darwin's family life and to the *Beagle* voyage were given at the same time to Down House, now under the care of the Royal College of Surgeons to which the fossil mammals collected on the voyage were given by Darwin in 1836. Since 1961 this nucleus has attracted further accretions and I consider myself fortunate that I drew Miss Allan's attention to this material, because I had formed the stubborn and uninstructed conviction that Charles Darwin's researches into plant distribution, structures, movements and feeding would repay informed study by someone with her knowledge and skill in conveying to the general reader the logic and excitement of scientific inquiry.

In 1875 Charles enlisted his son Frank – who, like his uncle Erasmus, after graduating as a doctor was disinclined for practice – as his assistant in research in plants. Frank later wrote his father's life and letters in three substantial volumes. Because of unique familiarity with his father's working methods and a shared experience of the experimental material, Frank dealt with the botanical studies in a group of chapters towards the end of the third volume. The *Life and Letters* have long been out of print, and the later treatment of botanical matters in *More Letters* is not designed to be read without prior knowledge of the former.

Darwin's plants have accordingly become isolated from the general accounts of his life. In the public mind the Galapagos Islands mean giant tortoises, marine lizards and oddly peculiar and diverse finches: the equally extraordinary plant life has not received the attention its strangeness merits. Then, when Darwin was writing up the natural history of the *Beagle*, Professor Henslow was incapable of identifying these plants, and so the botany of the voyage was not included. The plants had to await Joseph Hooker who some seven years later described Darwin's collections from the southerly regions of South America in his rare and expensive *Flora Antarctica*. For his paper on the Galapagos flora he was mainly indebted to Darwin whose collections were by far the largest and most complete ever made in the Archipelago.

Miss Allan has disentangled Darwin's contributions from these specialised sources, supplemented by research into unpublished letters exchanged with Hooker, as well as their joint listings of plant populations of the extreme southern localities, Darwin's world-wide and voluminous correspondence with other leading botanists and his own scattered notes. In short, she has integrated Darwin's plants into Darwin's life to fascinating effect, wearing her scholarship with reticence and tact.

By the end of 1838, two years after the return of the *Beagle*, Darwin saw clearly the overall plan of the work he had to do. Lengthy periods of severe debility stultified his programme, but his drive towards achievement was merciless. Relief was achieved by new observations of plants, when hypotheses could be constructed and subjected to the test of experiment. Too much speculation without due seasoning of experimental observation (the real ground for his suspicion of Lamarck and Herbert Spencer) was always anathema to Darwin, and the contemplation of the elegant contrivances of flower-parts, stems and leaves, seems to have provided a stable outcome from the racing of his over-active brain. The solace of clear-cut results completed the therapy.

It is now evident that the seven major works he wrote on plants, the *Origin* and the *Descent of Man* are closely inter-related elements in the total activity of a supremely gifted man. This account of Darwin as a working botanist reveals his grasp of detail, essential to him as well as to our understanding of him, without which his revolutionary speculative vision, compressed into the *Origin*, would never have triumphed.

Mea Allan has served him well, and we, her readers, should be grateful.

Sydney Smith

Introduction

It is perhaps no coincidence that the earliest portrait we have of Charles Darwin shows him at the age of six clasping a pot of flowers. Plants were to become the great absorbing interest of his life: he was to spend more time studying them than anything else. Through them he was to reveal some of nature's most wonderful secrets; the struggle for existence that lies behind the quiet beauty of a primrose, the drama of how flowers—by the use of intricate mechanisms evolved in a continuous process of adaptation—make insects and birds, the winds and even the currents of the seas, serve their purpose in the grand design of carrying on their race. In his hands plants were to provide the key to Evolution.

We think of him as the author of the *Origin of Species*, the book that revolutionised man's attitude to the world in which he lived, giving him reality in exchange for a myth, a global history that was living and natural and therefore understandable. A revolution is never a comfortable thing. It implies destruction. But if the *Origin* tore apart old ways of thinking, it set men free. They now had a theory, a truth on which to build, a tool.

Darwin's work on Natural Selection has vaguely been thought solely to concern the descent, by successful modification, of animals and birds. The fact that for the testing of his theory he sought information not only from such experts in the field as horse-breeders and pigeon-fanciers but also horticulturists, has never been remarked as significant. Nor has this aspect been followed through to his subsequent work on plants, except in the one or two essays written for the centennials of his birth and of the publication of the *Origin*.

Animals and plants are frequently bracketed in the *Origin*. But animals were not so easy to observe as plants; and plants were quick and responsive material for experiment; they grew in his garden and in his hothouse and greenhouse; they were all around, in hedgerow, field and meadow. Through friends and correspondents he was able to obtain seeds or living specimens of any plant he required. When observing the behaviour of a particular kind, there was always some pot or jar by him. In them he could see Evolution at work, right on his study table.

After publishing the *Origin* he went on to extend and develop his Natural Selection theory in seven major works explaining how plants live and grow, outwit their rivals, and survive to repeat their kind. Hence it has been said, by one of his grand-daughters, that he 'came into botany by the back-door'. But botany to Darwin was no mere collecting and drying of herbarium specimens and giving them a name. The living bodies of plants were what interested him, born of parents as man was, finding light and moisture and food, overcoming obstacles in the path of their progress. Darwin was the man who said 'Why?' and 'How?' He was known to go out in a gale to find out what was happening to a White Bryony. Though whipped about by the wind the bryony was still clinging safely to its tree. *Why* wasn't it blown off? *How* did it manage to cling on? The answer was that its coiled tendrils acted like elastic springs, giving to the wind, pulling itself back again.

'It has always pleased me to exalt plants in the scale of organised beings,' Darwin said in his *Autobiography*. His son Francis, whom he called Frank, described how he used to stand before a plant for a considerable time and touch it affectionately. He wrote: 'He had great delight in the beauty of flowers—for instance in the mass of Azaleas which generally stood in the drawing-room. I think he sometimes fused together his admiration of the structure of a flower and of its intrinsic beauty; for instance, in the case of the big pendulous pink and white flowers of Dielytra. In the same way he had an affection, half artistic, half botanical, for the little blue Lobelia. In admiring flowers, he would often laugh at the dingy high-art colours, and contrast them with the bright tints of nature. I used to like to hear him admire the beauty of a flower; it was a kind of gratitude to the flower itself, and of personal love for its delicate form and colour. I seem to remember him gently touching a flower he delighted in; it was the same simple admiration that a child might have.'

Perhaps it was because he loved plants so much that Charles Darwin was able to see them as complete beings, fellow-inhabitants of the world. He viewed them against the background of their environment—and opened up the new field of study, ecology. In observing the different forms of primrose and loosestrife flowers, some with a short style (the tube down which pollen is conveyed to the ovules) and some with a long style, and likening them to the 'males and females of ordinary unisexual animals', he discovered that 'legitimate' crosses produced a normal crop of seed and 'illegitimate' unions impaired fertility—thus throwing an important light on hybridisation. His researches into the phenomena of spiralling plant-movements, from the turning to the soil and the turning to the light of the first shoots emerging from the seed, to the complex movements of climbing plants, are echoed in present-day researches in the field of tropic responses—one result of which has been the discovery of plant-growth hormones. He conceived the idea of 'gemmules' as the carriers of inheritance—and thereby anticipated the twentieth-century conception of the gene and the science of genetics. The identity of many plants inhabiting the summits of mountains separated from each other by hundreds of miles of lowland, he explained was the result of plant-migrations during the cold period before the Glacial Age—in this anticipating the celebrated memoir of Edward Forbes, but he went farther than Forbes in applying the theory to transtropical migrations.

These are over-simplifications of complex matters which entailed years of persevering observation and experiment: it took eleven years of such labours to produce *The Effects of Cross- and Self-Fertilisation*. There are few departments of plant biology that Darwin did not enrich, either by original discovery or by so re-thinking a concept that it became workable material for future scientists. His genius was the happy combination of three things: his patience in ferreting out the meaning of each detail of a plant; his synoptic view of how and where these details fitted into the whole; and his philosophic powers of inference and deduction by which he was able to establish truths which have stood the test of time.

We might ask why Darwin wrote these books. Purely as a scientist probing scientific problems in the realm of plant biology? On the worth of his botanical work alone, present-day evaluation recognises him as the man who has given us a greater understanding of plants than any other, and who has still much to teach us. As a biologist he will always be among the few great of any century. But Darwin had no thought of reserving his ideas for the academic few. He addressed himself to those who would find his ideas of practical use in their work. From early days he was writing articles for the *Gardeners' Chronicle*, then as today not only a scientific but a practical horticultural journal read by leading professional gardeners and nurserymen. Thus in 1843 he contributed 'Double Flowers—their Origin', in 1844 'Manures, and Steeping Seeds', in the same year 'Variegated Leaves'. But before then, in the first volume of the *Gardeners' Chronicle*, published in 1841, eighteen years before the *Origin*, he was writing on 'Humble Bees' and their habit of cutting holes at the base of the corolla to obtain nectar without expending time and energy crawling down the tube of the flower, having witnessed hundreds of such 'felonious' attacks. He became a regular contributor on such useful topics as the 'Vitality of Seeds', and 'The Agency of Bees in the Fertilisation of Papilionaceous Flowers and on the Crossing of Kidney Beans', and, as he worked out each new theory, even contributing what he called 'tidbits' about it. A new book by Charles Darwin was an event to be reviewed at length; for in these, as in his papers, Darwin was writing for practising men. 'Under a practical point of view agriculturalists and horticulturists may learn something from the conclusions at which we have arrived.' 'Florists may learn . . . that they have the power of fixing each fleeting variety of colour . . .'

Darwin did not regard himself as a botanist. In his second letter to Joseph Dalton Hooker, who became his botanical mentor and closest friend, he classed himself as a 'Botanical ignoramus'. Writing to the Marquis de Saporta, he regretted that 'I have never been properly grounded in Botany, and have only studied special points.' To Asa Gray, Professor of Natural History at Harvard University: 'I perceive how presumptuous it is in me, not a botanist, to make even the most trifling suggestion to such a botanist as yourself.' (The suggestion was his publishing on the subject of American alpine plants.) On the other hand, there were these 'special points'. In his book on *The Fertilisation of Orchids*, when describing how *Listera ovata*, the Twayblade, was pollinated, he wrote that 'The structure and action of the rostellum has been the subject of a valuable paper in the "Philosophical Transactions" by Dr

Hooker, who has described minutely and of course correctly its curious structure.' He added: 'He did not, however, attend to the part that insects play in the fertilisation of the flowers.' He went on: 'C. K. Sprengel well knew the importance of insect-agency but he misunderstood both the structure and the action of the rostellum.' It was in the special points that Darwin excelled, reading in them a philosophy that went far beyond the bounds of Botany. He saw botany always as growing plants, saw them, too, in the wider context of the evolution of living matter down the stream of time. They were, after all, as he pointed out, the senior inhabitants of Earth, since plants were here before animals made their appearance, and long before man.

So the subject is of first importance to all of us. For neither man nor animal, nor even the insects, can live without plants and their flowers; grass, rice and corn, trees that give shade and shelter, vegetation that even in its death provides food for the unending cycle.

1

The Origins of Charles Darwin

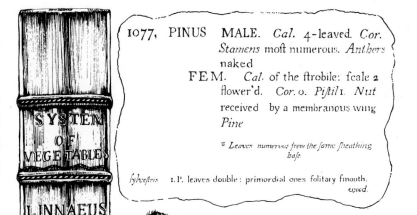

1077, PINUS MALE. *Cal.* 4-leaved. *Cor.*
Stamens moſt numerous. *Anthers*
naked
FEM. *Cal.* of the ſtrobile: ſcale **2**
flower'd. *Cor.* o. *Piſtil* 1. *Nut*
received by a membranous wing
Pine

* *Leaves numerous from the ſame ſheathing*
baſe.

ſylveſtris 1. P. leaves double : primordial ones ſolitary ſmooth.
wood.

*The mind of the young Charles Darwin wakens
to the wonders of plant life—*

*As a boy : 'I tried to make out the names of
plants,' he recalled in his* Autobiography. *He
did so with the help of Linnaeus's* Systema
Vegetabilium *which his grandfather, Erasmus
Darwin, had published in English in 1783 under
the imprint of the Botanical Society at Lichfield
which he founded.*

*Erasmus Darwin, a successful doctor, was an
authority on Linnaeus, inventor of the binomial
system of nomenclature in which the first word
stands for the genus, and the second the species
of the genus to which the plant belongs.*

*At his day school Charles 'smattered in botany',
and when he was eleven his father gave him his
own copy of John Berkenhout's* Botanical
Lexicon. *So when his elder brother wrote asking
him to 'look in English Systema Vegetab & copy
me out the specific description of Pinus
sylvestris', Charles was no stranger to botany.*

THE EVOLUTION OF Charles Darwin as a botanist seems, on the face of things, easy to trace.

His grandfather was Erasmus Darwin, a colossus who took everything in his stride, from being a busy general practitioner (achieving such a high reputation that he was repeatedly offered, though he always declined, the appointment of physician to George III) to being the brilliant inventor of, among other things, a speaking machine, a horizontal windmill, a rocket-motor, an office copying machine, a steam turbine, an automated water closet, a wire-drawn ferry and a canal lift. All of them worked: they were practical. He was the author of the massive two-volumed *Zoonomia* (a medical book in which he pioneered the humane instead of the brutal treatment of mental illness) and, with other books, of *Phytologia* (a 600-page survey of vegetable life, covering plant nutrition, the biological control of insects, ideas for sewage farms and a system of artesian wells, complete with engineering drawings for new drill ploughs and water pumps). He produced scientific theories to explain everything he saw. A brilliant meteorologist, he was the first to discover the reason for cloud-formation and the existence of cold and warm fronts, and he forecast correctly the composition of the outer atmosphere. He was a founder with Matthew Boulton of the Lunar Society of Birmingham, a body unique in the history of science. This was a group of philosophers and practical talented men whose achievements provided a driving force behind the industrial revolution. Its meetings (always held about the time of the full moon) were entirely informal. No minutes were kept. Its strength lay in the personal friendships of its members, each of whom helped the other. Thus Erasmus Darwin brought together James Watt and Boulton (the chief manufacturer of England and Father of Birmingham)—and the steam engine was born. Dr Joseph Priestley, Josiah Wedgwood, James Kerr, Dr William Small and Richard Lovell Edgeworth were all members of the Society who became eminent.

Not least was Erasmus Darwin's interest in botany. He was a great authority on the Linnaean system of classification, and he not only created a botanic garden but wrote an encyclopaedic poem of that name which became a best-seller. *Zoonomia* had been hailed as 'perhaps the most original work ever composed by mortal Man'. *The Botanic Garden* brought him extravagant praise acclaiming him as a poet greater than Milton. It was published in two parts: 'The Economy of Vegetation' and 'The Loves of the Plants', though Part II was published first because it was lighter and more likely to appeal to the public. In it Erasmus Darwin told the story of the sex-life of plants, disguising truths of nature in even more revealing metaphor, as when, exposing the love-life of *Collinsonia*, he made two males woo one female (knowing like a good Linnaean that *Collinsonia* belonged to the class Diandria, having two stamens, and the group or order Monogynia, having one pistil), in Linnaeus's own words *Mariti duo in eodem conjugio*, two husbands in the same marriage. He had even more fun with *Genista*, where 'Ten fond brothers woo the haughty maid', these being ten stamens arising from two bases, as if from two mothers, and united by their filaments. Linnaeus's Sexual System of classification scandalised his critics, who could not equate such 'loathsome harlotry' as several males (stamens) to one female (pistil) with the Creator of the Vegetable Kingdom. Such eroticism coming from the witty and learned pen of Dr Darwin delightfully shocked his eighteenth century readers, and probably taught them much about plant morphology.

Erasmus Darwin (1731–1802), famous doctor, poet, inventor, wit and gardener, who propounded many scientific theories including one concerning evolution

Erasmus had greatly admired Linnaeus's system, which 'superseded all others by its concise and elegant arrangement'. It also had the merit of 'shewing the great analogy between plants and animals'. Erasmus but drew a closer analogy, between plants and people.

He died in 1802. Two years later his second long poem was published. This again is of interest to us, for in *The Temple of Nature* he showed his remarkable vision by tracing the development of organic life to its culmination in man. Modern scientists have no quarrel with his supposition that earliest life was born in the ocean. He wrote:

Organic Life beneath the shoreless waves
Was born and nurs'd in Ocean's pearly caves;
First forms minute, unseen by spheric glass
Move on the mud, or pierce the watery mass;
These, as successive generations bloom
New powers acquire, and larger limbs assume,
Whence countless groups of vegetation spring,
And breathing realms of fin, and feet, and wing.

It was a poetic rendering of a theme in *Zoonomia* where he had imagined the earliest ancestor of all warm-blooded animals might have been 'one living filament' and directed attention to evidences of change in the life of an individual animal, sometimes of metamorphoses as in 'the butterfly with painted wings from the crawling caterpillar' or of 'the respiring frog from the sub-natent tadpole', and, over a longer period of time, variations 'by artificial or accidental cultivation, as in horses, which we have exercised for the different purposes of strength or swiftness, in carrying burthens or in running races'. Over still longer periods of time developments left useless rudimentary organs in their wake. He discussed sexual selection in the animal world, the male's desire for exclusive possession of the female and the tusks and horns provided for combat with rivals, the result being 'that the strongest and most active animal should propagate the species, which should thence become improved'. He explained how each animal is adapted for procuring its own type of food: the hard noses of swine for rooting, the trunks of elephants for reaching to the branches of trees, the hard beak of a parrot for cracking nuts.

This was Evolution. But Erasmus Darwin was not the first evolutionist—nor the last.

Next in the line was Erasmus Darwin's son Robert Waring, father of Charles. But before we speak of him, there was the brother of Erasmus, also Robert Waring, the author of *Principia Botanica* which, it is interesting to note, was published in 1787, exactly a century after the publication of Sir Isaac Newton's *Principia Mathematica*.

This Darwin was a botanical systematist. He also was deeply influenced by the work of Linnaeus, and from him we learn that his brother Erasmus 'suggests improvements on the Linnaean System to make it "more natural".' The meaning of 'natural' in this context denotes the evolution of a race or genetically related group of animals or plants. Robert did not get so far as Erasmus, and his book, though running into three editions, was ingenuous rather than scientific or prophetic. But Charles, his great-nephew, wrote appreciatively of his work, as containing 'many curious notes on biology—a subject wholly neglected in the last century.'

Erasmus Darwin had a great influence on his son, the other Robert Waring who was Charles's father. He too was a doctor and a successful one, with a large practice in and around Shrewsbury. Charles wrote of him: 'His chief mental characteristics were his powers of observation and his sympathy, neither of which I have ever seen exceeded or even equalled.'

He hated extravagance but could be supremely generous. One day a small manufacturer in Shrewsbury came to him and said that he would be bankrupt unless he could borrow £10,000 immediately, but that he was unable to give any legal security. Listening to his story and feeling intuitively that the man could be trusted he advanced the sum, a very large one for a young doctor. Sure enough, he was repaid.

Erasmus Darwin's son Robert Waring (1766–1848), also a famous doctor, whose second son, born in 1809, was to become one of the greatest thinkers of all time

But the most remarkable power Robert Darwin possessed was his ability to read not only a person's character but even the thoughts of someone he met briefly for the first time. Sometimes this power seemed almost supernatural, and he was skilled at making what Charles called 'good guesses'. Lord Shelburne, later the first Marquis of Lansdowne and famed for his knowledge of European affairs, came to consult him and afterwards harangued him on the state of Holland. Robert had studied medicine at Leyden, and there one day while on a country walk with a friend had been taken to the house of an elderly clergyman who was married to an Englishwoman. He was very hungry, but there was little for luncheon except cheese, which he could never eat. Grieved at this, Mrs A—— assured him that it was excellent cheese, sent to her from Bowood, the seat of Lord Shelburne. Robert wondered why cheese should have been sent to her from Bowood, but thought no more about it until the day Lord

Shelburne talked about Holland. He then said, 'I should think from what I saw of the Rev. Mr A——, that he was a very able man and well acquainted with the state of Holland.' At this, the Earl looked startled, and immediately changed the conversation. But next morning he called again, to say that it was of the utmost importance to know how he had discovered that the Rev. Mr A—— was the source of his information. On hearing Robert Darwin's explanation, he was so struck with his diplomatic guess-work that he never forgot him. Years later when Charles wished to become a member of the Athenaeum Club, the Earl, unasked, proposed him and got him elected. 'A queer concatenation of events,' Darwin remarked, 'that my father not eating cheese half-a-century before in Holland led to my election as a member of the Athenaeum!'

Robert Darwin's hobbies were pets and plants. He reared birds and animals, and his pigeons were well known in the town and far beyond for their beauty, variety and tameness. With an interest in botany almost as great as his father's he stocked the garden at The Mount—the family house he built overlooking the River Severn—with the choicest flowers, shrubs and trees, and like all good gardeners he was pleased to give roots and cuttings to others. Writing in 1808 to her brother Josiah Wedgwood, Mrs Darwin told him:

> The Dr sends you by to-morrow's Coach some suckers of the white Poplar, and as they have good roots, he has no doubts of their growing. If you want more say so, and they shall be sent. It is the common White Poplar. It is become so fashionable a tree that Lady Bromley has sent for some cuttings for Baroness Howe, to decorate Pope's Villa at Twickenham, as all his favourite trees have been cut down.

Charles himself was a gardener and had his own plot. After he left home in 1825 his sister Caroline wrote: 'It made me feel quite melancholy the other day looking at your old garden and the flowers coming up which you used to be so happy working.'

The garden meant much to them all, and Caroline was the garden correspondent. In another letter to Charles, written in February 1826, she told him:

> We have been very busy in the flower garden, planting sweet peas &c. I flatter myself it will look much gayer than it did last, that I know you will think it may easily do. I have remembered your admiration of the Hollyhocks at Maer & have been buying some, so that we will at least not be outdone in the flower. We are going to have pipes laid to have a supply of water in the flower garden, so next summer your good nature will not be so often taxed with 'Charles it is very hot' (very hot indeed you unthinkingly answer). 'Dear Bobby, the ground is so dry that the pans of water you brought half an hour ago did hardly any good, would you bring one more?' Papa pays frequent visits to the garden to see a *Leucojum vernum* which is now in blow and rather a rare plant.

As Robert Darwin had married a daughter of the famous potter (her brother was Josiah II), it was natural that his home should be 'a shrine of Wedgwood's art', in vases, plaques, figures, and other lovely ornaments. A special piece of ware he designed was made for him at Etruria (which name Erasmus Darwin had suggested for the magnificent works built in the late 1760s, because he thought that Wedgwood had rediscovered 'a species of non-vitreous encaustic painting' previously known only to the ancient Etruscans). This was a nursery lamp: it sold widely, but in Shrewsbury Robert Darwin confined its sale to some dealers in earthenware by the name of Cook who were poor and needed help. Like his father, Erasmus, he had a flair for invention.

He married Susannah Wedgwood in 1796, and on the 12th of February 1809 Charles Robert was born, fifth of their family of six children. His childhood was uneventful, and he records that 'my earliest recollection goes back only to when I was a few months over four years old, when we went to near Abergele for sea-bathing, and I recollect some events and places with some little distinctness.' His mother died in July 1817, when he was eight years old. Of a gentle sympathetic nature, she seems to have left little impression upon him. When he came to write his *Autobiography* he could remember hardly anything about her 'except her death-bed, her black velvet gown and her curiously constructed work-table.'

Charles Darwin at the age of six with his sister Catherine

In the spring of 1817 he was sent to a day-school in Shrewsbury kept by the Rev. G. Case, minister of the Unitarian chapel in the High Street, where he stayed for a year. He was much slower at learning than his three-years-younger sister Catherine. But already his interest in natural history, and more especially in collecting, was well developed. 'I tried to make out the names of plants, and collected all sorts of things, shells, seals, franks, coins, and minerals.' In his autobiographical notes, the details of this are slightly different. 'I remember I took great delight at school in fishing for newts in the quarry pool. I had thus young formed a strong taste for collecting, chiefly seals, franks, etc., but also pebbles and minerals—one which was given me by some boy decided this taste. I believe shortly after this, or before, I had smattered in botany, and certainly when at Mr Case's School I was very fond of gardening, and invented some great falsehoods about being able to colour crocuses as I liked.' For Charles was sometimes naughty. He recalled: 'One little event during this year has fixed itself very firmly in my mind, and I hope that it has done so from my conscience having been afterwards sorely troubled by it; it is curious as showing that apparently I was interested at this early age in the variability of plants! I told another little boy (I believe it was Leighton, who afterwards became a well-known lichenologist and botanist), that I could produce variously coloured polyanthuses and primroses by watering them with certain coloured fluids, which was of course a monstrous fable, and had never been tried by me.'

The Rev. William Allport Leighton, who published a *Flora of Shropshire* and other works including the *Lichen Flora of Great Britain*, later told Darwin's son Francis that he remembered his naughty schoolfellow bringing a flower to school and saying that his mother had taught him how by looking at the inside of the blossom the name of the plant could be discovered. 'This,' said Leighton, 'greatly roused my attention and curiosity, and I inquired of him repeatedly how this could be done?' But his lesson was naturally enough not transmissible.

Darwin confessed that as a little boy he was 'much given to inventing deliberate falsehoods, and this was always done for the sake of causing excitement. For instance, I once gathered much valuable fruit from my father's trees and hid it in the shrubbery, and then ran in breathless haste to spread the news that I had discovered a hoard of stolen fruit.' But he could say in his own favour that he was as a boy humane, though 'I owed this entirely to the instruction and example of my sisters. I doubt indeed whether humanity is a natural or innate quality.' He was fond of collecting eggs, but never took more than a single egg from a nest except on one occasion when he took them all, not for their value but 'from a sort of bravado'. His passion for collecting, 'which leads a man to be a systematic naturalist, a virtuoso, or a miser, was very strong in me, and was clearly innate, as none of my sisters or brother ever had this taste.'

On the 10th of August 1818 Charles entered Dr Samuel Butler's great school where his brother Erasmus also was, and there he remained for seven years till midsummer 1825 when he was sixteen years old. He was a boarder, which pleased him, as he 'had the great advantage of living the life of a true schoolboy', though as the school was hardly more than a mile from The Mount he very often ran home in the longer intervals between the callings-over and before locking up at night. 'This, I think, was in many ways advantageous to me,' he wrote in retrospect, 'by keeping up home affections and interests.' He remembered that he often had to run, and because he was a good runner he was usually back in time. 'But when in doubt I prayed earnestly to God to help me, and I well remember that I attributed my success to the prayers and not my quick running, and marvelled how generally I was aided.'

Nevertheless, Dr Butler's school was educationally a blank to him. 'Nothing,' wrote Darwin, 'could have been worse for the development of my mind.' The curriculum was strictly classical, except for a little ancient geography and history. The lessons of the previous day had to be learnt by heart, and Charles easily memorised forty or fifty lines of Virgil or Homer during morning chapel—and as promptly forgot them. But he worked conscientiously and never used cribs. He greatly admired the Odes of Horace, his sole pleasure in the classics. Out of school he found his pleasures in natural history. A letter from his sister Catherine asked: 'How does Mineralogy, Botany, Chemistry and Entomology go on?' In 1820, when he was eleven, his father had given him his own copy of *Clavis Anglica Linguae Botanicae; or A Botanical Lexicon* by John Berkenhout, in which 'The Terms of Botany, particularly those occurring in the Works of Linnaeus, and other modern Writers' were 'Applied, Derived, Explained, Contrasted, and Exemplified'.

Erasmus, towards the end of his school career, turned the garden tool-house into a science laboratory. It was properly equipped, and Charles was allowed to help 'Philos', as he then called his brother, 'as a servant' with most of his experiments. 'He made all the gases and many compounds, and I read with care several books on chemistry, such as Henry and Parkes' "Chemical Catechism".' Charles always regarded this as the best part of his education, for it showed him in practice the meaning of experimental science. But it earned him the school nickname of 'Gas' and the public rebuke of his headmaster who accused him of wasting his time on 'such useless subjects' and called him a *poco curante*, which to Charles, not understanding what he meant, seemed a fearful reproach.

Looking back on his school life and at his character during this period, he thought that the only qualities promising well for the future were 'strong and diversified tastes, much zeal for whatever interested me, and a keen pleasure in understanding any complex subject or thing. I was taught Euclid by a private tutor and I distinctly remember the intense satisfaction which the clear geometrical proofs gave me. I remember with equal distinctness the delight which my uncle gave me (the father of Francis Galton) by explaining the principle of the vernier of a barometer.' Apart from science he was fond of books and used to sit for hours, generally in a window in the thick school walls, reading the historical plays of Shakespeare. He also read poetry such as Thomson's 'Seasons' and the recently published poems of Byron and Scott. Later in life he wholly lost, to his great regret, all pleasure in Shakespeare and poetry of any kind. Longer-lasting was his vivid delight in scenery, first awakened in 1822 during a riding tour on the borders of Wales. A wish to travel in remote countries was given him early in his schooldays by a book called *Wonders of the World*, though he often disputed with other boys the veracity of some of its statements. Towards the end of his schooldays he became passionately fond of shooting. 'I do not believe that any one could have shown more zeal for the most holy cause than I did for shooting birds. How well I remember killing my first snipe, and my excitement was so great that I had much difficulty in reloading my gun from the trembling of my hands.'

It is interesting, considering his later passion for geology, that he collected minerals 'with much zeal but quite unscientifically—all that I cared about was a new-*named* mineral, and I hardly attempted to classify them.' He studied insects 'with some little care'. Ten years old when he went for a holiday to Plas Edwards on the Welsh coast, he was surprised to see a large black and scarlet Hemipterous insect, many moths of the Zygaena family, the Sphinges, and a *Cicindela* beetle, none of which were to be found in Shropshire. 'I almost made up my mind to begin collecting all the insects I could find dead, for on consulting my sister, I concluded that it was not right to kill insects for the sake of making a collection.' After reading White's *Selborne* he took to bird-watching, and made notes on their habits. 'In my simplicity,' he recalled, 'I remember wondering why every gentleman did not become an ornithologist.'

Erasmus went up to Cambridge in 1822 leaving Charles still at school and in charge of the tool-house laboratory. Indeed, he had given it over to him. 'Dear Bobby,' he wrote in November, 'I think it would be an improvement in the Lab to have some more shelves fixed up. The places I have thought are—' and here he drew a sketch with details. 'I have ordered a small goniometer (an instrument to measure angles) so that we shall be able to separate the different crystals in your Lab.' Charles had written to ask if he could procure some minerals he wanted. 'I daresay I shall be able to obtain some specimens of rock for you, for Prof Sedgwick said that at the Gog Magog hills (about 4 miles distant) there were a vast number of specimens which I shall certainly some day go and explore.' Adam Sedgwick was to have a profound influence on Charles Darwin's life.

In another letter Erasmus mentioned that 'Prof. Henslow (on Mineralogy) has twice shewn us an experiment of a test of Arsenic by burning it with a blow pipe . . . The lectures are very entertaining, & this is his first course so that he will be improved by the time you come up.'

John Stevens Henslow, professor of botany by the time Charles did go up to Cambridge, was to have an even greater influence on him.

Meanwhile, the two brothers were working at a bit of botany on their own. Philos wrote: 'Apropos to Linnaeus I wish you would look in English Systema Vegetab & copy me out the specific description of Pinus sylvestris.'

Charles was sixteen when his father decided that he was doing no good at school. Though he was for his age 'neither high nor low in it', later reflection brought Charles Darwin to the conclusion that he was considered by his masters and by his father as a very ordinary boy, even below the normal standard of intellect. To his deep mortification his father said to him: 'You care for nothing but shooting, dogs, and rat-catching, and you will be a disgrace to yourself and all your family.' Charles, who worshipped his father, 'the kindest man I ever knew and whose memory I love with all my heart', thought he must have been angry and somewhat unjust when he used such words.

There was talk first of all of sending him to Cambridge, to join Erasmus. The decision was cancelled: he was to go instead to Edinburgh University. In February 1825 Erasmus wrote to his brother: 'I don't know whether to be sorry or glad about your Edinburgh plans. I think it is ten thousand pities if you do not come to Cambridge (wh. seems to be the case) & I shall venture to add it is a pity you leave school so soon, but to this latter doctrine you will hardly give credit.'

Charles did not question his father in removing him from school at so young an age. As his daughter Henrietta wrote in the Life of their mother: 'What his father did or thought was for him absolutely true, right, and wise.'

It was now decided to send Erasmus, too, to Edinburgh, there to complete his medical studies in the greatest school of its day, Charles to commence them. For Robert Darwin thought he saw a doctor in his younger son also. Erasmus was delighted at the change of plan. 'It will be very pleasant our being together. We shall be as cosy as possible & I almost think that when you have arrived at the dignity of a "Varsity" man that I shall leave licking you. We shall have some good amusements in scheming out our plans next summer.'

But that summer was a busy one for Charles. His father put him to the test, allowing him to attend some of his poorer patients, chiefly children and women. In the *Autobiography* we read: 'I wrote down as full an account as I could of the case with all the symptoms and read them aloud to my father, who suggested further enquiries and advised me what medicines to give, which I made up myself. At one time I had at least a dozen patients, and I felt a keen interest in the work.' Robert Darwin thought his trust not misplaced. 'My father who was by far the best judge of character whom I ever knew declared that I should make a successful physician,—meaning by this one would get many patients. He maintained that the chief element of success was exciting confidence; but what he saw in me that convinced him that I should create confidence I know not.'

It was fortunate for Charles Darwin that he started his university career at Edinburgh. The time was not ripe for Cambridge.

2

First Discoveries

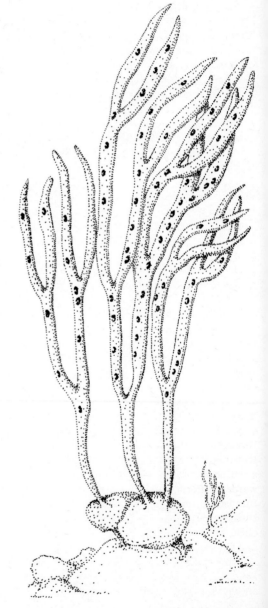

*When Charles was sixteen he became a medical
student at Edinburgh University. His spare time
was spent in natural history pursuits. He made
friends with the Newhaven fishermen and
sometimes went out with them when they trawled
for oysters, in this way collecting marine
specimens.*

*Botanists thought that the small black globules
on the seaweed* Fucus lorius *were its buds.
Charles discovered that they were the egg-cases
of the leech* Pontobdella muricata. *He read a
paper on them at the Plinian Society.*

   Fucus lorius, *also spelt* loreus, *is now
renamed* Himanthalia lorea.

THE TWO BROTHERS found comfortable lodgings at 11 Lothian Street, conveniently less than two hundred yards from the University gates. Their landlady, a Mrs Mackay, was 'a nice clean old body—exceedingly civil and attentive,' as Charles wrote to his father. Their rooms were only four flights of steps from the ground floor, which was 'very moderate' compared with other lodgings he and Erasmus had inspected. They had 'two very nice and *light* bedrooms and a sitting-room', for which they were grateful, since light rooms were scarce articles in Edinburgh, most of them being 'little holes in which there is neither air nor light'. On Saturday the 22nd of October Charles registered as a student, signing the matriculation book 'Charles Darwin—Shropshire'. He and Erasmus were not long in 'paying in their pound' for the use of the library, which they did the following Tuesday, and the records show that they took out more books than was usual among the students. Next day Charles attended his first classes—in *Materia medica* at 8 a.m., chemistry at 10, and anatomy at 1 o'clock. He was one of 902 medical students in that academic year, and one of the 250 who had come from England.

Whatever he expected of Edinburgh, and he came as a serious student, he was bitterly disappointed—that is, in the lectures, for apart from those by Dr Thomas Charles Hope on chemistry, which covered pharmacy ('I like both him and his lectures *very* much,' he told his sister Caroline), he found them intolerably dull. The lectures by Dr Andrew Duncan, junior, 'at 8 o'clock on a winter's morning are something fearful to remember,' he wrote in his *Autobiography*, and to his sister: 'Dr Duncan is so very learned that his wisdom has left no room for his sense. The Materia Medica as given by him cannot be translated into any word expressive enough of its stupidity', while 'Doctor —— made his lectures on human anatomy as dull as he was himself.' This was Alexander Monro *tertius* who did not sustain the great reputation made by his grandfather and his father who had preceded him in the chair of anatomy. 'I dislike him and his lectures so much,' Charles wrote, 'that I cannot speak with decency about them.' Human anatomy disgusted him. Better were the twice-weekly clinical lectures on the patients in the Royal Infirmary, which he enjoyed. He had a 'perpetual ticket' for the hospital and regularly visited the medical wards. Different was surgery. It was twenty years before Sir James Simpson's introduction of chloroform, and scenes in the operating theatre could be horrific. Charles saw two very bad operations, one on a child, but rushed away before they were completed. He never went again, and the two cases haunted him for years.

Life in Edinburgh became more interesting in his second year. Erasmus was now in London and Charles sought out new companions among his fellow-students. One was William Francis Ainsworth, later surgeon and geologist to the Euphrates expedition. He was a Wernerian geologist, adhering to the Neptunist theory of Abraham Gottlob Werner who had believed that all rocks were deposited as precipitates from water (as against the Vulcanists who thought they were volcanic in origin). Charles found him superficial and glib of tongue. More congenial was John Coldstream who was interested in marine zoology. 'Prim, formal, highly religious and most kind-hearted,' Charles described him. William Kay, William Alexander Browne and George Fyfe were other friends. The last two and Coldstream proposed him for membership of the Plinian Society, a body entirely after his heart, for here he was able to indulge his passion for natural history. He was elected a member on the 28th of November 1826, and a week later chosen as one of the five members of its Council.

The Society met every Tuesday evening in a basement room in the University and had about one hundred and fifty members, though the number present at the meetings was usually not more than twenty-five. The minute-book shows that Charles attended all but one of the nineteen meetings held from the date of his election to the 3rd of April 1827 when he left Edinburgh. He found them stimulating, and the papers read at the last five meetings in 1826 show how wide was the range of subjects discussed. They included the alleged oviposition of the cuckoo in the nests of other birds; extra-uterine gestation; the results of the analysis of the Cheltenham waters; oceanic and atmospheric currents; the anatomy of expression; instinct, and the various purposes to which the formation of a vacuum is applied in the animal kingdom. In January 1827 two of the subjects before the Society were 'The Sap-vessels of the *Solanum tuberosum*, founded on the commencement of a series of experiments in this department of physiological botany', a paper presented by Allen Thomson, later to become a distinguished embryologist; and one read by Ainsworth on the principles of natural classification including some considerations of specific characters. In February there was a paper on a peculiar variety in the shape of the leaf of *Laurus nobilis*. In March Allen Thomson described the circulation of the sap in vegetables. Charles is noted as having participated in the discussions on four of the evenings. It would have been particularly interesting to know what he said on the principles of natural classification, especially on specific characters, when he spoke on that subject.

The west block of the University was occupied by the Natural History Museum founded by Robert Jameson. It was notable for the excellent preservation of its specimens and their scientific arrangement, and for its large collections of birds and fossils. So extensive and important was the entire museum collection that it was reputed to be second only to that of the British Museum. Two experienced naturalists who worked there were Robert Edmund Grant who became Professor of Comparative Anatomy and Zoology at London University in 1827, and William MacGillivray, later Professor of Natural History at Aberdeen. Charles came to know them well. He sometimes went with Grant on walks, and learnt much from him. He was sixteen years older than Charles, who thought him 'dry and formal in manner but with much enthusiasm beneath this outer crust.' In March 1827 they visited the Black Rocks at Leith, and there Charles found a large *Cyclopterus lumpus*, the common Lump Fish. Its length from snout to tail was 23½ inches, its girth 19½ inches, he recorded in his diary on the 16th. 'Dissected it with Dr Grant.'

He made friends with the skippers of the Newhaven dredgers and asked them to let him have anything interesting they found. On the 19th one of them presented him with some specimens of *Flustra carbasea*, the Sea Mat. Studying them under the microscope he made what he called 'one interesting little discovery'. It was that the so-called ova of the *Flustra* had the power of independent movement by means of cilia, hair-like processes which they lashed about; and that the eggs were in fact larvae. He made a second discovery, as the minute-book of the Plinian Society recorded on the 27th.

> Mr Darwin communicated to the Society two discoveries which he had made:
> 1. That the ova of the Flustra possess organs of motion.
> 2. That the small black globular body hitherto mistaken for the young Fucus lorius is in reality the ovum of the Pontobdella muricata.
> At the request of the Society he promised to draw up an account of the facts and to lay it, together with the specimens, before the Society at their next meeting.
> Dr Grant detailed a number of facts regarding the Natural History of the Flustra.

A minute of the next meeting records the presentation of

> A specimen of the Pontobdella muricata, ⎱ By Mr Darwin
> with its Ova and young ones   ⎰

Charles had also made friends with some of the Newhaven fishermen and he sometimes went out with them when they trawled for oysters. This was how he got the specimens which were the subject of his second paper at the March meeting. Exhibiting them to the Plinian members on the 3rd of April he pointed out that the small black globular bodies (the fishermen called them Peppercorns) which had been supposed to be the young state of the brown seaweed were in fact the egg-cases of the wormlike leech.

Further to this, Grant prepared a description of the egg-cases, or cocoons, for the *Edinburgh Journal of Science*, in which he stated that 'the merit of having first ascertained them to belong to that animal is due to my zealous young friend Mr Charles Darwin of Shrewsbury, who kindly presented me with specimens of the Ova exhibiting the animal in different stages of maturity.'

The Wernerian Society, founded, like the Plinian Society, by Robert Jameson, was another source of information about natural history subjects. Professor Grant, who was a member, occasionally took him to its meetings, and there Charles heard John James Audubon give two lectures on North American birds, though 'sneering somewhat unjustly' at Charles Waterton, the naturalist, traveller, and author of *Wanderings in South America*. Living in Edinburgh was a negro who had travelled with Waterton and gained his livelihood by stuffing birds. Charles paid him for lessons in taxidermy and often used to visit him, for he found him a pleasant and intelligent man.

As a member of the Royal Medical Society of Edinburgh he attended their meetings quite regularly, though as the subjects were exclusively medical he did not much care about them. 'Much rubbish was talked there, but there were some good speakers,' was his later comment. There was the great occasion when he went with Leonard Horner, the geologist and educationalist, to a meeting of the Royal Society of Edinburgh. Sir Walter Scott who was in the chair as president apologised to the meeting as not feeling fitted for such a position. Charles looked at the famous poet and at the whole scene with some awe and reverence (if the occasion was before the 23rd of February 1827, Scott had not yet divulged his authorship of the Waverley Novels), writing of it in 1876: 'I think it was owing to this visit during my youth and to my having attended the Royal Medical Society, that I felt the honour of being elected a few years ago an honorary member of both these Societies, more than any other similar honour. If I had been told at that time that I should one day have been thus honoured, I declare that I should have thought it as ridiculous and improbable as if I had been told that I should be elected King of England.'

Edinburgh boasted a chair of natural history, of which Robert Jameson had been for twenty-two years regius professor. The subject then included geology and zoology, and during his second year Charles attended these lectures and found them incredibly dull. 'The sole effect they produced on me was the determination never as long as I live to read a book on Geology or in any way to study the science.'

He had looked forward to geology. At home an old man by the name of Cotton had pointed out to him the large boulder in Shrewsbury called the Bell Stone. He knew a great deal about rocks and had told Charles that there was no stone of the same kind nearer than Cumberland or Scotland, solemnly assuring him that the world would come to an end before anyone would be able to explain how the Bell Stone came to Shrewsbury. He was telling this to the boy who later in life was to solve many problems in connection with erratic boulders! Meanwhile the mystery of the Bell Stone made a deep impression on him, but Professor Jameson continued to be a severe disappointment. In a field lecture one day at Salisbury Crags he discoursed on a trap dyke, declaring that it was a fissure filled with sediment from above, and adding with a sneer that there were men who maintained that it had been injected from beneath in a molten condition. Charles was astounded. Volcanic rocks were all around them: the trap itself had margins of amygdaloid, an igneous rock, and the strata on each side were indurated by volcanic pressure. 'When I think of this lecture, I do not wonder that I determined never to attend to Geology,' he wrote in retrospect.

We sometimes live to eat our words.

Vacations during the two years at Edinburgh were, as he wrote, wholly given up to amusements, though he kept up his reading and always had some interesting book in hand. In the June of 1826 he went for a long walking tour to North Wales with Nathan Hubbersty, the assistant master at Shrewsbury School. They accomplished thirty miles most days, and climbed Snowdon. In October he went with his sister Caroline on a week's riding tour, again in North Wales, a servant with saddle-bags carrying their clothes. Caroline was a delightful companion with her quick appreciation of beauty and the changing scenes of travel. He had another trip with her in April 1827, after a tour from Edinburgh of Dundee, St Andrews, Stirling, Glasgow, Belfast and Dublin. The entry in his diary reads: 'Then London & Paris with Uncle Jo.' This was Josiah Wedgwood the younger, whom they usually called Uncle Jos. His daughters Fanny and Emma had been staying for the last eight months with their aunt Jessie Sismondi at Geneva. Jos was going out to fetch them home, and Charles was to travel as far as Paris and then return. On the passage to Dieppe he made a very hearty dinner on roast beef, 'though not quite well'— a foreshadowing of the years of the *Beagle*. This was the only time Charles Darwin ever set foot on the Continent.

Best of all were the visits to Uncle Jos at Maer in Staffordshire, especially for the shooting in the autumn. Charles used to place his shooting boots open by his bedside at night, so as not to lose even half a minute in putting them on in the morning. He recorded every bird he shot throughout the season, keeping a tally by making a knot in a piece of string tied to a button-hole.

A few miles away from Shrewsbury was Woodhouse, the home of the Owens who were all close friends of the Darwins. Sarah the eldest daughter was a 'prodigious friend' of Charles's sister Susan. They were a big family and one day while shooting there two of them played a trick on Charles. They had noticed the satisfaction with which he tied each fresh knot, and after a while one of them cried out, 'You must not count that bird, for I fired at the same time!' Wickedly they kept this up for some hours before telling him the joke, though it was no joke to poor Charles who knew he had shot a large number of birds but did not know how many and could not add them to his precious list. 'How I did enjoy shooting,' he recalled, 'but I think that I must have been half-consciously ashamed of my zeal, for I tried to persuade myself that shooting was almost an intellectual employment; it required so much skill to judge where to find most game and to hunt the dogs well.'

Apart from the shooting, he enjoyed any visit to Maer where there was a sense of freedom absent at The Mount in a household dominated by the opinions of Dr Darwin. The country was pleasant for walking and riding, and in the evenings there was 'much very agreeable conversation, not so personal as it generally is in large family parties, together with music.' In the summer the whole family used to sit on the steps of the portico looking over the flower garden, with the steep wooded bank opposite the house reflected in the lake, and here and there a fish rising or a water-bird paddling about. Nothing ever left a more vivid picture on his mind than these evenings at Maer, and he was attached to and greatly revered Uncle Jos who was a silent and reserved man. But sometimes he would talk openly with Charles in whose opinion 'he was the very type of an upright man with the clearest judgement', adding: 'I do not believe that any power on earth could have made him swerve an inch from what he considered the right course.'

This clear judgement of his uncle Josiah Wedgwood was to mark the turning-point in Charles Darwin's life.

Meanwhile there was the question of his career. He left Edinburgh University in April 1827 without completing his medical studies. Not long after he had gone there he had become convinced that his father would leave him enough property to live on comfortably, and according to his own confession this belief checked any strenuous effort to learn medicine. It had not taken much intuition on Dr Darwin's part to know how Charles felt, if he had not long ago learnt from Charles's letters to his sisters that he disliked the thought of being a physician. But the hardworking Dr Darwin was not one to encourage idleness. He now proposed that Charles should become a clergyman, as an antidote perhaps to his son's turning into the 'idle sporting man' which seemed his probable destination. With the history of Darwin liberal thought—Erasmus Darwin's evolutionary ideas being regarded as subversive of religion, Robert classifying himself as a freethinker without any belief in the supernatural—this would otherwise have been a ludicrous suggestion, and Charles himself had scruples about declaring his belief in all the dogma of the Church of England, though in some ways the thought of becoming a country clergyman appealed to him. He asked for time to consider, and meanwhile carefully read Pearson on the Creed and a few other books on divinity, and as he did not then in the least doubt the strict and literal truth of every word in the Bible he soon persuaded himself that the Creed must be fully accepted. 'It never struck me,' he wrote about this decision, 'how illogical it was to say that I believed in what I could not understand and what is in fact unintelligible. I might have said with entire truth that I had no wish to dispute any dogma; but I was never such a fool as to feel and say "credo quia incredibile".'

Preparations were made. He was to go to one of the English universities and take a degree. Before then some cramming was necessary, for he had not opened a classical book since leaving school and found to his dismay that in the two intervening years he had forgotten almost everything he had learnt, including some of the Greek letters. His father engaged a tutor for him and the resilient Charles soon recovered his lost knowledge, easily translating again his Homer and Greek Testament. He came into residence at Christ's College, Cambridge, on Christmas Day, 1827, having been admitted a pensioner on the 15th of October.

Cambridge was to occupy him for three years. But as at Edinburgh and school Charles regarded his time there as 'wasted, as far as the academical studies were concerned'. He attempted mathematics and in the summer of 1828 went to Barmouth with George Ash Butterton as his private tutor ('a very dull man'), who was only four years older. He got on very slowly. The work was repugnant to him, chiefly because he was unable to see any meaning in the early steps in algebra. In later years he deeply regretted that impatience had stopped him from going far enough to understand something of the great leading principles of mathematics, for it seemed to him that men thus endowed had an extra sense, though he did not think he would ever have succeeded beyond a very low grade. In classics he did nothing except attend a few compulsory college lectures. During the second year he had to work for a month or two to pass the Little Go, which he did easily at the end of March 1829. Again, in his last year he worked 'with some earnestness' for his B.A., brushing up his classics together with a little algebra, and Euclid which gave him as much pleasure as it had in schooldays. It was also necessary to get up Paley's *Evidences of Christianity* and his *Moral Philosophy*. This Charles did so thoroughly that he could have written out the whole of the *Evidences* from memory, if not in Paley's clear language, which was one of the things about him Charles admired. The logic of this book and of his *Natural Theology* gave him as much delight as did Euclid. He was charmed and convinced by Paley's long line of argumentation. In fact the careful study of his works was the only part of his Cambridge course which he then felt, and continued to believe, was of the least use to him. (He was proud of a personal link with his hero, for his rooms in college were those traditionally occupied by Paley—in the middle staircase of the first court, on the right-hand side.) So by writing a good paper on Paley, by doing Euclid well, and by not failing miserably in classics, he gained a good place among the men going in for the ordinary degree, and was placed tenth in a class list of 178 successful candidates.

The pattern of his out-of-study hours again followed school and Edinburgh. This time botany took the place of marine zoology, as that subject had replaced the chemistry he and Erasmus had practised in the garden tool-house.

But he made time for play. Christ's in his day was a pleasant, fairly quiet college, but 'with some tendency towards horsiness'. With his love of shooting and hunting and, when the season for these pursuits was over, riding across country, he got into a sporting set which included some 'dissipated low-minded young men'. They often used to dine together, and they sometimes drank too much, with singing and card-playing after-wards. Looking back to these times from his chair as an old man, he knew that he ought to feel ashamed of days and evenings thus spent, but as some of his friends were very pleasant and they were all in the highest spirits he could only recall them with pleasure. It was probably all innocent fun, for when Francis Darwin came to edit his father's *Life and Letters* and interviewed some of his contemporaries, he gathered that he had exaggerated the Bacchanalian nature of these parties.

He had many other friends of widely different tastes. One was Charles Whitley who was afterwards senior wrangler and an honorary canon of

Durham. He inoculated Charles with a taste for pictures and good engravings, and this liking for art—though not natural to him, as he confessed—lasted for several years, many of the pictures in the National Gallery in London giving him much pleasure, the *Lazarus* of Sebastiano del Piombo exciting in him a sense of sublimity.

He also got into a musical set through his warm-hearted friend John Maurice Herbert, later county court judge of Cardiff and the Monmouth circuit. From associating with these men and hearing them play, he acquired a real love of music and often used to time his week-day walks so as to hear the anthem sung in King's College Chapel. This gave him such intense pleasure that his backbone would sometimes shiver. Occasionally he even hired the choristers to sing in his rooms. The odd thing was that he was utterly destitute of a musical ear, so that he could not discern a discord, keep time or even hum a tune. In later years this became a family joke, but despite his inability to tell one tune from another music never failed to charm him.

His second cousin William Darwin Fox (they shared the same great-grandfather, Robert Darwin of Elston) was also at Christ's College, and from him Charles imbibed a passion for collecting beetles. As proof of his zeal he told the story of how one day in tearing some old bark off a tree he saw two rare beetles and seized one in each hand. He then saw a third which was new to him and which he could not bear to lose, so he popped the one he held in his right hand into his mouth. Alas, it ejected an acrid fluid that burnt his tongue and he was forced to spit it out. He lost the beetle, as well as the third one. But he went on collecting, in the winter employing a labourer to scrape moss off old trees and bring it to him in a large bag, and also to collect the rubbish at the bottom of the barges bringing reeds from the fens. In this way he got some rare species. Triumph came on the day he opened James Francis Stephens's *Illustrations of British Entomology* and read the magic words 'Captured by C. Darwin, Esq.' Beetle-collecting left an indelible impression. Even when he was an old man he could still remember the exact appearance of certain posts, old trees and banks where he had made a good capture.

He became friendly with Albert Way of Trinity, a future archaeologist, and with Harry Stephen Meysey Thompson of the same college, afterwards an M.P. and a leading agriculturist, and went out entomologising with them. Once, when they were not around, he wrote to Fox saying: 'I am dying by inches from not having any body to talk to about insects.'

He did not then know that his knowledge of insects was to be so useful in the years ahead.

It was soon after Charles came up to Cambridge that Fox introduced him to the Rev. John Stevens Henslow, the man who was to change the entire course of his life. Erasmus, by his description of Henslow's 'very entertaining' lectures, had whetted Charles's interest in him, for Edinburgh had satiated and sickened him of lectures. Professor of Mineralogy from 1822–27, Henslow in 1825 had taken on the professorship

There were three men who most influenced the career of the young Charles Darwin. The first was John Stevens Henslow, professor of botany at Cambridge, who recommended him as naturalist to H.M.S. *Beagle*

of botany as well, a post he was to hold for the rest of his days.

Before he took over, botany at Cambridge had dwindled almost to extinction. In fact his aged predecessor Professor Thomas Martyn had been unable to give his course of lectures for twenty-nine years. The time was propitious for a resurgence of interest in botany. Martyn had been a firm adherent of the 'artificial' Linnaean classification, and this was now being shaken by the 'natural' system advocated by de Jussieu. John Edward Gray, botany lecturer at the Borough School of Medicine in London, had published (under his father's name) a *Natural Arrangement of British Plants* and taught the views of de Jussieu at various other places in London. This was in 1821, in the same year that William Jackson Hooker, Regius Professor of Botany at Glasgow, published his *Flora Scotica*, Part I of which, devoted to flowering plants only, was arranged according to the Linnaean system, Part II including flowering plants with cryptogams and other orders being arranged according to the Natural System. This was the first Flora using the new system.

Henslow was an all-round naturalist, but he first pursued geology with Adam Sedgwick, Woodwardian Professor of Geology at Cambridge from 1818, and with him established the Cambridge Philosophical

Society, of which he was elected secretary. Conchology and entomology were his next interests. The bivalve *Cyclas henslowiana* was named in his honour, and his collection of insects was presented to the Philosophical Society he helped to found. On the death of Dr E. D. Clarke he offered himself for the professorship of mineralogy, devoting himself to the study of chemistry as well. He was only twenty-six years of age and still B.A. when elected to that chair. On taking the chair of botany he 'knew very little, indeed, about botany,' as he later confessed, adding, however, that he 'probably knew as much of the subject as any other resident in Cambridge.' His views on botany were unorthodox. No slave to its systematic side, he was much more interested in what we would now call ecology, the relationship of a plant with its habitat; and the relationship of the plants of one country with the same species in other countries. 'For,' he wrote in 1827, 'to ascertain the geographical distribution of a well-known species is a point of vastly superior interest to the mere acquisition of a rare specimen.'

One of the first things Henslow did on assuming the chair was to plan for a new botanic garden. The one he inherited was in the heart of the town, the soil was bad, and there was no possibility of extending its five acres. Pronouncing it 'utterly unsuited to the demands of modern science' he sought the advice of William Jackson Hooker on acreage, the extent of the Glasgow greenhouses, and what should be the salary of a curator or head gardener. Not until 1831 was his dream realised, when forty acres between the Trumpington and Hills Roads were purchased. Operations were held up till a lease expired in 1844, and the ground then had to be cleared, so that the removal of the plants to their new home did not begin till 1846, replanting being completed by the end of 1852. But Henslow lived to see the Cambridge Botanic Garden a mecca for botanists from all over the world.

Knowing from his brother that Henslow was a man who knew every branch of science, Charles was 'accordingly prepared to reverence him'. He was not disappointed, and afterwards told Fox that meeting him 'was the luckiest day of my life.'

During term Henslow kept open house every Friday evening, when he welcomed all undergraduates sharing his keen interest in natural history, as well as older members of the University. Fox often went, and through him Charles received an invitation, thereafter regularly visiting his friendly house. To enter it was at once to feel completely at ease, and this in the presence of the man who 'knew everything', according to one of the other students who went there. Charles, even on that first evening, read him as a man of utter sincerity and kindness. Though awed by the wide scope of his knowledge as he led or joined in a discussion, which he did brilliantly, Charles noticed that he never paraded it, and his manner to the older and distinguished visitors and to the youngest student was exactly the same: to each he showed the same winning courtesy.

Charles left with the determination to attend his lectures on botany.

3

The Man who Walked with Henslow

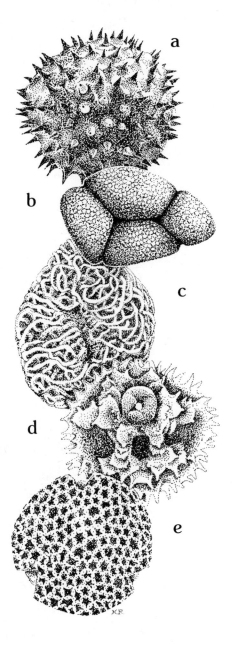

a

b

c

d

e

Charles went up to Cambridge in 1828, giving up medicine for divinity. Though botany was not a required subject he attended the lectures of John Stevens Henslow.

Examining plants under a microscope he found that the pollen grains of Orchis morio, *the Green-winged Orchid, were wedge-shaped, and later that the pollen grains of other flowers had each their different shape. One day he was to write a book on orchids and the fantastical mechanisms of their flowers.*

Henslow, recommending Charles as naturalist to the Beagle *voyage, told him that he was 'the best qualified person I know of . . . not on the supposition of yr. being a finished naturalist, but as amply qualified for collecting, observing & noting anything new to be noted in Natural History.'*

The fascinating shapes of pollen grains:
(a) Cotton (*Gossypium* species)
(b) Green-winged Orchid (*Orchis morio*)
(c) Storksbill (*Erodium moschatum*)
(d) Dandelion (*Taraxacum officinale*)
(e) Meadow Cranesbill (*Geranium pratense*)

MEDICAL STUDENTS OF those days studied botany in order to gain a knowledge of the plants which were part of the *materia medica*. For a divinity student, as Charles was, the subject was outside the curriculum. But he enrolled during each of his three years at Cambridge, and attended every lecture.

Under previous lecturers, botany at Cambridge had been merely instructional, to the students a dutiful cramming of facts. Henslow taught them differently. To learn by their own discovery was his method, so each student educated himself by examining and recording plant structures first seen by his own dissections, and each had a round wooden plate for dissecting upon. They worked with living specimens, particularly flowers. Henslow also used large drawings to illustrate the parts of plants, with such visual aids hammering home the lesson, and these he prepared himself. By the time his students arrived at examination stage they were well able to describe the organs of plants systematically, and explain their relations, uses and meanings, both from a physiological and classificatory point of view, thus proving, as Joseph Hooker wrote, 'that they had used their eyes, hands and heads, as well as their books.' The phenomena of plant-life, such as the different colours of flowers, were studied; the laws of phyllotaxy governing the way leaves are arranged on a stem or the scales on a pine cone; hybridisation; variation in leaves and plant 'monsters' or those departing in form or structure from the normal type of the species. A special study was the geographical distribution of plants and the effects on them of their new environment.

Nor were his students kept in the lecture room. Two or three times in each term he took his class on an excursion: either a long walk to the habitat of some rare plant, or in a barge down the river to the fens, or in coaches to some distant place like Gamlingay to see the wild Lily of the Valley or catch on the heath there the rare Natterjack Toad. These excusions made an enduring and joyful impression on Charles's mind. Henslow on such occasions was always in boyish good spirits, and laughed as heartily as a boy at the misadventures of those who chased the splendid Swallowtail butterflies across the boggy and treacherous fens. He used to pause every now and then to lecture on some plant or other object: he always had something interesting to say about each insect, shell or fossil collected on the day's foray. After it was over they trooped into the nearest inn to dine, 'and most jovial we then were,' Charles Darwin remembered.

It is clear from the notebooks he kept, which he began in Edinburgh (they had marbled covers and a brown leather spine), that he took a serious interest in botany. His notes at all times were brief, sometimes containing little more than the dates of the principal events of his life, the history of a year being compressed into a page or less. But in his

Cambridge notebook he devoted five pages to botany. There are careful notes on the experiments he was doing on pollen grains: under the microscope he examined those, for instance, of *Orchis morio*, the Green-winged Orchid, noting that they were green in colour and wedge-shaped, and fastened at their narrow ends one to the other by a highly elastic thread. The pollen grains of a geranium were yellow and round. He also examined the epidermis or outer skin of the geranium petals and found them very curious, consisting of six- or occasionally seven-sided cells. Then came a leaping thought, typical of Darwin in his later years, brilliantly conjectural for a nineteen-year-old: 'It would be curious experiment to put box with *boiled* earth on top of house & see how soon any plants would come there.'

It was as though, already, he were poised on the brink of discovery.

He found Henslow the best of teachers, with a gift for drawing conclusions from long-continued minute observations. His judgment was excellent and his temper imperturbable, though he could be roused to wrath and prompt action in the face of cruelty. In a memoir compiled at his death by Leonard Jenyns, Charles Darwin wrote of him:

> No man could be better formed to win the entire confidence of the young and to encourage them in their pursuits. He would receive with interest the most trifling observation in any branch of natural history; and however absurd a blunder one might make, he pointed it out so clearly and kindly, that one left him in no way disheartened, but only determined to be more accurate the next time.

He gave an instance of this in his *Autobiography*.

> Whilst examining some pollen-grains on a damp surface I saw the tubes exserted, and instantly rushed off to communicate my surprising discovery to him. Now I do not suppose any other Professor of Botany could have helped laughing at my coming in such a hurry to make such a communication. But he agreed how interesting this phenomenon was, and explained its meaning, but made me clearly understand how well it was known; so I left him not in the least mortified, but well pleased at having discovered for myself so remarkable a fact, but determined not to be in such a hurry again to communicate my discoveries.

To Fox he wrote in November 1830: 'Henslowe is my tutor & a most *admirable* one he makes, the hour with him is the pleasantest in the whole day. I think he is quite the most perfect man I ever met with.'

As time went on at Cambridge he got to know Henslow very well and felt himself accepted as a friend. Very often there was an invitation to join his family at dinner in the evening, though, as he said to Fox: 'You have not told me half enough what you think of Mrs Henslowe. She is a devilish odd woman. I am always frightened whenever I speak to her & yet I cannot help liking her.'

During the latter half of his time at University Charles took long walks with him most days. The dons called him 'the man who walks with

Henslow'. They talked on every subject, including religion. Henslow was strictly orthodox and once told him that he would be grieved if a single word of the Thirty-nine Articles was altered. Besides his University work he was curate of St Mary-the-Less, the exquisite little church next to Peterhouse in Trumpington Street, where the light coming from windows of pale green glass makes you feel that you are under the fall of a sea-wave poised over shimmering sands.

Henslow went on to do great things for industry and for agriculture. Through his knowledge of geology he discovered nodules in the cliffs at Felixstowe, which were the petrified droppings of extinct reptiles and fishes. They contained, as he suspected, phosphate of lime, and went into commerce as coprolite, a valuable fertiliser. Abortive attempts had been made in some counties to find coal. He pointed out the value of geology in intimating where coal was possible and also where it was impossible. He noticed that farmers stripped off the leaves of the mangold-wurzel, under the mistaken impression that it was the roots which made their 'bulbs'. Henslow pointed out that this was a wasteful practice because it was the leaves that made them. He did much work on the diseases of wheat.

Meanwhile he was Charles Darwin's great teacher of botany.

During his last year at Cambridge Charles read with profound interest Alexander von Humboldt's *Personal Narrative*. This and Sir John Herschel's *Introduction to the Study of Natural Philosophy* stirred up in him 'a burning zeal to add even the most humble contribution to the noble structure of Natural Science'. Humboldt had a golden pen. His descriptions of exotic nature and scenery so entranced Charles that he copied out long passages about Tenerife. Henslow and his brother-in-law Leonard Jenyns sometimes went on longer excursions into the country with some of their friends. Among them were Marmaduke Ramsay, tutor of Jesus College, and Richard Dawes, the future Dean of Hereford who became famous for his education of the poor. These were men of standing, and Charles was proud that he was always allowed to join them. On one of these excursions he read aloud Humboldt's glories of Tenerife, and some of the party at once declared they must go there, though Charles thought they were only half in earnest. He himself was quite in earnest.

He could not stop thinking, talking, and dreaming of the Canaries, and even hoped Henslow might come with him. He read and re-read Humboldt, '& I am sure nothing will prevent us seeing the great Dragon Tree,' he wrote from Shrewsbury. This was the colossal *Dracaena draco* standing in the garden of Dr Franqui, its circumference being nearly 48 feet. He went to London to find out the cost of a passage. It was £20, and ships touched and returned from June to February. He began learning Spanish, 'working like a tiger at it'.

He passed his finals in January 1831, later writing to Fox: 'I do not know why the degree should make one so miserable—both before and afterwards. I recollect you were sufficiently wretched before & I can assure you I am now.' He did not know what about, but believed it was 'a beautiful provision of nature' to offset leaving so pleasant a place as Cambridge where he had so many friends ('Henslow among the foremost').

There was still the academic year to finish, but he looked forward to the spring term, 'walking & botanising with Henslow'.

A 'most magnificent anonymous present of a microscope' had come to him—from some graduated Cambridge man, he supposed. He only wished he knew who, just to feel obliged to him. 'Did you ever hear of such a delightful piece of luck?' He did not quote the note that came with it: *If Mr Darwin will accept the accompanying Coddington's Microscope, it will give peculiar gratification to one who has long doubted whether Mr Darwin's talents or his sincerity be the more worthy of admiration, and who hopes that the instrument may in some measure facilitate those researches which he has so fondly and so successfully prosecuted.*

But his plans for the future were not at all settled. As for his Canary scheme, his other friends most sincerely wished him there, he plagued them so with talking about it.

Henslow took pity on his unsettled state and persuaded him to begin the serious study of geology. So at Shrewsbury Charles examined sections and coloured a map of parts around the town, and he bought himself a clinometer. 'It cost 25s made of wood, but the lid was plate of brass graduated. Cary did not approve of a bar for the plumb; so that I had *heavy* ball instead.' (Cary was presumably the son of William Cary, the instrument-maker.) He added in this letter to Henslow: 'I put all the tables in my bedroom at every conceivable angle & direction. I will venture to say I have measured them as accurately as any Geologist going could do.' But he suspected that the first expedition he took, clinometer and hammer in hand, would send him back very little wiser and a good deal more puzzled than when he started. 'As yet I have only indulged in hypotheses but they are such powerful ones, that I suppose, if they were put into action but for one day, the world would come to an end.' He was never so serious that he could not poke fun at himself.

The geological map of Shrewsbury was an exercise preparatory to going on a tour with Adam Sedgwick, to whom Henslow had introduced him, and it was due only to Henslow's persuasions that Sedgwick consented to take him, for in all his time at Cambridge Charles had never attended even one of his eloquent and popular lectures. In previous years Sedgwick had geologised in Scotland, the Alps, Cumberland and South Wales. He was now anxious to extend his studies into North Wales. Shrewsbury was a half-way house, and he stayed the night at The Mount, before the two of them set off on Friday the 5th of August in Sedgwick's gig.

On the evening before, Charles, thinking to interest him, told him about a large worn tropical volute shell which a labourer had found in an old gravel pit near the town. To his surprise Sedgwick at once said that it must have been thrown away by someone, adding that if it really had been embedded in the pit this would be the greatest misfortune for geology, as it would overthrow all that was known about the superficial deposits of the Midland counties. Charles was utterly astonished at Sedgwick's not being delighted at so wonderful a fact as a tropical shell being found near the surface in the middle of England. Instead, he received a lecture about drawing conclusions from hearsay and from isolated facts. But there was a lesson in the reproof. 'Nothing before,' he wrote in his *Autobiography*, 'had ever made me fully realise, though I had read various scientific books, that science consists in grouping facts so that general laws or conclusions may be drawn from them.'

Next morning they started for Llangollen, Ruthven, Conway, Bangor and Capel Curig, at Llangollen meeting the British Ordnance draughtsman Robert Dawson who was making topographical maps. He gave them some geological notes, and next day they went north towards St Asaph and Abergele. To make a geologist of him Sedgwick often sent Charles on a line parallel with his, telling him to bring back specimens of the rocks and mark the stratification on a map. Thus at Capel Curig where they said goodbye to each other Charles began a lone twenty-seven-mile compass traverse south-south-west across the rugged Snowdonian and Merioneth mountains to Barmouth.

The two, of course, kept notes. Charles's filled twenty pages, and show not only a thorough knowledge of geology but are deductive and speculative. He drew empirical generalisations, and related broad areas of one discipline to another, an example being the observation that the character of the vegetation depended on rock types. The great Merioneth anticlyne between Maentwrog, Harlech, Barmouth and Dolgellau is called the Harlech Dome. Charles Darwin's description of its rocks was the first to be done of the region in any detail.

Sedgwick was one of the most outstanding and respected geologists in England. His methods of surveying the field, of meticulous note-taking, thorough collecting and cataloguing of specimens, were invaluable teaching. But both he and his pupil failed to observe at Cwm Idwal glacial grooves in the rocks.

At Barmouth were some Cambridge friends who were reading there. Robert Lowe, later Viscount Sherbrooke, has described how Darwin arrived, carrying with him in addition to his other burdens a hammer weighing fourteen pounds. He remembered him as being full of modesty, and that he saw a something in him which marked him out as superior to anyone he ever met, adding: 'The proof of which I gave of this was somewhat canine in nature, I followed him. I walked twenty-two miles with him when he went away, a thing which I never did for anyone else before or since.'

Charles was home at the end of August and looking forward to the shooting at Maer. 'For at that time,' as he wrote, 'I should have thought myself mad to give up the first days of partridge-shooting for geology or any other science.'

A letter was awaiting him, marked *To be forwarded or opened if absent.* It was from Henslow and was dated '24 Aug. 1831'.

My dear Darwin,
    Before I enter upon the immediate business of this letter, let us condole together upon the loss of our inestimable friend poor Ramsay

of whose death you have undoubtedly heard long before this: I will not now dwell upon this painful subject as I shall hope to see you shortly fully expecting that you will eagerly catch at the offer which is likely to be made to you of a trip to Terra del Fuego & home by the East Indies—I have been asked by Peacock who will read & forward this to you from London to recommend him a naturalist as companion to Capt Fitzroy employed by Government to survey the S. extremity of America. I have stated that I consider you to be the best qualified person I know of who is likely to undertake such a situation—I state this not on the supposition of yr. being a *finished* Naturalist, but as amply qualified for collecting, observing, & noting anything new to be noted in Natural History. Peacock has the appointment at his disposal & if he cannot find a man willing to take the office, the opportunity will probably be lost. Capt. F. wants a man (I understand) more as a companion than a mere collector & would not take anyone however good a Naturalist who was not recommended to him likewise as a *gentleman*. Particulars of salary etc I know nothing. The Voyage is to last 2 yrs & if you take plenty of Books with you, anything you please may be done—You will have ample opportunities at command—in short I suppose there never was a finer chance for a man of zeal & spirit. Capt. F. is a young man—What I wish you to do is instantly to come to Town and consult with Peacock (at No 7 Suffolk Street Pall Mall East or else at the University Club) & learn further particulars—Don't put on any modest doubts or fears about your disqualifications for I assure you I think you are the very man they are in search of—do conceive yourself to be tapped on the Shoulder by your Bum-Bailiff & affect[e] friend

J. S. Henslow

(turn over)

The expedn. is to sail on 25 Sept: (at earliest) so there is no time to be lost

Charles's entry about this in his diary was the brief note: 'Refused offer of Voyage.'

A painful decision lay behind it. He was instantly eager to accept the offer, only to be met by his father's strong objections. Categorically they were these, as we know from the list Charles made of them:

1. Disreputable to my character as a Clergyman hereafter.
2. A wild scheme.
3. That they must have offered to many others before me the place of Naturalist.
4. And from it not being accepted there must be some serious objection to the vessel or expedition.
5. That I should never settle down to a steady life hereafter.
6. That my accommodations would be most uncomfortable.
7. That you, that is, Dr Darwin, should consider it as again changing my profession.
8. That it would be a useless undertaking.

Charles considered his father's objections perfectly valid, as he told Henslow in his reply, including one he had not put on his list. This was *'the shortness of his time'* (underlined by Charles), which indicated a piece of moral blackmail on the part of Dr Darwin, still vigorous at sixty-five and with another seventeen active years before him. Charles added: 'But if it had not been for my Father, I would have taken all risks.'

The Peacock mentioned by Henslow was the Rev. George Peacock of Trinity College, Lowndean Professor of Astronomy and a friend of Captain Francis Beaufort, R.N., Hydrographer to the Royal Navy. Peacock wrote that he had received Henslow's letter too late in the evening to forward, but that in the interval he had seen Captain Beaufort at the Admiralty, who entirely approved of his offer. 'I trust that you will accept it as an opportunity which should not be lost,' he said, '& I look forward with great interest to the benefit which our collections of natural history may receive from your labours.'

The second man who most influenced young Darwin's career was Robert FitzRoy, captain of the *Beagle*. He chose Charles as his naturalist—despite the shape of his nose!

He described Captain FitzRoy as a public-spirited and zealous officer, of delightful manners and greatly beloved by all his brother officers. 'You may be sure therefore of having a very pleasant companion who will enter heartily into all your views.' The ship would sail about the end of September and he must lose no time in making known his acceptance to Captain Beaufort and the Admiralty Lords. In common with Henslow he felt the greatest anxiety that he should go, and hoped that no other arrangements were likely to interfere. He joined Henslow's wish that Charles should come up to London to see Captain Fitzroy and to complete his arrangements. The Admiralty were not disposed to give a salary, although they would furnish him with an official appointment and every accommodation. If, however, a salary should be required he was inclined to think that it would be granted. He also mentioned that the expedition was entirely for scientific purposes '& the ship will generally wait your leisure for researches in natural history etc.'

In his reply to Henslow Charles told him that he had written to Mr Peacock and asked him to communicate with Captain FitzRoy, adding: 'Even if I was to go my Father disliking would take away all energy, & I should want a good stock of that.—Again I must thank you; it adds a little to the heavy but pleasant load of gratitude which I owe to you.—'

That seemed to be that. But there was one alleviating hope. Dr Darwin while giving such strong advice against going 'does not decidedly refuse me', and he had added the words fortunate for Charles: 'If you can find any man of common sense, who advises you to go, I will give my consent.' But so much weighed against it that Charles felt the day lost. He wrote his refusal and next morning went to Maer to be ready for the 1st of September, telling his Uncle Jos all that had happened.

He was out shooting when his uncle sent for him. He offered to drive him over to Shrewsbury and talk with his father. Charles thought they had better pave the way by first writing to him, and accordingly Josiah sat down and carefully penned his objections to his brother-in-law's objections, itemising the list and couching it all sensibly and straight-forwardly; perfectly, moreover, rounding off his arguments, as in his comments on No. 5, which read:

> You are a much better judge of Charles's character than I can be. If on comparing this mode of spending the next few years with the way he will probably spend them if he does not accept this offer, you think him more likely to be rendered unsteady & unable to settle, it is undoubtedly a weighty objection—Is it not the case that sailors are prone to settle in domestic and quiet habits.

Charles started off his own letter by fearing that he was again going to distress his father, but that after consideration he thought he would again excuse his stating his opinions on the offer of the voyage. He begged for a decided answer—Yes or No. If No, he would never mention the subject again. If Yes, he would go directly to Henslow and consult with him, and then come to Shrewsbury. The Yes coming after the No shows in which direction Charles's hopes were winging.

The letters were sent on the 31st of August by public carrier, called 'the car', and Charles asked for an answer the following day by the same means. Uncle Jos, however, thought that they might as well receive Dr Darwin's verdict in person, doubtless thinking that he could add any necessary persuasions. They drove over to Shrewsbury on the 1st of September.

It was Josiah Wedgwood, Charles Darwin's uncle, who turned the scales in favour of the voyage

It was all ironed out in the smoothest way possible. 'As my uncle thought it would be wise in me to accept the offer, and as my father always maintained that he was one of the most sensible men in the world, he at once consented in the kindest manner,' was how, in his *Autobiography*, Charles Darwin recorded the tremendous moment.

One thought nagged at him. The expense. No salary on the voyage meant an extra burden on his father, and he had been rather extravagant at Cambridge ('The Governor has given me a £200 note to pay my debts & I must be more economical,' he had written to Fox in May). To console his father he said to him that he would be 'deuced clever to spend more than my allowance whilst on board the Beagle', to which Dr Darwin answered with a smile: 'But they tell me you are very clever.'

There was one more hitch. From his brother's rooms at 17 Spring Gardens, London, Charles wrote to Fox: 'On 2^{d} started for Cam^{idge}: then again from a discouraging letter from my captain I again gave it up. But yesterday everything was smoother:—' (He had seen FitzRoy) '& I think it most probable I shall go.'

William Darwin Fox was then curate at Epperstone, Nottingham.

A letter to Henslow written on the 5th of September started off:

My dear Sir,
Gloria in excelsis, is the most moderate beginning I can think of.

The ship was to leave on the 10th of October, and he was to sail in her! As Naturalist.

From now on it was all breathless haste to make preparations for a voyage that was to last not two years but three (in fact it lasted for five). At Cambridge he learnt the answer to No. 3 on the list of his father's objections: three people had already been offered the post. The first was Leonard Jenyns who, as Charles wrote to his sister Susan, 'was so near accepting it that he packed his clothes. But having two livings he did not think it right to leave them—to the great regret of all his family.' Then: 'Henslow himself was not very far from accepting it, for Mrs Henslow most generously & without being asked gave her consent, but she looked so miserable that Henslow at once settled the point.' The third was a friend of FitzRoy, and this was the reason for the Captain's discouraging letter. But again the situation was saved. Charles learnt of this at his first meeting with FitzRoy, and hastened to tell Henslow on the 5th of September: 'What has induced Cap. Fitzroy to take a better view of the case is; that Mr Chester, who was going as a friend, cannot go: so that I shall have his place in every respect.'

The way was now clear. Charles wrote in high glee: 'Things are more prosperous than I should have thought possible—Cap Fitzroy is every-thing that is delightful, if I was to praise half so much as I feel inclined, you would say it was absurd only once seeing him.—I think he really wishes to have me.—He offers me to mess with him & he will take care I have such room as is possible.—But about the cases he says I must limit myself: but then he thinks like a sailor about size: Cap. Beaufort says I shall be upon the boards & then it will only cost me like other officers.' Which was another point that eased his mind. Furthermore: 'Cap Fitzroy has a good stock of books, many of which were in my list, & rifles etc. So that the outfit will be much less expensive than I supposed.' He added: 'You cannot imagine anything more pleasant, kind & open than Cap. Fitzroy's manners were to me.—I am sure it will be my fault if we do not suit.'

But in FitzRoy's mind, which could be idiosyncratic, there was something that did not suit him about Darwin. Little did Charles then know that even at the interview, when all seemed happily resolved, FitzRoy was sitting sizing him up on the principles of physiognomy propounded by Johann Kaspar Lavater. It was the shape of Darwin's nose that gave him misgivings. According to Lavater this shape of nose was not compatible with the energy and determination required to withstand hardship, as hardship there would be on the long voyage ahead.

Aloud, FitzRoy voiced his fears that nothing would be so miserable for him as having someone who was uncomfortable, as in a small vessel they would be thrown together. He thought it his duty to state everything from the worst point of view, and Charles learnt that if he messed with him he must live poorly, with no wine and the plainest of dinners. He was asked not to make up his mind quite yet, though FitzRoy added that he thought the voyage would have much more pleasure than pain for him. It all sounded uncertain. It was not even sure yet whether they would indeed go right round the world, so that on the following day, the 6th, when Charles wrote to his cousin to say that it was most probable that he would be going, he had to add the caution: 'But it is not certain, so do not mention it to anybody.' He similarly warned his family.

But there was really no doubt in Charles Darwin's own mind. To Susan he confided: 'I feel a predestination I shall start.' He was making lists of everything, and asked her: 'Tell Nancy [their old nannie] to make me soon 12 instead of 8 shirts, tell Edward to send me up in my carpet bag . . . my slippers, a pair of lightish walking shoes—My Spanish book . . . my new microscope . . . my geological compass . . . a little book, if I have got it in my bedroom—*Taxidermy* . . .' He must have a rain-gauge, he told Henslow, and would he see about an iron net for shells? He hoped Henslow would excuse all the trouble he was giving him. Even on the day of the doubtful interview, he had written to Susan that he thought he would go on Sunday to Plymouth to see the vessel. His diary recorded for that day, the 11th of September: 'Went with Capt. FitzRoy in steamer to Plymouth to see the Beagle.' They arrived on Wednesday evening, and Charles wrote that he had scarcely ever spent three pleasanter days.

Next morning in the dockyard at Devonport he looked down at the little *Beagle* that was to be his home for the next five years.

The Plants of the *Beagle*

i. The Pampas and the Land of Fire

The Brazilian forest gave Charles Darwin his first taste of the grandeur of the tropics. It was nothing more nor less than 'a view in the Arabian Nights', with the advantage of reality. 'I collected a great number of brilliantly coloured flowers, enough to make a florist go wild.'

Between the sea and a wilderness of salt lagoons were the scorching plains where cactuses were growing and other succulent plants of the most fantastic shapes. They were storage tanks for their own water supply, their spines protecting them against predators.

But above all it was the 'wonderful beautiful flowering parasites' that impressed him. He collected hard, anything new that came his way.

In the forest he noted the lianas binding the trees like ropes and binding each other. 'Twiners entwining twiners—tresses like hair,' he wrote. They were to be the subject of another book.

S HE WAS ONLY NINETY feet in length, and when Charles saw her propped up with timbers in the dry dock without her masts or bulkheads, he thought H.M.S. *Beagle* looked more like a wreck than a vessel commissioned to go round the world. After her previous voyage much of her timber was found to be rotten, and before any sailing could be done not only repairs but improvements were to be made. Belonging to the class of ten-gun brigs (nicknamed 'coffins' in the navy from their behaviour in severe gales), she was to have her upper decks raised, increasing her tonnage from 235 to 242 tons burthen, which would also make her safer in heavy weather.

The *Beagle* emerged rigged as a barque, and to young Darwin she looked 'most beautiful' when at last she lay at the quayside refitted and ready for the voyage. 'Even a landsman must admire her, *we* think her the most perfect vessel ever turned out of the Dock yard,' he wrote enthusiastically to Henslow. 'One thing is certain, no vessel has ever been fitted out so expensively & with so much care.—Everything that can be made so is of Mahogany, & nothing can exceed the neatness & beauty of all the accommodation.' The corner of the poop cabin which was to be his own private property was, however, 'most wofully small'. He had just room to turn round in, and that was all. He was six feet tall, and when it came to slinging his hammock (which Captain FitzRoy showed him how to do) he had to remove the drawer where he kept his clothes, so as to gain an extra twelve or so inches for the foot-clews. When one thinks that the ship's company numbered sixty-seven, and there were, besides, three strange passengers with an immense amount of luggage, no wonder space was limited, though Charles was to find to his great surprise that a ship 'is singularly comfortable for all kinds of work.—Everything is so close at hand, being cramped makes one so methodical, that in the end I have been a gainer.' Certainly neatness and method were good training for a future scientist.

Robert FitzRoy had a passion for exactness, and the chronometers he brought on board were his special pride. 'No vessel,' Charles declared, 'ever left England with such a set of chronometers, viz 24, all very good ones.' These were for the work on which the ship would be engaged, the completing of the survey begun by the *Beagle* and her companion *Adventure* five years earlier. After working along the coasts of Brazil and the Argentine, and round the gale-swept Horn to Chile, they were to make a chain of longitudinal measurements around the world. On the three-day trip from London to Plymouth, FitzRoy had set aside all doubts of Darwin. He told him that the chronometers would be his special care.

The ship was to sail on the 10th of October. On Monday the 19th of September he went by the mail coach to Cambridge to say goodbye to Henslow, going on to Shrewsbury by the 'Wonder' to take farewell of his family, on the 2nd of October leaving home for London. But October passed. Then the sailing date was fixed for the following month, and Charles wrote to FitzRoy: 'What a glorious day the 4th of November will be for me! My second life will then commence, and it shall be as a birthday for the rest of my life.'

November passed. He kept himself busy, learning all he could and getting to know his shipmates. The officers from being a collection of unknown faces became individuals. He thought them 'a very intelligent, active, determined set of young fellows', if rather rough, and Henslow upbraided him for this, telling him to look for the real and sterling worth he was sure to find under many a rough surface. John Lort Stokes was FitzRoy's assistant surveyor. His father had commanded the *Beagle* before FitzRoy, and in 1841 he was himself to command her. Darwin watched him prepare the astronomical house for observations on the dipping needle, and assisted Captain FitzRoy in reading the various angles. He dined with the Kings: Midshipman Philip Gidley King and his father Captain Philip Parker King who had commanded the *Adventure*. He now had the regular job every morning of taking and comparing the differences in the barometers. The surgeon was Robert Maccormick who in 1839 was to sail with James Clark Ross to the Antarctic ('My friend the Doctor is an ass, but we jog along very amicably.').

Monday the 21st of November was a great day, when he carried all his books and instruments aboard—and fell into a panic on the old problem, want of room. Where and how was he to stow them? FitzRoy showed him, 'such an effectual & goodnatured contriver, that the very drawers enlarge on his appearance & all difficulties smooth away.'

H.M.S. *Beagle* in Sydney Harbour

On the 23rd the *Beagle* was loosed from her moorings and moved about a mile to Barnet Pool where she was to remain till the day of sailing. Erasmus arrived on the 2nd of December, they had a happy few days together, and when the *Beagle* sailed on the morning of the 10th he came with her as far as the breakwater where he left the ship and where Charles Darwin's misery of sea-sickness began, with the *Beagle* pitching bows-under in a heavy gale. They put back to their anchorage at Barnet Pool the following day. The next attempt was made on the 21st of December. The morning was calm and the sun shone red through a mist. Ill-luck met them while tacking round Drake's Island where they stuck fast on a rock, fortunately clearing it in half an hour. Then when they were only eleven miles from the Lizard another south-west gale drove them back to the Pool. It was not until the 27th of December that the sails of the *Beagle* filled with the longed-for east wind and they finally left the shores of England.

When FitzRoy had written to Captain Beaufort saying that he liked much what he had seen of Darwin and requesting that he 'apply for him to accompany me as Naturalist', it might be asked what he expected of him. George Peacock had recommended him as a 'savant', and we know that in geology he had been well grounded by Sedgwick, apart from his own early interest in rocks and minerals; that in zoology he had learnt to dissect marine animals with Grant at Edinburgh and made two original discoveries about them; also that from his young days he had taken more than a boyhood interest in birds and insects, at Cambridge developing his love of entomology to the extent of knowing both the common and rare species. We know that collecting of any kind had always been a passion with him, particularly in natural history, and we know that he collected plants and that in John Stevens Henslow he had one of the greatest botany teachers. When Darwin later wrote that he was a 'botanical ignoramus' he was comparing himself with such systematic giants as Joseph Dalton Hooker who became the head of Kew and the most honoured botanist of all time, and with Asa Gray who led the field in America. Enough for the Naturalist to the *Beagle* expedition that if he was what Henslow called 'not a *finished* Naturalist', what he lacked in experience (and he was only twenty-two) he made up for by being what Henslow also called 'the best qualified person I know of who is likely to undertake such a situation' and 'amply qualified for collecting, observing, & noting anything new to be noted in Natural History'. What was more, he had an abiding curiosity that already sent shafts of inquiry turning the wheels of his mind, and already he had the habit of outgrowing his contemporaries.

But there was another thing that Captain FitzRoy expected of his naturalist. For him the voyage had a second purpose. A deeply religious man with a fixed belief in the literal truth of the Bible, and particularly the book of Genesis, he was looking forward to discoveries being made in natural history which would prove that record to be true absolutely. How horrified he would have been, could he have foreseen that his trusted young naturalist would one day disprove the special acts of creation in which he believed so implicitly. How ironic that it was he who gave Charles Darwin the post that assured for him five years of the most ideal training possible, in a course that gave him an almost cosmic yet intimate view of the world and its beings past and present, fitting him as nothing else could have done to turn him into one of those 'false philosophers' against whom FitzRoy inveighed so bitterly.

Meanwhile there was no inkling of this. During the delays before sailing, exasperating to Darwin as to himself, he had written to Beaufort: 'Darwin is a very sensible hard-working man, and a very pleasant messmate. I never saw a "shore-going fellow" come into the ways of a ship so soon and so thoroughly as Darwin.' And after only a few weeks at sea: 'Darwin is a regular Trump.' On his part Charles hero-worshipped the man as his 'beau ideal of a Captain', as he wrote home to Susan. Though little older than himself, the twenty-six-year-old FitzRoy was already known as a brilliant navigator.

Heading for the Canary Islands the *Beagle* ploughed her way through stormy seas that kept Charles in his hammock. It was not until the 6th of January when they sailed into the harbour at Santa Cruz that he was well enough to go on deck. Having steeped himself in Humboldt's *Personal Narrative* (and now possessing his own glorious copy, a parting present from Henslow) he was picturing all the delights of fresh fruits in beautiful valleys when a shore boat came alongside and a small pale man announced that because of a cholera scare a twelve-day quarantine was in force. They could not wait so long, so it was 'Up jib' and away for St Jago in the Cape Verde Islands.

Darwin's brief included marine invertebrates, and on the nine days to their first landfall the net he put out to drag astern caught such quantities of minute creatures that he was kept fully occupied. John Clements Wickham, the First Lieutenant, was responsible for order and tidiness on board, and when specimens began to spread across Darwin's table and pile up in bottles and boxes he would curse and revile the 'Flycatcher' or 'Philosopher', as Darwin was nicknamed. 'If I were skipper, I would soon have you and all your damned mess out of the place!' he would growl. For all that, 'Wickham is a glorious fellow,' thought Darwin. The good Henslow had offered to receive his boxes and keep all his specimens in good order until his return.

They anchored at Porto Praya on the 16th of January. Viewed from the sea the place looked utterly sterile, so that over a wide sweep of the lava plains scarcely a green leaf could be seen. Volcanic fires of a past age and the scorching tropical sun had made the soil unfit for vegetation. Darwin went ashore with a party to announce their arrival to the *Governador*, and before returning to the ship walked across the town and came to a deep valley, and here for the first time he met the glories of tropical vegetation. Tamarinds, banana trees and palms flourished at his feet. Wild flowers were growing riotously everywhere. From Humboldt's description he had expected a great deal, so much that he was afraid of being disappointed. 'How utterly vain such fear is, none can tell but those who have experienced what I today have,' he wrote ecstatically. It was not only the graceful forms of the plants or the richness of their colours, it was the numberless and confused associations that rushed together on his mind. He returned to the shore, hearing the notes of unknown birds and seeing new insects fluttering about still newer flowers. 'It has been for me a glorious day, like giving to a blind man eyes, he is overwhelmed with what he sees & cannot justly comprehend it. Such are my feelings and such may they remain.' On the following day he went out collecting—selected specimens of rocks, marine animals, and plants, after which he 'returned to the ship heavily laden with my rich harvest, & have all evening been busily employed in examining its produce.'

His first letter home was full of the expedition. He recommended his father to procure some tropical plants, for the pleasure they would give him. Some time later the reply came. 'I got a Banana tree, it flourishes so as to promise to fill the hothouse. I sit under it and think of you in similar shade.'

They spent three weeks in the Cape Verde Islands, every day going on some expedition—to Quail Island, inland on a trip with Maccormick following a broad river which by good luck brought them to the celebrated Baobab trees. Darwin had read about them: some were supposed to be 6,000 years old. With two other officers he went on a riding expedition to Ribeira Grande, nine miles west of Praya, and to Santa Domingo. By the end of their stay he had collected some forty flowering plants. Among them were many lush tropical things, odd fruits like the Lablab or Hyacinth Bean (*Dolichos lablab*) and two forms of the Bladder Cherry (*Physalis alkekengi*), with enchanting shrubs like Barbados Pride (*Caesalpinia pulcherrima*) whose long stamens hung like a lady's eardrops. There were beautiful convolvulus, and at Santa Antonia a charming little plant carpeting the rocks with snowy flowers, *Paronychia gorgonocoma* whose European cousins we grow in our rock gardens.

But geology still carried the day, which was right and proper for the development of Darwin's theory of evolution, for it gave him his first clue to the past. He wrote to Henslow that St Jago's geology was, he believed, quite new, and that there were some facts that would interest Mr Lyell. The first volume of Charles Lyell's *Principles of Geology* was another of his indispensable reference books (recommended to him by Henslow who had advised him to read it but on no account to believe it). And now here he was making his own geological discoveries. A stream of lava had formerly flowed over the sea bed of triturated recent shells and corals. These it had baked into a hard white rock and the whole had then been upheaved. But the line of white rock revealed that later there had been subsidence around the crater, which had since been in action and poured forth fresh lava. This was a new and important fact, and it dawned on Charles that he might perhaps write a book on the geology of the various countries they would visit. That evening at dinner he confided his hopes to FitzRoy who at once encouraged him to make careful and copious notes of all he observed. The result, in 1842, 1844 and 1846, was three volumes: *Coral Reefs*, *Volcanic Islands*, and *Geological Observations on South America*.

On their way to Bahia they touched at St Paul's Rock ('a strangely bedevilled rock', Darwin called it.) 'This is a serpentine formation,' he declared. 'Is it not the only island in the Atlantic which is not *Volcanic*?' His assumption was correct. The remarkable conical hill on Fernando Noronha, their next call, was indeed volcanic. It was about 1,000 feet high, and a stay on the island of a few hours gave him time to climb it. Half-way up, great masses of the columnar rocks were shaded by laurel-like trees and ornamented by others covered with showers of a pink-flowered creeper. This was the way of the climbing plants on Fernando. They flung their garlands over every tree, even over the beautiful magnolias, in the forest that covered the entire island so thickly intertwining that it required great exertion even to crawl along. There are nine plants from Fernando Noronha preserved in Darwin's herbarium, which is at the Botany School, Cambridge. One was new. Henslow named it *Pisonia darwinii*. Its flowers were inconspicuous. As Darwin wrote: 'We had no gaudy birds, no humming birds, no large flowers.' He looked forward to Brazil where they would next land: there he would see the real grandeur of the tropics.

About 9 o'clock on the morning of the 28th of February they were near to its coast, looking at a broken line of trees and vegetation all of a brilliant green, and when they landed at Bahia even Humboldt's glorious descriptions fell far short of reality. 'The delight I experienced,' Darwin wrote, 'bewilders the mind; if the eye attempts to follow the flight of a gaudy butter-fly, it is arrested by some strange tree or fruit; if watching an insect one forgets it in the stranger flower it is crawling over; if turning to admire the scenery, the individual character of the foreground fixes the attention. The mind is a chaos of delight . . .'

Next day he was writing: 'Delight is however a weak term for such transports of pleasure: I have been wandering by myself in a Brazilian forest: amongst the multitude it is hard to say what set of objects is most striking; the general luxuriance of the vegetation bears the victory, the elegance of the grasses, the novelty of the parasitical plants, the beauty of the flowers, the glossy green of the foliage, all tend to this end.' A paradoxical mixture of sound and silence pervaded the shady parts of the wood, the noise from the insects being so loud that in the evening it could be heard in the ship anchored several hundred yards from the shore. Yet within the depths of the forest a universal stillness reigned. Next day again he could 'only add raptures to the former raptures'. He walked a few miles into the interior, and each new valley he came upon was more beautiful than the last. 'I collected a great number of brilliantly coloured flowers, enough to make a florist go wild.' The Brazilian scenery was nothing more nor less than 'a view in the Arabian Nights, with the advantage of reality.'

Among the plants he collected were the elegant *Maranta porteana* with its leaves bright green on the upper side and striped with white, rich purple beneath—a new discovery; two ageratums of celestial-blue flowers; two shrubby desmodiums whose flowers come purple, blue, rose, and white; the showy *Polygala paniculata*; two different pavonias whose heads of flowers were like delicate sea-anemones; *Lantana fucata* of rosy flowers in evergreen leaves; three euphorbias: in all more than fifty plants, with climbers and grasses such as *Eleusina indica* of tufted spikes, a rust-coloured millet and one of the Greek-named olyras.

Geology at this moment was far from his thoughts, though on the 14th of March he found some interesting geological structures and spent some pleasant hours in wandering on the beach.

At the end of the month they arrived at the Abrolhos, a group of five small rocky islands uninhabited but for vast numbers of birds. Two parties landed after breakfast and Darwin immediately began an attack on the rocks, insects and plants, adding nine more specimens to his herbarium. One was outstanding, an iresine whose inconspicuous flowers were compensated by handsome striking foliage.

Then it was Rio de Janeiro and its conical sugar-loaf hill. As the *Beagle* was returning to Bahia to correct an error in longitudinal readings, Darwin and Augustus Earle the ship's artist decided to explore the neighbourhood during the weeks she would be away. They found lodgings in a delightful house at Botofogo, a village some miles along the coast, and here, landing by the dinghy, an accident very nearly lost Darwin his equipment when huge waves swamped the boat and he saw to his dismay his precious books, instruments, gun-cases and everything else he possessed tossing amid the surf. Fortunately everything was retrieved and nothing completely spoilt.

He lost no time in finding a way of reaching the interior, and luck was with him. Patrick Lennon, an Irishman, was going to visit his coffee plantations on the Rio Macae, about 150 miles north of the capital. They set out on horseback on the 8th of April, a party of six. The day was powerfully hot and as they rode through the woods everything was motionless except for large and brilliant butterflies lazily fluttering about.

From the start of the voyage Darwin had been keeping records of what most impressed him, jotting these down in small pocket notebooks. On the 9th came his first excitement in plants. 'Started about $\frac{1}{2}$ after six and passed over scorching plains—cactuses and other succulent plants: on the stunted and decaying trees beautiful parasites—orchids with a delicious smell.' The plains lay between the sea and a wilderness of salt lagoons where egrets and cranes made a beautiful picture as they fished. The succulent plants were of the most fantastic shapes, but above all it was the 'wonderful beautiful flowering parasites' that impressed him. He collected hard, anything new that came his way: 'a frog and several Planorbis, Helix and Puccinea'. He saw more than a hundred buzzards in a flock. Then it was into the boundless forest, riding mile after mile in such torpid heat that he began to feel feverish and sick. By the end of the day he was really ill, though he still noted punctiliously: 'During the morning C. Frio appearing from refraction like inverted tumblers. Gneiss dipping to the south (and then the north)'. After a troubled sleep he 'nearly cured' himself by eating cinnamon and drinking port wine, but he was glad when evening brought them to Socego, the house of Signor Manoel Joaquem da Figuireda, father-in-law of one of the party who had come to visit him. His *fazenda* consisted of a piece of cleared ground where mandioca—the source of tapioca—was grown, sugar cane, rice and beans. It was still the days of slave labour and more than a hundred were kept here. One morning when Darwin rose before daylight to enjoy the stillness of the forest, the silence was suddenly broken by a Catholic morning hymn raised by all these blacks. The effect, he wrote, was sublime.

All the same, he abominated slave labour. FitzRoy defended the system. At Bahia, he told Darwin, he had visited a great slave owner who had called up many of his slaves and asked them whether they were happy, and whether they wished to be free, and all had answered no. Darwin asked whether he thought that the answer of slaves in the presence of their master was worth anything, at which FitzRoy flew into one of his rages, saying that as Darwin doubted his word they could no longer live together. But after a few hours he sent an officer with an apology and the request that he continue to mess with him. The two were to have several serious quarrels on the voyage, but each had an affection and respect for the other, and sometimes it was FitzRoy's magnanimity that healed the breach, and sometimes it was Darwin's steady commonsense.

They stayed at Socego for some days, and Darwin liked nothing better than to wander by himself in the forest. Henslow had sent him an engraving depicting one of these tropical forests, but Darwin thought it underrated rather than exaggerated the luxuriance. 'Nothing but the reality can give any idea how wonderful, how magnificent the scene is,' he told him. He liked to sit down and eat his lunch on one of the rotten trees that age had flung across the floor of the forest. The strange white boles of the living trees never failed to move him, the one brightness in a gloom so dense that only a glimmer penetrated from above, their lofty height reaching far upwards to where, unseen by him, their tops shone dazzling in the tropical sun. He noted the lianas binding the trees like ropes and binding each other. 'Twiners entwining twiners—tresses like hair,' he wrote. This was no mere poetry but scientific observation; and it is interesting, in view of the fact that one day he was to write a book about them, that he was so early captivated by climbing plants. One he measured. 'A creeper circum: 1 ft 4.' These scenes in the primeval Brazilian forest were to remain vividly in his memory as the most moving experience of the whole voyage.

Various expeditions took up his time until the *Beagle* returned to pick him up at the end of June: a long walk to the Gavia mountain where liliaceous plants were luxuriant, a ride to Tijeuka to see the waterfalls where he delighted in the wealth of ferns, and again and again into the depths of the forest. So much was new that he found one hour's collecting sometimes kept him fully employed for the rest of the day.

But he was troubled, and at Montevideo he unburdened himself to Henslow. First, in sending his specimens, he was afraid that Henslow would think the collection very small. 'But I have not been idle, and you

Southern portion of
South America,
showing Charles
Darwin's inland
expeditions

must recollect what a very small show hundreds of species make.' The box
he was sending contained a good many geological specimens: he had tried
to get samples of every variety of rock. 'If you think it worth your while
to examine any of them I shall be very glad of some mineralogical
information about them, especially on any numbers between 1 and 254.'

The orchid *Catasetum tridentatum* (*macrocarpum*). It was Darwin who solved the mystery of its three sexes

Calceolaria darwinii which Charles Darwin found in Fuegia in 1832. It is now a rock garden plant

He had made a duplicate catalogue. As to his plants, he was ashamed and despondent, feeling that it was useless to make a collection when he knew nothing about them. 'It is positively distressing to walk in the glorious forest amidst such treasures and feel they are thrown away upon one,' he wrote despairingly. Yet, he thought, his collection from the Abrolhos was interesting, and he suspected that it contained nearly their whole flowering vegetation. The same might be said of the collection he had made at St Jago, if only in that case because the flora was sparse. Thus he wrote, imploring Henslow's guidance. It took many months for letters to catch up with the ship, and he was to wait long to be told that his plants would be eagerly awaited by the botanists.

They arrived at Montevideo on the 26th of July, and after being imprisoned in the ship for a fortnight Darwin was glad to land. The scenery was very uninteresting: there was scarcely a house, an enclosed piece of ground or even a tree to give it an air of cheerfulness. But he found a charm in the freedom of walking over boundless plains of turf; and when he began to look closer he saw that the bright green sward, grazed short by the cattle, was full of dwarf flowers. There was one that looked like an English daisy: he greeted it as if it were an old friend. And, he wrote in his Journal, what would a florist say to whole tracts so thickly covered by *Verbena melindres* that it made a blazing carpet of the most glorious scarlet? (Alas, he could not claim this plant as his own, for it had been discovered in 1827 by the Scottish landscape gardener John Tweedie who now lived in Buenos Aires.)

To this city they moved up-river the following day. FitzRoy had heard of some interesting old charts there and wished to see them. The vastness of the river amazed Darwin: it was only just possible to see the north and south shores. Returning a few days later they found fighting going on. Although Uruguay had been created as a buffer state in 1828, Argentina and Brazil were still contesting possession, and it was not until the 13th of August that it was deemed safe to walk in the country. There was little time left for exploring, for with his surveying work finished here FitzRoy was anxious to press on southward.

They made their way down the coast in fine weather, arriving at Bahia Blanca at the beginning of September, there getting themselves entangled in the midst of shoals and mud banks and being rescued by a schooner which piloted them to safety in a sheltered bay. Her captain and part-owner was a Mr Harris who had two smaller schooners in the Rio Negro, farther south, and FitzRoy now conceived the plan of hiring them to help with the surveying, which would shorten their stay on the east coast. The news was hailed with joy by everybody. While this work was going on, Darwin was able to spend the rest of September and up to mid-October exploring.

The country was an undulating sandy plain covered with coarse herbage: he guessed that in summer it must be a desert. It was now the southern spring and all the flowers were in bud. He had a field day on the 14th of September when he added twenty plants to his collection, and it was interesting that the shrubs were evergreens, able to store moisture in their leaves against drought. The attractive *Schinus dependens* had bright green foliage and hanging flowers of yellowish-white which would turn to black berries. *Colletia longispina* told of another plant contrivance, this time against foraging animals, as did the bristly *Margyricarpus setosus* whose green flowers would be followed by most handsome white berries. There was a pretty little alpine, *Draba patagonica*, cushions of leaves covered with brilliant golden flowers, and among climbers was *Lathyrus tomentosus*, a cousin of the garden Sweet Pea. Before leaving Bahia Blanca he found three more species of pea. In later years when he came to write *The Movements and Habits of Climbing Plants* he worked out the fascinating life-history of these legumes, in this as in many other experiments extending his evolutionary theory into the world of plants.

By the time the *Beagle* left Bahia Blanca he had added nearly eighty plants to his collection. Among them were some beautiful grasses, one a delicate festuca whose name was *Vulpia tenella*; the Sweet Grass, *Melica papilionum*, and the Canary Grass *Phalaris* from which canary seed is obtained. It was now the height of spring: the birds were laying their eggs and the flowers were in full bloom. In places the ground was covered with the pink flowers of a wood sorrel, those of a wild pea—one of his lathyruses, and a dwarf geranium.

Meanwhile there was nothing extraordinary about the morning of Saturday the 22nd of September 1832. It dawned calm, bright and clear, where yesterday had been windy, though this was not unusual—the weather was mixed. FitzRoy suggested a cruise about the bay, and Darwin and Sulivan the second lieutenant (later Admiral Sir James Sulivan) fell in with the idea. Not that there was much scenery to look at: water and sky indistinctly separated by a ribbon of mud banks. They landed on Punta Alta, about ten miles from the ship, and here Darwin found some rocks, writing in his Journal: 'These are the first I have seen, & are very interesting from containing numerous shells and the bones of large animals.' The following day he returned to look for more fossils, '& to my great joy, I found the head of some large animal, imbedded in a soft rock. It took me nearly three hours to get it out. As far as I am able to judge, it is allied to the Rhinoceros.'

Farther on, was a bed of reddish clay 'containing much fewer shells—but armadillo,' as he wrote in his notebook. Armadilloes were plentiful in the neighbourhood. On a hunting expedition he had been able to examine one at close quarters, for the Gauchos had caught it and roasted it for their dinner. So he had no difficulty in identifying the extinct giant with its osseous compartmented coat. He was greatly excited, and deeply impressed, feeling as if he were looking back over his shoulder to the beginning of things.

They sailed north again for the Rio Plata, and at Montevideo he caught up with news from England in letters that were five months old. Among his mail was the recently published second volume of the

Principles of Geology in which Charles Lyell rejected Lamarck's theory that in past ages the world contained animals and plants entirely different from those of modern times. Lamarck had no proof of this: Lyell had evidence in the mammals and dicotyledons (plants of the highest development) he had found in ancient coal measures. Charles Darwin now had proof also—in his armadillo.

The *Beagle* had been absent from England for almost a year when on the 16th of December 1832 they reached the stormy shores of Tierra del Fuego and the Strait of Magellan. Charles was longing to see the southern limits of the great continent, but his first impression was of a land of valleys and trees which might have been pretty but for the gloomy sky and storm-haze that hung over them. Far to the south was a chain of lofty mountains whose summits glittered with snow.

On the 19th of January three whale-boats and the yawl left the ship with the dual purpose of surveying the Beagle Channel and returning to their native land the three passengers they had been carrying: York Minster, Jemmy Button, and a nine-year-old girl, Fuegia Basket. These were three Fuegians whom FitzRoy had picked up on the previous voyage, given their whimsical names (Jemmy had been purchased for the price of one pearl button) and for the past year been educating in England at his own expense. Like the fabled Indian princess Pocahontas they had been presented to the King and Queen. Now they were being returned to their homes to spread Christianity and civilisation among their countrymen. So FitzRoy fondly hoped. He detailed a shore party to find their relatives, and soon a group of savages appeared, one old man delivering himself of a long harangue inviting the three to stay with them. Poor Jemmy who had become quite a dandy, always wearing kid gloves and distressed if his polished boots were dirtied, was obviously painfully ashamed of his ragged and half-naked people, and by now he hardly understood a word of his own language. Darwin wondered what would become of him.

Next day he attempted to penetrate some way into the interior. There was no level ground and the hills were so thickly wooded as to be quite impassable. He therefore followed the course of a mountain torrent and was well rewarded by the grandeur of a ravine whose gloomy depths were in keeping with the violence of nature. In every direction were irregular masses of rock and uptorn trees, and for a moment the entangled mass of the thriving and the fallen reminded him of Brazil. There was a difference. 'For,' as he wrote, 'in these still solitudes Death, instead of Life, seemed the predominant spirit.'

The trees were mainly of one species, *Nothofagus betuloides*, the birch-leaved evergreen beech whose foliage was of a peculiar brownish-green tinged with yellow, so that the whole landscape was of this sad and sombre colour. In a letter to Henslow he mentioned a 'parasitical bush' he had found growing on these beeches. It was a species of *Myzodendron (brachystachyum)*, a remarkable genus and sole representative of the family. Darwin's specimen is preserved in the Kew Herbarium. Another discovery at Tierra del Fuego was a fungus which the Fuegians ate uncooked: it was named *Cyttaria darwinii* by the Rev. Miles Joseph

(*top left*) In Tierra del Fuego, the Land of Fire, Charles Darwin found a 'parasitical bush' (*Myzodendron brachystachyum*) growing on the Southern Beech trees

(*top right*) Another inhabitant of the Southern Beech (*Nothofagus betuloides*) was a new fungus, to be named *Cyttaria darwinii*. It was eaten raw by the Fuegians

(*bottom left*) *Acaena magellanica*, a discovery in Good Success Bay in the Strait of Magellan, is now a garden plant

(*bottom right*) He found *Chlorea magellanica* on Elizabeth Island in the eastern part of the Strait where, he wrote, the floras of Fuegia and Patagonia are blended

Berkeley, the great mycologist who dealt with all Darwin's fungi from the voyage. Beautifully preserved specimens are at Cambridge.

Darwin was glad the following day when, taking the same route, he left the gloomy trees behind him and came out into the open. Here between the forest and the line of perpetual snow that reached to the glittering white summits he was excited to find a band of peat carpeted with minute alpines. Some were evergreens like the little *Acaena magellanica* of russet and purple flower-heads; a diminutive myrtle with coin-like leaves, and the pretty creeping *Pernettya pumila* of solitary white flowers resembling a fairy campanula. Two had leaves as small as a cress (one was *Cardamine geranifolia*, the other *C. glacialis*), and there was a forget-me-not of just the size for the garden of a doll's house, with a little buttercup whose leaflets were borne in threes like a clover (it was later named *Ranunculus biternatus*). The tiny *Gunnera magellanica* formed carpets of lilliputian rhubarb leaves: its outsize cousin *Gunnera manicata* which we know as a handsome decoration at pond edges was not discovered until 1867. There was a beautiful new gentian, two charming violas, *Senecio candicans* of white woolly leaves, and two insectivorous plants—one a butterwort, the other a sundew. Again later, Darwin was to write a book about these insect-eating plants. With grasses, other alpines, and shrubs that included two barberries (*Berberis buxifolia* and *B. ilicifolia*) his collection amounted to some forty plants.

He took a professional pride in drying and pressing his specimens as Henslow had taught him to do, but in April had to write dolefully to Cambridge that 'nearly all the paper for drying plants is spoiled and half of this curious collection.' One of Tierra's murderous gales had sprung up, a sea stove in one of the boats, and there was so much water on the decks that every place was afloat—including the poop cabin, the paper and the plants. To everybody Tierra del Fuego was a detestable place.

Yet it was here that Darwin renewed the pledge he had made when leaving the shores of England, now deciding, as he wrote in his *Autobiography*, 'that I could not employ my life better than in adding a little to natural science.'

And this was in Good Success Bay.

It was now the Falklands. They arrived at Port Louis, the most easterly point, early in the morning of the 1st of March, to be greeted by the news that England had taken possession of the Islands. For some time uninhabited, Argentina had recently claimed them. Britain sent the *Clio*, and the Union Jack was now flying.

Darwin found East Falkland Island very dreary. The land, low and undulating with stony peaks and bare ridges, was almost completely covered by a brown wiry grass, and he could find few other plants. The largest bush had daisy flowers of yellowish-white and was scarcely as tall as English gorse. There was no moss and not a single tree—which surprised him, seeing that Fuegia was covered with them. The soil was peaty, yet no alpines were to be found. The difference between the East Falklands and the Fuegian group was striking.

In April they sailed back to the South American mainland and

Darwin was landed at Maldonado at the entrance of the Rio Plata. Here he was to spend ten weeks. He procured lodgings at the house of a well-known old lady called Donna Francisca and next day rode a few miles around the town. The country was similar to what he had seen at Montevideo but was rather more hilly. There was the same grass plain with its beautiful flowers and birds, the same hedges of cactus, and the same entire absence of trees. It seemed that the level Pampas lands were unfavourable to the growth of trees. He set himself the task of discovering why. Could it be the force of the winds or the kind of drainage? No such reason, however, was apparent: the rocky mountains around Maldonado afforded protection; there were various kinds of soil; streamlets of water were common in nearly every valley, and the clayey nature of the earth seemed adapted to retain moisture. It was known that the presence of woodland was generally determined by the annual amount of rainfall, and in this province abundant rain fell during the winter, while the summer, though dry, was not excessively so. This was a conundrum he continued to puzzle over during the long years of the voyage, adding pieces of information as he studied forest and desert lands, soils, rain-bearing winds and even geological formations, in the effort working out the fascinating story of geographical distribution—as when he compared the sparse vegetation of the East Falklands with the densely afforested Fuegia; and then Fuegia with the West Falklands, writing: 'Both the direction of the heavy gales of winds and of the currents of the sea are favourable to the transport of seeds from Tierra del Fuego, as is shown by the canoes and trunks of trees drifted from that country and frequently thrown on the shores of western Falkland. Hence perhaps it is that there are many plants in common to the two countries.'

Later, writing to Joseph Hooker, he was to declare that the geographical distribution of plants was 'that almost keystone of the laws of creation.'

4

The Plants of the *Beagle*

ii. Rivers and Mountains

In Patagonia Darwin saw plants growing 140 miles inland which seemed to him out of place. 'Are they not Cordilleras plants crawling downwards?' he asked.

One was Cruckshanksia glacialis, *a Chilean plant which he numbered 2042 in his collection.*

The following year, when he climbed the Cordilleras, he was able to visualise the formation of prehistoric South America. The plains were once inland seas; straits had joined the Pacific and Atlantic. The plants of Chile and Patagonia proved his point. They were 'absolutely the same.'

At 7,000 feet where no tree could grow he found a grove of silicified trees which told him they had once been submerged by the sea, and the land then heaved up to make these mountains. Thus he worked out, through vegetation, the origin of a continent.

T HE *BEAGLE* RETURNED from her surveying on the 24th of May, and
Charles was joyful when FitzRoy told him that he hoped to double
Cape Horn the following summer. 'My heart exults whenever I think of all
my glorious prospects of the future,' he wrote, and: 'My heart has revelled
with delight to hear the orders for getting 12 months provisions ready for
our next visit.'

They started their cruise to the Rio Negro on the evening of the 24th of
July, and with them went the schooners, one of which FitzRoy had now
bought (hoping that the Admiralty would recoup him), renaming her
Adventure in memory of the *Beagle*'s first sister-ship. At the mouth of the
Negro Darwin landed to explore more of the countryside while the *Beagle*
went off on another trip. He was accompanied by Syms Covington
('fiddler and boy to Poop Cabin') who was now his personal servant. The
times were so dangerous that FitzRoy had forbidden him to go anywhere
alone or unarmed, and as they rode up to Patagones they passed the ruins
of some fine *estancias* destroyed by the Indians.

It was Darwin's intention to make a traverse from Patagones to
Bahia Blanca and on to Buenos Aires. Between lay territory where
Indians roved, and he had to admit that there was justice in their
hostility, for every year the white men were annexing more of their land
for grazing cattle. The Argentine Government had now sent General Juan
Manuel Rosas to exterminate them, and this with a villainous army of
half-caste negroes, Indians and Spaniards he was actively doing, leaving
in his wake a chain of armed outposts. Rosas was now at Bahia Blanca,
and Darwin had to get his permission to travel inland. He set off on the 6th
of August with Syms Covington, Harris (of the little schooners), a guide,
five Gauchos and a troop of horses.

At first, a great flat plain offered little that was interesting in the
way of plants: it was almost entirely covered with prickly bushes. But
growing near Salinas, a large salt lake ('far more salt than sea', he noted),
were sea plants. On the following day they saw their first tree. A striking
object on the plain, it was reverenced by the Indians as an altar of their
god Walleechu. Being winter it had no leaves but on its many branches
hung strings (threads pulled from their ponchos), *yerba*, meat and other
offerings. Around it were the bleached bones of horses slaughtered as
sacrifices. That night was the first that Darwin spent under the open sky.
He found the Gaucho life exhilarating—'to be able at any moment to pull
up your horse and say, "Here we will pass the night." ' Saddle-cloths were
their bedding. On one occasion the cloths were frozen stiff by morning
after a night of heavy dew.

As they approached the Rio Colorado the plain gave way to turf
spangled with tall clover flowers. Willows marked the course of the river,
and crossing it they found the encampment of General Rosas who readily

gave the necessary passport and order for Government post-horses. After a brief stay they set off for Bahia Blanca where Darwin had a rendezvous with Captain FitzRoy who, he hoped, would consent to his going on by land to Buenos Aires.

While waiting for the *Beagle* to arrive he went to his old haunt of Punta Alta where in September of the previous year, 1832, he had discovered the fossil bones of his armadillo. Now, carefully digging around, he found the place a perfect catacomb of monsters of extinct races. They were all embedded in stratified gravel and reddish mud, and he came to the conclusion that the carcases of these animals, inhabitants of the surrounding countries in that far distant past, must have been carried down the rivers—then emptying separately into a vast bay, now united in the one great stream of the Plata—and their skeletons entombed in the estuary mud which was then tranquilly accumulating.

First came parts of three heads and other bones belonging to a megatherium, largest of the ground sloths and at twenty feet long bigger than an elephant, then bones of a megalonyx, another ground sloth. Skeletons of these two edentates had already been discovered, and in 1804 Georges Cuvier, the brilliant colleague of Jean Baptiste de Lamarck, had correctly described them. Both had been huge woolly animals of grotesque shape—bear-like bodies with tapering snouts like a camel—when they lived here some 350,000 years ago.

Darwin's third tremendous find was the nearly perfect skeleton of a scelidotherium, a type of ant-eater as large as a rhinoceros. This was an original discovery. Find after find came out of the past, the ninth and last the remains of a toxodon, a huge swamp-wallowing animal with hooves. All enormous creatures, Darwin wondered what their food had been. The teeth of the megatheroids indicated that they had probably lived on the leaves and twigs of trees, while the colossal breadth and weight of their hindquarters, together with their thick tails and enormous heels, told him that the trees must have been strongly rooted to have withstood them as they reached up with their powerful arms and great claws. The general assumption amongst naturalists was that large animals required a luxuriant vegetation. Darwin thought this was completely false, arguing that in any book of travels describing the southern parts of Africa there were allusions on almost every page either to the desert character of the country or to the numbers of large animals inhabiting it. There was also the camel, an animal of no mean bulk, which was the emblem of the desert and lived on very little. He decided that, so far as the quantity of vegetation was concerned, the great quadrupeds of the later Tertiary epochs might, even on the steppes of Siberia, have lived on the spot where their remains were now found. But, he pointed out, 'I do not here speak of the *kind* of vegetation necessary for their support; because, as there is evidence of physical changes, and as the animals have become extinct, so it may be supposed that the species of plants have likewise been changed.'

Charles Darwin was well on his way to thinking out his Theory.

On the 8th of September, after a grand reunion with his shipmates and with FitzRoy's blessing on his enterprise, Darwin and Syms Covington set off with a guide on a 400-mile expedition where each *posta* in the slender chain was under threat from the Indians. So at each, after changing horses, one or two soldiers convoyed them to the next.

It was remarkable how the vegetation changed after they crossed the Salado river. From coarse herbage on one side they came to a carpet of fine green grass on the other. Darwin thought it must be something to do with the soil, though the townspeople of the Guardia del Monte assured him that it was due to the grazing and manuring by the huge herds of cattle. Near the Guardia he found the southern limit of two European plants. One was the fennel, the other the Cardoon, *Cynara cardunculus*. North, in the neighbourhood of Montevideo and other towns he had seen fennel blanketing ditch-banks, the thistle-like Cardoon covering square miles.

After a long day's ride over the rich green plain where here and there was a solitary *estancia* and its one ombu tree, *Phytolacca dioica* of evergreen leaves and buttressed trunk, they came to a post-house whose owner refused to let them stay unless they could produce a proper passport. There were so many robbers about that he would trust nobody. When, however, he read the magic words 'El Naturalista Don Carlos', the title conferred on Darwin by General Rosas, smiles and civility were as unbounded as previous suspicions, though Darwin suspected that the man had not the least idea what a naturalist might be.

Buenos Aires greeted them like an English spring, the outskirts of the city burgeoning with peach and willow trees all throwing out their fresh green leaves. They rode to the house of a friendly merchant by the name of Edward Lumb, and there for five days rejoiced in the comforts of an English home before setting out on an excursion to Santa Fé, nearly three hundred miles away on the banks of the Parana. Beyond Areco the *estancias* were few and far between, there being little good pasture. Most of the land was acrid clover or choked by giant thistles which when they were fully grown (to the height of a man) were the hiding places of bands of robbers—as Darwin learnt when he asked at a house whether robbers were numerous and received the reply: 'The thistles are not up yet.' However, he wrote himself a cautionary note: 'For future Pistol in hand; not leave Guide.'

The luxuriant gardens of Corunda made it one of the prettiest villages he had seen in South America; but from here to Santa Fé, his guide warned him, the road was notorious for marauding Indians who were able to pounce on luckless travellers from their hiding places in open woodlands of prickly mimosa trees. Darwin was glad to reach Santa Fé and was surprised that it had a much warmer climate, although between here and Buenos Aires there was a difference of only three degrees latitude. The brilliantly coloured birds and flowers reminded him of Brazil. There were orange trees and immense ombus, and he was delighted to find a number of new cacti and other plants to add to his collection. Some were in seed. He was always careful to take seeds when he could get them, thinking of new beauties to be added to English gardens. Alas, his plant hunting was cut short when he was confined for two days to his bed by an overpowering headache. A good-natured old woman came to help him, but as her remedies smacked horribly of witch-doctoring he preferred to keep his headache.

It was an attack of fever, and after crossing the Parana river, a tortuous passage of four hours of winding about its various branches, he was glad to take refuge in the house of an old Catalonian to whom he had a letter of introduction. His intention had been to cross the province of Entre Rios and return to Buenos Aires by way of the Banda Oriental. But a walk to the Barranca left him feeling so weak that he decided to shorten the expedition, and thinking that the *Beagle* was sailing earlier than she actually did he made up his mind to go by river. Darwin always reckoned the size of a river by comparison with his Shrewsbury Severn. The Parana was 'generally as broad as the Severn and much deeper and more rapid'.

It should have been a speedy passage, but the *balandra* or one-masted vessel had a nervous navigator. No sooner had they started than the weather blew up and he decided to moor the boat to the branch of a tree growing on one of the many muddy islands which were thick jungles of willows and other trees tangled with creepers, thickets in which capybaras and jaguars found a retreat. Deciding to explore the island, Darwin had not got very far when he came upon fresh jaguar tracks. Reluctantly he went back to the boat. A gale delayed them another day, this time not so annoyingly, for he discovered that the river cliffs were a mine of important bones.

The voyage to the mouth of the Parana took eight long days, and at Las Conchas, a village just outside Buenos Aires, they left the *balandra* with the intention of riding into the city. On landing, rebel soldiers surrounded them and he found to his dismay that he and Covington were virtually prisoners. Violent revolution had broken out and the ports were under an embargo. He could neither return to the boat nor go into Buenos Aires. In a fright now that the *Beagle* would sail without them he argued his way to the presence of the rebel commander who finally agreed to give him safe conduct to the bridge, but not his servant, their guide, or their horses. How Darwin tricked his way past the sentinels by showing them an out-of-date passport, and then bribed a man to smuggle Syms Covington into Buenos Aires, made exciting hearing for his shipmates when on the 4th of November, after a stormy voyage in a crowded packet bound for Montevideo, he saw with relief the spars of the *Beagle* and at last climbed aboard.

The ship was not to sail for another month. Captain FitzRoy had collected so much fresh information with the help of his two auxiliary schooners that he needed time to tabulate the new knowledge in his charts. Darwin seized the opportunity to ride to the River Uruguay and its tributary the Negro, a fine river of blue water almost as large as its namesake farther south. Geologically the trip was interesting; botanically there were at first more immense beds of the thistle and Cardoon he had seen on the Pampas lands. Then came good coarse grass growing as high as his horse's belly. Yet there were square miles without a single head of cattle. He thought that if well stocked the province of Banda Oriental could support an astonishing number of animals, and of course history has proved him right.

He stayed for a few days with a very hospitable Englishman to whom his friend Edward Lumb had given him an introduction, at this *estancia* and others learning the life of the country with its cattle and trade in hides. All the way along he had been collecting seeds, and had bundles of them to despatch to Henslow. He had noticed that among the unfamiliar plants were ones he knew in England as friends of the hedgerow and meadow, or enemies of the garden, which must by some means or other have found their way to South America. A thought occurred to him, and writing to Henslow on the 24th of November he said: 'It will be a Botanical problem to find out to what country the weeds belong. It might be curious to observe whether European weeds have undergone any change by their residence in this country.'

This was an early fruit of Henslow's teaching in the realm of geographical distribution, of which Darwin was to become a master. It was more: it was an early excursion into the realm of the modification of species under different conditions of existence.

On the 6th of December they sailed from the River Plate, never again to enter its muddy waters. The next seventeen days were a magic part of the voyage. The weather was fine and calm, and one evening when they were about ten miles out from the Bay of San Blas they found themselves in the midst of a cloud of butterflies 'in banks or flocks of countless myriads, extending as far as the eye could range.' Even with a telescope it was not possible to see anything but butterflies. The seamen called out that it was 'snowing butterflies', and so it seemed. Then on one very dark night the sea presented a wonderful, most beautiful spectacle. A fresh breeze was blowing and every part of the moving waters flashed with phosphorescent light. Before her bows the ship drove two billows of liquid fire and her wake was a milky train. The crest of each wave was bright. It was just as if the whole ocean were alight with white flames.

Ahead, Tierra del Fuego awaited them. They were to pick up the coastal survey at Port St Julien, and crew and officers went about their ship duties with high hopes that in a few more months they would have finished charting the interminable east coast and the hated regions of Tierra del Fuego. Its name meant the Land of Fire, from the camp fires of the local Indians, called the Canoe People, by which they kept themselves warm throughout the night. Ships sailing in the dark saw these fires all along the coast. As to the natives, to all on board the *Beagle* it was a desolate land of tempestuous storm, cold and wet.

They put in at Port Desire, as the *Adventure* was found not to sail well and her sails were to be altered. As always, Darwin took the opportunity to go ashore. The yawl under the command of Edward Main Chaffers, the *Beagle*'s master, was being sent to survey the head of the creek.

Darwin always found the first landing in any new country interesting, especially when it was of a different character. Here, two and three hundred feet above some masses of red porphyry rock, stretched a level plain cut up into valleys. In the week ashore he found six adesmias—shrubs with pea-flowers—nine different grasses, a broom, a glorious fern (*Polystichum adiantiforme*) with fronds up to three feet long, a geranium, and a calceolaria which here was a hardy rock plant but in England was to be grown in greenhouses, and the Evening Primrose *Oenothera dentata*. Four plants were new and were named after him: *Panagyrum darwinii* which had daisy-like flowers; *Chiliotrichum darwinii*, another composite; the shrubby *Baccharis darwinii*, and the cactus-like *Opuntia darwinii* which Henslow described in the *Magazine of Zoology and Botany*, Volume I, page 466. Darwin found it growing on the arid gravelly Patagonian plain, not far inland. Its solitary yellow flowers grew in spirals up its fleshy stem, and lest we should think, even now, that Darwin knew little about botany, or was not yet interested in plant behaviour, let me quote what he did. He found that the opuntia 'was remarkable by the irritability of the stamens, when I inserted either a piece of stick or the end of my finger in the plant. The segments of the perianth also closed on the pistil, but more slowly than the stamens.' He was simulating the touch of a pollinating insect. Like Darwin's fungi it was bottled in spirits of wine, as were his cacti, the only way of preserving succulent plants. It is preserved at Cambridge.

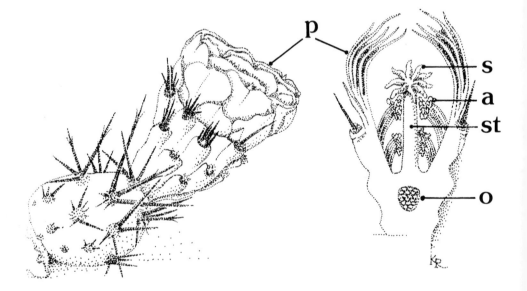

Darwin's first early experiment was on the *Opuntia* named for him by Henslow. Poking its flower with a straw he simulated the touch of a pollinating insect

But just as interesting to Darwin as the unfamiliar and new plants, was finding the same Thrift that grew by the sea at home, and the little creeping *Veronica peregrina* which had certainly made a long peregrination. Now widespread in Europe and first noted in England about the year 1680, the speedwell's original home was North America.

The *Adventure* not being ready on the 4th of January, FitzRoy decided to run down to Port St Julien, about 110 miles to the south, to survey some of the coast between. The terrain was the same table-land as at Port Desire and yielded little of interest, some seventeen plants including two grasses, and nothing that was new or interesting. On the 22nd of January 1834 the *Beagle* and *Adventure* stood out to sea, four days later passing the white cliffs of Cape Virgins and entering the Strait of Magellan, the stretch of difficult water between the South American mainland and Tierra del Fuego, nearly 400 miles long and varying in width from 2½ miles to 17 miles. After days of beating against strong westerly gales they anchored in St Gregory Bay where Darwin was amazed at the height to which the tide rose—between 40 and 50 feet. Who, he asked, could wonder at the dread felt by the early navigators of this Strait?

Elizabeth Island gave him *Anemone decapetala* which, however, he had seen at Cape Negro the year before. But now there was a real find in the yellow 'tobacco-pouch' flower we all know with chestnut spots on its lower lip. It was named *Calceolaria darwinii*.

Thinking to make a base at Lando Bay they anchored there, but as no good drinking water could be found they sailed on to Port Famine. Although scenically the place was thoroughly uninteresting Darwin's attention was aroused when he studied the plants, finding them an intermixture of Patagonian and Fuegian. The climate was intermediate and many of the plants were common to both.

He could never resist climbing a mountain. At Port Famine he left the ship at four in the morning for Mount Tarn, the highest point at 2,600 feet, finding at 2,000 feet *Cerastium arvense*, the Field Mouse-ear, a little plant with narrow downy leaves and white flowers, widespread in England. It was one more wanderer noted by Darwin.

From the 14th of February to the 21st a complete survey was made of the east coast of Fuegia. They landed only once, in a large wild bay where the country resembled Patagonia in its openness and absence of trees. But it was pretty and park-like. Ponsonby Sound where they anchored on the 4th of March was entirely different. Rising abruptly from the water's edge to 3,000 feet were snow-clad mountains densely clothed with forest on their lower slopes and breaking the sky-line with a series of sharp peaks. They looked very grand and splendid when the afternoon sun turned their snow to rosy-red. One, the highest at 7,000 feet, was unnamed. Captain FitzRoy made his mess-mate a present of it, calling it Mount Darwin.

They sailed down to Woolya, Jemmy Button's territory. It was a populous part of the country and they were followed by seven canoes. Arriving at the spot where they had left their three passengers nearly fifteen months before, they were alarmed to see some Fuegians arming themselves with bows and arrows. Then a canoe came towards them with a flag hanging out. In it was poor Jemmy Button—thin, pale, and without a remnant of clothes except for a piece of blanket round his waist. His hair hung over his shoulders and he was so ashamed of himself that he turned his back to the ship. Aboard he told them that York Minster had robbed

Cerro de Campana or
Bell Mountain in the
Cordillera. A plaque
on its summit records
Darwin's climb

him of everything he possessed and run away with Fuegia Basket. But he
made light of his troubles: he had got a young and nice-looking squaw and
had no wish to return to England. He lighted a farewell signal fire as the
ship stood out of the Sound on her course to East Falkland Island.

Here early one morning Darwin set out on an expedition with two
Gauchos and six horses. The country was uniformly the same: undulating
moorland covered with a light brown withered grass and a few very low
shrubs growing out of a light peaty soil. Lichens were abundant and he

was delighted with an aster whose 'pale most beautiful' flowers were of 'auricula purple'. Three other composites had flowers of 'tile red', 'deep orange brown' and 'beautiful vermilion red'. He collected fourteen plants, one with silvery leaves to be named by William Jackson Hooker of Glasgow *Senecio darwinii*. The return journey to the ship was in a rainstorm which quickly turned the peaty soil into treacherous bog. Darwin's horse fell at least a dozen times, and sometimes all six were floundering in the mud together. To complete their misery they took a short cut by crossing an arm of the sea which was up to the top of the horses' backs. The wind was blowing violently and the waves broke over them. Even the hardened Gauchos were not sorry to reach their homes.

After her months of hard sailing and buffeting, FitzRoy decided to beach the *Beagle* so that her hull could be examined. They sailed to the mouth of the Santa Cruz river, and there she was laid aground. On two long walks Darwin found the plants, birds and animals to be the same as in other parts of Patagonia. The country remained uninteresting with level plains of arid shingle supporting the same stunted and dwarfed plants, the valleys the same thorn-bearing bushes. Even the banks of the river and streamlets were scarcely enlivened by a brighter green. So that he wrote of it: 'The curse of sterility is on the land.'

Now came a longer trip, when FitzRoy decided to send three whale-boats up the Santa Cruz river as far as they could navigate. They were away for eighteen days, on the 4th of May reaching a point about 140 miles from the Atlantic and 60 from the nearest inlet of the Pacific, on the 29th of April climbing to higher land, hailing with joy the snowy summits of the Andes which they could see occasionally peeping through their dusky envelopes of clouds.

It had been a laborious track up-river. The descent was rapid. Leaving on the 5th of May they shot down at ten miles an hour, covering in one day what had cost them five and a half days going up. They arrived back at the mouth of the river on the 8th, to find the *Beagle* with her masts up, fresh-painted and looking as gay as a frigate. A little after noon they went aboard, everybody disappointed with the expedition, much time lost and hardly anything seen or gained—everybody except Darwin to whom the cruise had been one of tremendous discovery, producing evidence of the great modern formation of South America. Finding sea-shells in the bed of the river and on the steppe-like plains, he was able to prove that in past ages South America had been separated by a strait joining the Pacific and Atlantic oceans. His bold theory, which he worked out with a mass of geological observations, was later hailed by the geologists as the explanation of what had been an unanswerable mystery. The plants provided an impelling clue. Darwin always numbered each specimen, supplying notes of any he thought remarkable. Against No. 2042, *Cruckshanksia glacialis* (now named *Oreopolus glacialis*) which was a *Rubus* named in honour of Alexander Cruckshanks who had collected in the Andes, he wrote: 'Plants 140 miles up the river: character of country same as at coast. as these plants, I never saw to this Coast are they not Cordilleras plants crawling downwards.'

It was not an idle query, as later he was to see.

Back at Cape Virgins it was to hear that there had been trouble with the Gauchos who had ferociously murdered five of the English residents on Tierra del Fuego, and soon after they left Berkeley Sound a British man-o'-war came in. On board was mail from England. Darwin's letters from home were dated October and November of the previous year.

The days were short now for surveying and on the 8th of June with dark ragged clouds massing almost to the bases of the mountains they weighed anchor, two days later passing out of the Strait between the East and West Furies to meet the long swell of the Pacific. The 28th of June saw the *Beagle* safely at anchor in the port of Santo Carlos in the Island of Chiloe.

In gardens where shrubs are treasured for their special beauty you will be sure to find one with tiny holly-leaves that in spring covers itself with tight little golden flowers ripening to grape-like berries, purple and dusted with bloom. 'We were driven into Chiloe by some very bad weather,' Charles Darwin wrote to Henslow, his 'Father in Natural

Four Darwin plants figured in Hooker's *Flora Antarctica*

(*below left*) *Hamadryas tomentosa*

(*below right*) *Calceolaria darwinii*

(*far left*) *Muhlenbergia rariflora*

(*far right*) *Asterina darwinii*

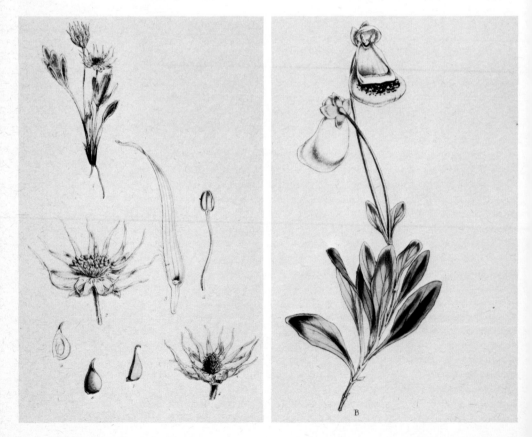

B

History'. It was there on this volcanic island off the coast of Chile that he made his find. Named *Berberis darwinii* by the future Director of Kew, William Hooker, 'Mr Darwin's Berberry' is now deservedly one of the most popular shrubs in cultivation.

Seen from a distance Chiloe bore a strong resemblance to Tierra del Fuego, hilly country clothed in forest. But the woods of 'the fragrant-leaved sassafras' and noble trees of Winter's Bark (*Drimys winteri*) were incomparably more beautiful. Instead of dusky uniformity there was the variety of tropical scenery, and except in Brazil Darwin had never seen so many elegant forms of flowers and leaves. On the high slopes of San Pedro he found again his old friend the Fuegian beech, though here they were 'poor stunted little trees, & at an elevation of little less than 1000 feet.' From their dwarfed appearance he thought this must be nearly their northern limit. Unless of course this was a new species.

The weather at first was delightful, but in a few days he was wondering if any part of the world was so rainy, and by mid-July they were all glad to leave, for at this time of year 'nothing but an amphibious animal' could tolerate its climate. No wonder such beautiful ferns grew there! He collected *Asplenium obtusatum*, the magnificent *Alsophila quadripinnata*, and *Polypodium ignaminia*. Late in the night of the 31st they anchored off Valparaiso. The morning sky was so clear and blue, the air so dry and the sun so bright, that all nature seemed sparkling.

To the joy of everyone the mail caught up with them at Valparaiso. There was a boxload for Darwin, in which were two letters from Henslow. He sat down immediately to answer them. The date was 24th July 1834.

At this spot Charles Darwin looked at the unfamiliar vegetation and asked: 'Are they not Cordilleras plants crawling downwards?'

You do not know how happy they have made me.—One is dated Dec. 12th, 1833, the other Jan. 15th of the *same year*! By what fatality it did not arrive sooner, I cannot conjecture: I regret it much: for it contained the information I most wanted about manner of packing etc etc: roots, with specimens of plants etc etc: this I suppose was written after the reception of first cargo of specimens.—Not having heard from you untill March of this year; I really began to think my collections were so poor, that you were puzzled what to say: the case is not quite on the opposite tack; for you are *guilty* of exciting all my vain feelings to a most comfortable pitch; if hard work will atone for these thoughts I vow it shall not be spared.

He added, thinking of his harvest in the Buenos Aires region: 'Have any of the B. Ayrean seeds produced plants?' The seeds and dried plants were being collected at Henslow's request for the Cambridge Botanic Garden herbarium which he was anxious to build up. The instructions Darwin had long been waiting for were in the following page of Henslow's January letter:

So far from being disappointed with the Box—I think you have done wonders—as I know you do not confine yourself to collecting, but are careful to describe. Most of the plants are very desirable to *me*. Avoid sending *scraps*. Make the specimens as perfect as you can, *root, flowers & leaves* & you can't do wrong. In large ferns & leaves fold them back upon themselves on *one* side of the specimen & they will get into a proper sized paper. *Don't* trouble yourself to stitch them—for the [they] really travel better without it, and a single label *per month* to

those of the same place is enough except you have plenty of spare time
or spare hands to write more.

He drew a sketch showing how a large leaf should be folded back at the
edges of one side.

Previously, in a letter dated 31 August 1833, which reached Darwin at
the East Falkland Islands, Henslow had written: 'The plants delight me
exceedingly, tho' I have not yet made them out—but with Hooker's work
and help I hope to do so before long.' He was referring to Glasgow's
professor of botany, Dr William J. Hooker (to be knighted in 1836 for his
services to botany). He had written to Hooker in January, telling him
that he had received some plants from his friend Darwin, including fungi
and a few algae, chiefly from Rio de Janeiro and Montevideo. 'I must try &
make them out somehow,' he wrote. In another letter: 'Your Flora will be
pre-eminently useful in making them out.' This was the 'work' he
referred to, being 'Contributions to a Flora of South America and the
Pacific Islands', a feature which Hooker and George Arnold Walker
Arnott ran in the *Botanical Miscellany* and its continuation *The Journal of
Botany*, and later in the *Companion to the Botanical Magazine*. Hooker's
correspondence with botanists and travellers was world-wide and full of
valuable information; his herbarium, already the largest in the world,
was teeming with plants new to science. To make these known he founded
all three journals.

By March 1834 Henslow at last got down to looking at Darwin's plants.
But he did not get very far, even referring to the *Botanical Miscellany*. 'I
see there are some among them which you have not noticed,' he told
Hooker. There was a very small *Ranunculus* he thought was new, some
fine specimens of *Anemone triternata*, and some unfamiliar crucifers.

From the East Falklands in March 1834, thinking all was well, Darwin
had replied to Henslow's August 1833 letter:

I am very glad the plants give you any pleasure, I do assure you I was
so ashamed of them, I had a great mind to throw them away; but if
they give you any pleasure I am indeed bound, and will pledge myself
to collect whenever we are in parts not often visited by Ships and
Collectors.—I collected all the plants, which were in flower on the
coasts of Patagonia at Port Desire & St Julian; also on the Eastern
parts of Tierra del Fuego, where the climates & features of T. del
Fuego and Patagonia are united. With them are as many seeds, as I
could find (you had better plant all the rubbish which I send, for some
of the seeds are very small).—The soil of Patagonia is *very* dry,
gravelly & light,—in East Tierra, it is gravelly—peaty & damp.

He was anxious that his plants should be grown in their natural
conditions.

With no homeward-bound ship at the East Falklands he had added:
'There is no opportunity of sending a cargo: I only send this, with the
seeds, some of which I hope may grow, & show the nature of the plants far
better than my Herbarium.' He was much more interested in living plants

than in dried specimens. But now, having Henslow's further letter he was confident that his herbarium might be of some scientific value.

While the *Beagle* and *Adventure* went on with their work of charting the waters around Valparaiso he was able to make plans to stay ashore. He asked Henslow to continue sending letters to him here, addressing them care of Richard Corfield, an old Shrewsbury school friend he was delighted to find living in the suburb of the Almendral. He took up residence with Corfield on the 2nd of September, in the letter to Henslow telling him that in the box he was sending 'there are three small parcels of seeds: the one in the oblong box I have labelled as coming from T. del Fuego. It comes from Chiloe: (Climate etc etc like T. del Fuego but considerably warmer)—I do not much expect, that any one seed will grow.'

Not all the seeds did survive, particularly those packed in the same box as skins treated with arsenical soap or camphor. But some did, and these are recorded on his herbarium sheets as having 'flowered 11 August 1836', 'flowered 24 Sept 1836', and so on.

He went exploring, finding the immediate neighbourhood of Valparaiso unproductive. During the long hot summer no rain fell, and consequently the vegetation was scanty. Farther afield he was rewarded. 'There are many very beautiful flowers,' he wrote in his *Journal*, 'and, as in most other dry climates, the plants and shrubs possess strong and peculiar odours, even one's clothes by brushing through them became scented.'

On the whale-boat trip up the Santa Cruz river he had glimpsed the glories of the Andes from the east side. Now viewing them from the other he planned to climb them. The expedition was to be of tremendous importance to geology. A piece of fossil botany was to supply the clue to his discovery. It happened in two stages: one when he set off on the 14th of August to explore the lower reaches and climb the 6,400-foot Campana or Bell Mountain; and more importantly when, after returning to Valparaiso in March 1835, he again penetrated the Andes, this time on a journey of over 500 miles.

His companion and guide was Mariano Gonzales. They travelled with four horses and two mules, occasionally sleeping at hospitable *haciendas* but more often under the open sky.

The first day's ride was northward along the coast to Quintero. Darwin had been told about great beds of shells which the local people burnt for lime. They were spread some yards above sea level. Up a few hundred feet the same old-looking shells were still plentiful, and climbing up he found some at 1,300 feet—proof of the elevation of the whole coast. Further proof came when he reached the summit of Bell Mountain and was able to see how the narrow strip of land between the Andes and the Pacific was traversed by mountains running parallel with the great Andes range, and that the succession of level plains between them must anciently have been inlets and deep bays of the sea. Darwin's historic ascent is commemorated in a plaque on the mountain's summit.

San Fernando, 120 miles from St Jago, was his farthest point south, and here he turned seaward again, near Rancagua passing through extensive woods of the Roblé or Chilean oak which he noted was different from the Roblé of Chiloe. This was its most northerly limit of growth.

He was not feeling well, but he struggled on, collecting as he went, until he reached Casa Blanca where he sent to Valparaiso for a carriage, thankfully reaching Richard Corfield's house the next day. Here he had to remain in bed until the end of October, counting it a grievous loss of time. During his absence changes took place in the affairs of the *Beagle* expedition. The *Adventure* had saved much time in the work of charting, and FitzRoy was hoping he could keep her. The Admiralty, however, refused to purchase her and FitzRoy perforce had to sell her. Dispirited as he was, he delayed the sailing of the *Beagle* till the 10th of November, by which time Darwin was well again.

They were not back at Valparaiso until the 11th of March in the following year, 1835, when Darwin had thoroughly explored Chiloe by going round it and crossing it in two directions. There was much to see. On the large island of Tanqui he noticed one day growing on the soft sandstone cliffs some very fine plants of what the natives called Pangi. It was *Gunnera chilensis* (*scabra*), which we know as one of the handsome species of the genus for growing by the side of a stream or pond. Its introduction date is given as 1849, and obviously Darwin did not make a herbarium specimen of it, for its leaves were gigantic: one was nearly eight feet in diameter with a circumference of twenty-four feet, its stalk being more than a yard long. A bold rocky hill near Punta Huantamó was covered by a plant locally called Chepones. Darwin believed it to be allied to the bromeliads. Its fruit, looking like an artichoke, was packed with seed-vessels containing a sweet pulp which the Chilotans used for making *chichi* or cider. In scrambling through beds of it everybody's hands were badly scratched by its prickly leaves. Darwin was amused when their Indian guide turned up his trousers, as more in need of protection than his own hard skin.

Running along the coast from Cape Tres Montes they found a harbour which Darwin thought might be of great use to vessels in distress navigating this dangerous coast. It could easily be recognised by a perfectly conical hill which quite beat the famous Sugar Loaf at Rio de Janeiro. He succeeded in reaching its 1,600-foot summit, a laborious undertaking, the ascent being so steep. But by using the trees as ladders he reached the top, with difficulty crawling through thickets of fuchsia covered with their beautiful scarlet flowers.

He found wild potatoes growing everywhere on the Chonos Islands near the beaches. The tubers were oval and generally quite small, but though they shrank when boiled they made good eating. He found a new sedge, to be named *Carex darwinii*, and in the woods the fragrant-leaved *Myrtus luma*. This beautiful tree with its bole of shining cinnamon is now one of the most ornamental of any in our gardens, though growing only in the mildest districts. It was one of Darwin's discoveries but his seeds must have perished, for its introduction date is given as 1843. There were ferns, rushes and a new shrub in *Lomatia ferruginea* which proved hardy in Cornwall—its shoots were coated with rust-coloured velvety down, and its flowers were golden-yellow and deep rose touched with scarlet. Yet another wonderful discovery was *Mitraria coccinea* of bright scarlet tubular flowers and seed-pods in the shape of a bishop's hat, an attractive evergreen we grow in sheltered gardens. Another Chonos plant named after Darwin was the nettle *Urtica darwinii*.

Darwin at this stage was interested as much in the situation of a plant as in the plant itself. Thus on finding in the central part of the Chonos Archipelago two little peat-forming plants (*Astelia pumila* and *Donatia magellanica*) which covered every patch of level ground, he remembered that he had seen the first in Tierra del Fuego where again it had formed an elastic peat. At Cape Tres Montes he found the Sea Pea (*Lathyrus maritimus*), and this was an example of a species restricted to the Northern Hemisphere, including local places in Britain, but occurring unexpectedly in southern Chile.

On the 4th of February (1835) they sailed for Valdivia, there on the 20th experiencing the worst earthquake that the oldest inhabitant could remember. The whole town was shaken violently, though none of the houses actually fell. To Darwin 'An earthquake like this at once destroys the oldest associations,' as he wrote. 'The world, the very emblem of all that is solid, moves beneath our feet like a crust over a fluid, one second of time conveys to the mind a strange idea of insecurity, which hours of reflection would never create.' Concepcion, which they reached a month later, had borne the brunt of the earthquake. The coast was strewn with timber and furniture as if a thousand great ships had been wrecked. Not a house was left standing. To add to the people's miseries a great wave had afterwards surged over the town.

To Darwin the most remarkable effect of the earthquake was the permanent elevation of the land. Around the Bay it was raised up two or three feet, as he saw by a rocky shoal which had been under the sea. Thirty miles away it was ten feet. Huge waves from previous earthquakes had carried shells on to land which was now 600 feet above sea level. At Valparaiso he had found shells at 1,300 feet.

They were back at Valparaiso on the 7th of March, and Darwin began preparations for crossing the Andes, setting out on the 18th with his companion Mariano Gonzales, and an *arriero* with ten mules and their *madrina*. This was a mare with a little bell round her neck who was a sort of stepmother to the troop, keeping the mules together at night, as well as by day when grazing. They started for the 12,000-foot Portillo Pass, with Mendoza as their first goal.

The Andes here consisted of two principal ridges, of which the Puquenas formed the division of the rivers and the republics of Chile and Mendoza. East of this was an undulating track with a gentle fall, and then the second line of the Portillo. Darwin first tackled the Puquenas, a tedious ascent rewarded by finding fossil shells on the highest ridge, before travelling across the intermediate tract to the foot of the range, on whose lofty heights he again found shells.

From the highest point he could see the vast plains which uninterruptedly extended to the Atlantic Ocean. This command of distance, because of the crystal clarity of the air, always astonished travellers.

He had been struck with the marked difference between the vegetation of the Chilean valleys and that on the eastern side. Coming down from the Portillo Pass he was struck by the similarity of the plants with those of Patagonia. We remember that on the whale-boat trip up the Santa Cruz river he had come upon alien plants and asked himself: 'Are they not Cordilleras plants crawling downwards?' Now he wrote: 'A great number of the plants ... were absolutely the same as, or most closely allied to, those of Patagonia.' The animals were characteristic of the Patagonian plains. 'We have likewise many of the same (to the eyes of a person who is not a botanist) thorny stunted bushes, withered grass, and dwarf plants.'

His observations were beginning to fit together like the pieces of a jigsaw puzzle.

Their journey was in a circle, from Mendoza turning westward and back to Valparaiso, back through a kindly scene where the lower slopes of the mountains were dotted over with the pale green Quillai or Soap Bark tree (*Quillaja saponaria*) and the magnificent chandelier cactus. The Uspallata Pass on the last leg was of a totally different geological construction to the Portillo. It consisted of different kinds of submarine lava alternating with volcanic sandstones and other sedimentary deposits, closely resembling some of the Tertiary beds on the shores of the Pacific. Where then was the certain proof that this part of the great Andes chain had also at some remote era risen slowly from the sea? For two days his geological hammer was busy.

He was in the central part of the range, up about 7,000 feet, when he saw on a bare slope some snow-white columns. They were a grove of petrified trees, still with their branches, eleven of them silicified and from thirty to forty of them of crystallised lime. Here was geological proof! Darwin could hardly believe his eyes, for he was gazing at the spot where 'a cluster of fine trees once waved their branches on the shores of the Atlantic, when that ocean (now driven back 700 miles) came to the foot of the Andes.' They had sprung from a volcanic soil which had been raised above the level of the sea, and subsequently this dry land with its upright trees had been let down into the depths of the ocean, to be covered by mud and by streams of submarine lava. Once more the subterranean forces had exerted themselves, and he now beheld the bed of that ocean forming a chain of mountains.

He chipped off some specimens of the petrified wood and sent them home. The trees were identified by Robert Brown, Keeper of the Botany Department of the British Museum, as 'belonging to the fir tribe, partly araucaria but with some curious points of affinity with the yew.'

4

The Plants of the *Beagle*

iii. The Galapagos and the Long Road Home

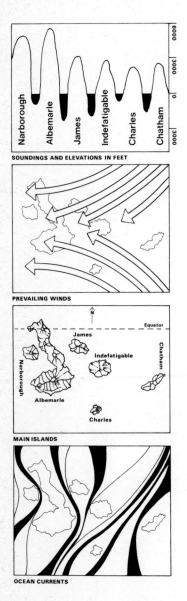

SOUNDINGS AND ELEVATIONS IN FEET

PREVAILING WINDS

MAIN ISLANDS

OCEAN CURRENTS

In the Galapagos Archipelago, the Enchanted Isles, Darwin found that each island had its own flora. The proportion of 100 new flowering plants out of 175 made it a distinct botanical province.

Why was this? Darwin was on a surveying ship and knew the ocean-depths between island and island, and also the strength of the capricious counter-currents that ran round them. All tended to isolate the islands from each other, and from the mainland of South America 500–600 miles away.

Yet the plants bore a resemblance to those of South America. These facts of modification helped Darwin form his evolutionary theory.

ON THE EVENING of the 4th of July 1835 Darwin said farewell to Mariano Gonzales with whom he had ridden so many leagues and discovered so much. Next morning he went to keep his rendezvous with the *Beagle* at Copiapó. In the two intervening months between this date and the 27th of April he had been exploring northern Chile while the ship went on with her charting.

They had set out from Valparaiso for Coquimbo where Darwin despatched a cargo of specimens to Cambridge, as always giving Henslow explicit directions.

In the B box, there are two Bags of Seeds, one ticket, Valleys of Cordilleras 5,000–10,000 ft high; the soil & climate exceedingly dry; soil very light & stony, extremes in temperature: the other chiefly from the dry sandy Traversia of Mendoza 3000 ft more or less.—If some of the bushes should grow but not be healthy, try a *slight* sprinkling of Salt & Salt-petre.—The plain is saliferous.—All the flowers in the Cordilleras appear to be Autumnal flowers.—They were all in blow & seed—many of them very pretty—I gathered them as I rode along on the hill sides: if they will but choose to come up, I have no doubt many would be great rarities.—In the Mendoza bag, there are the seeds or berrys of what appears to be a small potatoe plant with a whitish flower. They grow many leagues from where any habitation could ever have existed, owing to absence of water.—Amongst the Chonos dried plants, you will see a fine specimen of the wild Potatoe.—It must be a distinct species from that of the lower Cordilleras etc.—Perhaps, as with the Banana, distinct species are now not to be distinguished in their varieties, produced by cultivation.

Darwin claimed to be no artist but with this he sent a perfectly adequate sketch of his bush from the Traversia and a special note about it. 'Travelling Indians bring the seeds (vessels) from Bolivia & sell them at a high price for curing the toothache.' Henslow could not identify it from the *Botanical Miscellany* and sent it to Hooker. On it was Darwin's query: 'Can you inform me in what genus the seed-vessel belongs to?' It was labelled 'bright gamboge-yellow' and was curiously whorled like a shell. Or like the dropping of an extinct reptile? Henslow was to make his big discovery of coprolites in 1843 but must have found some fossil excrement before then, for Darwin made a joke of it. Perhaps his seed-vessel was Legumen coprolitiforme!

On the way to Coquimbo where the country became more and more barren, Darwin noted in his diary: 'It is rather curious the manner in which the Vegetation *knows* how much rain to expect.' This was altered in the first edition of his published *Journal* to: 'It is curious to observe how the seeds of the grass seem to know as if by an acquired instinct, what

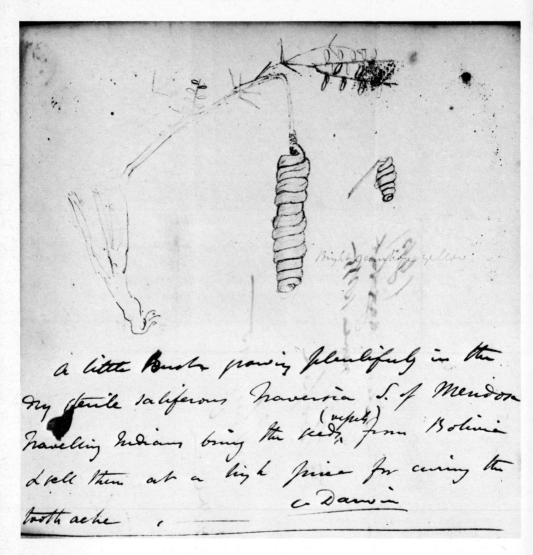

The capsule of the Toothache Tree uncurls when the seeds are ripe, in the process ripping them off and scattering them. Another plant-mechanism for the subject of Geographical Distribution

quantity of rain to expect.' In the second edition it was again altered. 'It is curious to observe how the seeds of the grass and other plants seem to accommodate themselves, as if by an acquired habit, to the quantity of rain which falls on different parts of this coast.' A sentence interesting as showing how Darwin's thoughts on Natural Selection developed.

After Coquimbo they had ridden to Guasco and thence to Copiapó, a distance along the shore northward of 420 miles but, with the inland excursions they took, many more.

Now the fabled Galapagos Islands lay ahead.

They anchored at Iquique on the coast of Peru on the 12th of July. The town had about a thousand inhabitants and stood on a little plain of sand at the foot of a great wall of rock rearing up two thousand feet. The place was utter desert, a light shower of rain falling only once in many years, so that the people had to import everything, water having to be brought in boats from Pisagua, some forty miles to the north. Here all the plants Darwin found on the rocky heights above the coast were a few cacti growing in the clefts, and strewing the loose mountain sand a curious greenish lichen lying on the surface quite unattached. Looking from a distance like a film of yellow-green grass it belonged to the genus *Cladonia* and somewhat resembled the Reindeer Moss, *C. rangiferina*, which Darwin must have seen growing on the high Scottish moors. Further inland on a forty-two-mile ride he saw only one other plant, a minute yellow lichen growing on the bones of some dead mules.

The next stop was in the Bay of Callao, the seaport of Lima, Peru's capital, where they stayed for six weeks. It was a small, dirty, badly-built town. The inhabitants, both there and at Lima, were of every mixture of European, Negro and Indian blood. They were a depraved and drunken set of people and even in Lima the air was loaded with foul smells from the heaps of filfth piled up in all directions—though the City of Kings must once have been a splendid town, Darwin thought, judging by the extraordinary number of its churches. He was glad to get out of the place and do some exploring.

The plain round the outskirts of Callao was sparingly covered with a coarse grass and nothing else; but on the hills near Lima the ground was carpeted with moss and beds of the beautiful yellow lily-like *Hymenocallis amancaes*. North of the town the climate became damper, and on the banks of the Guayaquil were the most luxuriant equatorial forests. These he heard about but did not visit: the country was in such a political turmoil that on the Anniversary of the Independence when High Mass was being celebrated, instead of each regiment displaying the Peruvian flag a black one with a death's head was unfurled—and this during the *Te Deum Laudamus*! This state of affairs was unfortunate for Darwin, as it prevented him from taking any excursions beyond the limits of the town. He spent most of his time on board, writing up his notes about Chile.

The *Beagle* sailed for the Galapagos on the 7th of September, and on the morning of the 17th they anchored at Chatham Island. The Spaniards who visited the Galapagos in 1535 had nicknamed them Las Islas Encantadas, the Enchanted Isles—not, however, for their beauty but because of the capricious counter-currents around their shores, which seemed alternately to attract and repel ships, as if by witchcraft. 'Nothing,' Darwin wrote, 'could be less inviting than the first appearance. A broken field of black basaltic lava, thrown into the most rugged waves, and crossed by great fissures, is everywhere covered by stunted, sun-burnt brushwood, which show little signs of life. The dry and parched surface, being heated by the noonday sun, gave to the air a close and sultry feeling, like that from a stove: we fancied even that the bushes smelt unpleasantly.'

He was to be ashore for only an hour on this first landing, for the ship was sailing round Chatham Island and there would be other opportunities for making longer excursions. Diligently he collected as many plants as he could in the time, but succeeded in getting only a few in flower (a herbarium specimen must have the flower if possible), and they were such wretched-looking, ugly little weeds that he thought them more fit for an arctic than an equatorial flora. From a short distance away the brushwood appeared to be as leafless as an English tree in winter. Indeed, so insignificant were leaf and bloom that it was some time before he discovered that almost every plant was not only in full leaf but that the greater number were in flower!

The commonest bushes were euphorbias and some had leaves so small that they could be measured only in botanical 'lines'. When you consider that a line is only one-twelfth of an inch long, and that the leaves of *Euphorbia recurva*, one of Darwin's bushes, were half a line long and two lines broad (and were few in number and brown in colour) no wonder he had difficulty in seeing them. An acacia and a great odd-looking cactus were the only trees that had any proper shape.

The *Beagle* anchored in several bays, and Darwin, accompanied as usual by Syms Covington, spent a day and a night ashore on a part of the island where black volcanic cones dotted the land, none higher than 100 feet. The day was glowing hot, and scrambling over the rough terrain and through tangled thickets was fatiguing work. But he was well repaid by a strange Cyclopean scene. As he was walking along he met two enormous tortoises, each of which must have weighed at least two hundred pounds. One was eating a piece of cactus. The other gave a deep hiss and drew in its head. 'These huge reptiles,' he wrote, 'surrounded by the black lava, the leafless shrubs, and large cacti, seemed to my fancy like some antedeluvian animals.' The few dull-coloured birds sitting quietly on the trees took no notice of him.

Charles Island was the only one inhabited. Five or six years before, more than two hundred people, mostly coloured, had been banished from Ecuador for political crimes and landed here to make the best of it. An Englishman by the name of Lawson was acting as Governor, and by chance he came down to the beach to visit a whaling vessel he was expecting. He offered to take Darwin to see the settlement, which was nearly in the centre of the island, four and a half miles inland and up about a thousand feet.

The first part of the road lay through a thicket of the same brushwood as on Chatham Island, and then as they climbed everything gradually became greener. Passing round the side of the highest hill they were refreshed by a fine southerly breeze, and suddenly the whole scene was as brilliantly green as England in springtime. Just below was the settlement. A flat area had been cleared and sweet potatoes and bananas were being cultivated. Scattered irregularly over it were the houses. Lawson introduced Darwin to some of the people, all of whom complained of their poverty, although food was to be had for the taking. Pigs and goats roamed the woods, and their main supply came from the helpless giant tortoises whose numbers, even then, were being greatly reduced by visiting ships when they were hunted down by the hundreds. It was these great chelonian reptiles which had given the Islands their name, from the Spanish *galápago*, a tortoise.

Although the smallest he visited, Charles Island was richest in plants. Not since leaving Brazil had Darwin seen so tropical a landscape, though there was a marked absence of the lofty, various and all-beautiful trees of that country, as he remembered them so vividly. Still, 'It will not easily be imagined how pleasant the change was from Peru & Northern Chili in walking in the pathways, to find *black mud* and on the trees to see mosses, ferns & Lichens & Parasitical plants adhering.' There were many different grasses. A wild Cotton tree was here only the size of a shrub. The largest tree of any kind was only one to two feet in diameter and had few leaves and crooked branches. It smelt like a balsam. Then he found not only a new species of plant but a new genus. There were two different species. Joseph Hooker was to name one *Galapagoa darwinii*, the other whose stems and leaves were of a lurid brown colour *G. fusca*.

Albemarle Island was the next to be visited. The *Beagle* was nearly becalmed between it and Narborough, and Darwin had plenty of time to study their coasts. Both were covered with frozen deluges of black naked lava which had flowed either over the rims of the great cauldrons, like pitch over the rim of the pot in which it had been boiled, or had burst forth from vents on their flanks. In their descent the streams of lava had spread over miles of the sea-coast.

Next morning he went exploring, finding the rocks abounding with great black lizards between three and four feet long. 'Imps of darkness', one of his shipmates called them. These were the famous marine iguanas of the Galapagos.

To a plant hunter the island was miserably sterile. But there was an interesting erigeron. In 1962 Gunnar Harling wrote in *Acta Horti Bergiani* that this (*Erigeron lancifolium*) ought to be placed in an independent new genus, *Darwiniothamnus*, with another species (*E. tenuifolium*) Darwin found on Charles and James Islands. 'It seems very fitting,' he said, 'to commemorate in this way Charles Darwin, who not only brought home the type material of both species of the genus but also through his large and excellent collections laid the foundations of our knowledge of the flora of the Galapagos Islands.' William Hooker's son Joseph, when he came to work on the collection, noted their distinctness, and although he was a 'splitter' and not a 'lumper' he decided to include them in *Erigeron*.

On the 8th of October they arrived at James Island which with Charles Island had been named for the Stuart kings. Darwin, Benjamin Bynoe (now Acting Surgeon) and three of the crew were to spend a week here while the *Beagle* went back to Chatham Island for water. Their walk inland was a long one. Not until they had gone about six miles and upwards to some 2,000 feet did the country show signs of green, for lower down the land was parched and the trees as usual nearly leafless, though here the same species attained a much greater size. Those up in the hills deserved, Darwin thought, the title of a wood. They were mainly of one particular tree, and this again was not only a new species but belonged to a new genus of the Compositae. Joseph Hooker named it *Scalesia darwinii*.

Four of the plants
Darwin found in the
Galapagos Islands

(*above left*)
Clerodendron molle

(*above right*) *Opuntia
galapageia*

(*far left*) *Phoradendron
henslovii*

(*far right*)
*Darwiniothamnus
lancifolius*

The biggest had a circumference of eight feet and several were of six feet.

But commonest was a tree with pale bright-green leaves covered with daisy-like flowers, also belonging to the Compositae family. This Darwin numbered 3294 in his collecting list. Hooker had an interesting comment to make on the 'very peculiar vegetable forms' of the Galapagos, citing eight tree-sized composites which grew there. There were also tree-sized clerodendrons and tree-sized 'snowberries' (*Chiococca*), all tropical in appearance.

Another thousand feet up they came to some springs, small but welcome, as the water was good and deliciously cold. Up here the clouds hung low for the greater part of the day. The vapour from them, condensed by the trees, dripped down like rain, which accounted for the brightly green and damp vegetation and muddy soil. 'The contrast to the sight & sensation of the body is very delightful after the glaring dry country beneath,' Darwin wrote. He enjoyed two days of collecting here. There were many plants, especially ferns, although the tree-fern was not among them. Nor were there any palms. He collected twenty-one James Island ferns, of which seven were new. Some were difficult, others impossible, for a herbarium sheet to accommodate. Of the one to which Darwin fixed his specimen of *Polypodium pleiosoros* Joseph Hooker wrote: 'Only the upper portion, about half a foot long, of an apparently very long frond exists in Mr Darwin's collection,' adding, 'and it does not accord with any described species or with any in the Hookerian herbarium.' (Writing this in 1849 he was referring to his father's vast herbarium, the largest and most complete in the world. By this time Sir William was at Kew as Director of the Royal Botanic Gardens, his herbarium occupying thirteen rooms of his private house. On his death in 1865 it was bought for

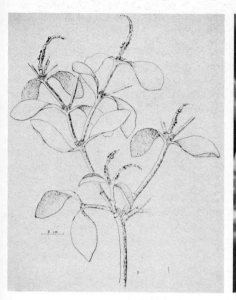 

Kew as a national memorial.) Another giant fern which was new was *Polypodium paleaceum* which had very long graceful fronds (only a two-foot portion being affixed), while *Adiantum parvulum* was small with notches cutting its fronds. A very beautiful fern of the same genus was named in honour of Darwin's 'Father in Botany' *Adiantum henslovianum*.

So wet was it in these upper reaches that there were large beds of a coarse cyperus, a water-loving plant and cousin of the Egyptian papyrus, where flocks of a small water-rail had their homes. The giant tortoises also liked this wet region: they were swarming round the springs to drink their fill. They had made pathways for travelling to and from them, and Darwin thought it very comical to see the huge creatures, with necks outstretched, deliberately pacing onwards. 'I think,' he noted, 'they march at the rate of 360 yards in an hour; perhaps four miles in the 24.' Fishermen in search of water merely had to follow their pathways, as Darwin and Syms Covington had done. In the low dry part of the island there were few of them, their place being taken by great yellow lizards which lived entirely on fruits and leaves. Darwin saw them crawling up the mimosa trees to get them. One of these trees was *Acacia tortuosa*. He took specimens of it. The yellow lizard, the Galapagos land iguana, also ate a succulent cactus which had yellow flowers and a branching cylindrical trunk beset with strong spines growing from it in star-like clusters. It grew as tall as ten feet and was common on rocky ground.

If the Galapagos were interesting for the 'antedeluvian' creatures, there was also a most singular group of finches whose beaks were nicely varied according to the work they had to do. Thus the one which fed on large seeds had a strong blunt beak for cracking them; the one which fed on smaller seeds a smaller blunt beak, another which fed on insects having a thin delicate beak. 'Seeing this gradation and diversity of structure in one small, intermittently related group of birds, one might really fancy that from an original paucity of birds in this archipelago, one species had been taken and modified for different ends.' So wrote Darwin in his *Journal of researches into the geology and natural history of the various countries visited by H.M.S. Beagle round the world.* It was published in 1839, twenty years before the *Origin of Species*.

But, he added, 'The botany of this group is fully as interesting as the zoology.'

He was careful to keep separate the collections he made on each island. It was fortunate he did so, for when he compared the herbarium sheets an astonishing fact emerged. It was that while many plants were common among the islands, each island had in addition plants of its own not found on any other island. There were also plants he recognised as having seen in South America, although there were not so many of these. This surprised him, considering that the Galapagos were only between 500 and 600 miles from that continent. There was no land bird common to the Islands and the American mainland, so the Galapagos were deprived of this very common means of transporting seeds and berries. As to the individual floras, it was not surprising that Charles Island, the only one inhabited, had the greatest number of non-native plants. Yet most of the islands were in sight of each other, Charles being fifty miles from the nearest part of Chatham and thirty-three from the nearest part of Albemarle, and so on, James Island only ten miles from the nearest part of Albemarle.

Again, neither the nature of the soil, nor the height of the land, nor the climate, nor the general character of the associated beings (animals, insects, and birds), and therefore their action upon each other, differed much. If there was any appreciable difference in their climates, it must be, Darwin argued, between the windward group (Charles and Chatham) and the leeward group, though there seemed to be no corresponding difference in the plants of these two halves of the archipelago.

The following masterly interpretation was how Darwin suggested that the puzzle could be solved:

The only light which I can throw on this remarkable difference in the inhabitants of the different islands, is, that very strong currents of the sea running in a westerly and W.N.W. direction must separate, as far as transportal by the sea is concerned, the southern islands from the northern ones; and between these northern islands a strong N.W. current was observed, which must effectually separate James and Albemarle Islands. As the archipelago is free to a most remarkable degree from gales of wind, neither the birds, insects, nor lighter seeds, would be blown from island to island. And lastly, the profound depth of the ocean between the islands, and their apparently recent (in a geological sense) volcanic origin, render it highly unlikely that they were ever united; and this, probably, is a far more important consideration than any other, with respect to the geographical distribution of their inhabitants. Reviewing the facts here given, one is astonished at the amount of creative force, if such an expression may be used, displayed on these small, barren, and rocky islands; and still more so at its diverse yet analogous action on points so near each other. I have said that the Galapagos Archipelago might be called a satellite attached to America, but it should rather be called a group of satellites, physically similar, organically distinct, yet intimately related to each other, and all related in a marked, though much lesser degree, to the great American continent.

This interpretation was stage two in Darwin's future working out of his theory of the origin of species.

After a month in the Galapagos the survey was completed, and on the 20th of October the ship's head was put towards Tahiti and they began the long passage of 3,200 miles across the Pacific. Day and night the trade winds blew, bearing them steadily onwards at the rate of 150 or 160 miles a day. By the 1st of November they had left behind them the gloomy region extending far out from the coast of South America, and every day the sun shone brightly out of a cloudless sky. So with studding-sails set on each side they pleasantly crossed the blue ocean.

Their first hints of a landfall were the increasing numbers of sea birds, especially two species of terns, and then the smudge, hardly more than a line, that became the level green surface of Dog or Doubtful Island. So insignificant was it on the domain of the wide all-powerful ocean that it seemed an intruder. By the evening of the 13th they had passed through the whole of the Tuamotu Archipelago, then sometimes called the Dangerous or Low Islands, and finding two not on the map they took their locations.

To Darwin most of the islands looked very uninteresting, merely a long brilliantly white beach capped by a low bright line of green vegetation. The width of the dry land was trifling: from the mast-head it was possible to see at Noon Island right across the smooth lagoon to the opposite side, a great stretch of water about ten miles wide. These were all coral islands, whose mysterious origin Darwin was to solve.

Tahiti lived up to all that he expected from the tales told of it by Captain Cook and Sir Joseph Banks: canoes to greet them, and crowds of men, women and children with laughing merry faces ready to befriend them. Behind the white sands and the strip of land covered with orchards growing tropical fruit, wild precipitous peaks beckoned. On Tuesday the 17th Darwin set off on a first climb. The mountains were split into fearsome ravines. He climbed up one of the ridges to a height of between two and three thousand feet, his notes reading: 'Wonderful view; Cordilleras nothing at all like it—ascended a fern-slope—excessively steep—middle of day—vertical sun—steaming hot—cascades in all parts, enormous precipices—columnar—covered with Lilies, Bananas and trees.' The ferns on the slope were almost exclusively dwarf ones, but at the highest point he reached were tree-ferns. It was the wealth of tropical fruits which most interested and impressed him. On a two-day excursion with the local missionary into the interior he was amazed at the prodigality of nature: the forests of banana trees and the fruit lying in heaps rotting on the ground, the brakes of wild sugar-cane, the wild arum whose roots when baked were good to eat and the young leaves better than spinach, the liliaceous plant called Tì which was as sweet as treacle and served them for dessert, the coconuts and pineapples.

But Tahiti had been well botanised, and shore-time in New Zealand, the next country they visited, was limited to seven days. Darwin squeezed in excursions, though finding on his first day that the country was impracticable for walking: the hills were thickly covered by tall ferns and a low bush growing like a cypress. When he tried the sea beach, left and right his walk was stopped short by creeks and deep rivers. At Waimate, about fifteen miles from the Bay of Islands, the British Resident offered to take him in his boat up a creek to where he would see a pretty waterfall. Some way up they left the boat to follow a well-beaten path bordered on each side by the tall fern which covered the whole countryside. After walking for some miles it was still uniformly clothed with it, although occasionally the banks of a river would be fringed with trees, and here and there on hillsides were clumps of woods. 'The sight of so much fern,' wrote Darwin, 'impresses the mind with an idea of useless sterility.' Later he learned that when it was ploughed in, the land became productive.

On another day he was taken to see a forest of the famous Kauri Pine. Measuring one of these noble trees he found it to be thirty-one feet in circumference. The trunks rose smooth and cylindrical for a height of up to ninety feet without branches. Among them grew the great ones, standing like gigantic columns. On the outskirts of the wood he saw plenty of *Phormium tenax*, the New Zealand flax, with its long sword-like leaves whose undersides were coated with strong silky fibres. Women workers scraped this off with a broken shell to make the beginnings of cloth. There were beautiful tree-ferns. In many places the common dock was a familiar weed, proof of the rascality of an Englishman who sold the seed as that of the tobacco plant.

They were all glad to leave New Zealand, finding it not a pleasant place. The natives lacked the charming simplicity of the Tahitians, and most of the English residents were 'the very refuse of Society'. Neither was the country itself attractive. But then, Darwin had seen little of it. His visit is commemorated in a charming little bush called *Hebe darwiniana*. 'An airy and elegant dwarf evergreen shrub with myriad of small pointed leaves', it is now an alpine treasure of our gardens. Thus a variegated variety of it is described by an alpine nurseryman of our own day, Joe Elliott. Its flowers are white and grow in racemes near the tops of the shoots in July and August.

Steering their course to Sydney they arrived at Port Jackson early in the morning of the 12th of January 1836, the straight line of yellowish cliff reminding Darwin of the coast of Patagonia. The harbour was fine and spacious, and beautiful villas and attractive cottages were scattered along the beach. Further inland were large stone houses. Sydney already had a population of 23,000 and hummed with industry and wealth. Darwin hired a man and two horses to take him to Bathurst, then a village and the centre of a great pastoral district, hoping in the hundred and twenty miles to get a general idea of the country. They were early days in the colony. He saw parties of convicts working in irons under the eyes of sentries with loaded guns.

The most remarkable feature of the landscape was the uniform character of the vegetation. Everywhere was open woodland, the trees nearly all eucalyptus. He visited the Blue Mountains, expecting to see a bold chain crossing the country and finding instead merely a sloping plain which rose almost imperceptibly to a height of 3,000 feet. There, too, was the eucalyptus. He visited a sheep farm, and went out kangaroo-hunting but did not see a single one. The English settlers had not been very particular in giving names to the Australian trees: hearing of oaks Darwin went to see them and found them to be casuarinas.

Then it was Tasmania, more green and cheerful to Darwin's eyes. He climbed Mount Wellington, failing on his first attempt because of an impenetrable wood. Next day he took a guide with him, but he was a stupid fellow who led him by the south or damp side. Here the vegetation was luxuriant, and dead trees and branches made the labour of climbing almost as great as in Tierra del Fuego or Chiloe. It cost them five and a half hours of hard climbing to reach the summit. In many parts the gum trees grew to a great size and formed noble forests. In some of the damper ravines tree-ferns grew to enormous heights: Darwin saw one about twenty-five feet high to the base of the fronds. It was six feet in girth. The foliage of these trees, forming elegant parasols, created a gloomy shade.

When they came to St George's Sound on the 6th of March there were eight days for exploring. The country was dull, the poor sandy soil covered either with thin low brushwood and wiry grass or a forest of stunted trees. But in the open parts he found great numbers of Grass trees, *Xanthorrhoea preissii*, a remarkable plant looking like a palm but sprouting a tuft of coarse grass instead of the crown of noble leaves. Darwin left the shores of Australia without sorrow or regret.

The Southern Keeling or Cocos Islands were everybody's dream of an earthly paradise: the low circular reef with its line of huge breakers crashing on to it, the groups of elegant coconut trees dotted here and there, and, once through the single opening, coming from the heaving blue ocean into a lagoon several miles wide, emerald green water so still and clear that you could look down and see the dark bands of the living coral.

Keeling had about a hundred inhabitants, most of them runaway Malay slaves, with a few English residents. All were dependent on the coconut tree, its oil and nuts, for their living, which was not very prosperous. The soil was merely rounded fragments of coral, and apart from the coconut there were only five or six other kinds of trees. One was the Cabbage tree, *Andira inermis*, which grew to about thirty feet and had, Darwin was told, panicles of purple flowers among its pinnate evergreen leaves. It was one of the 'rain trees', frequently attacked by such insects as cicadas, which so injured them that the sap exuded in drips and the legend arose that rain continually fell from them. The woods were as thick as a jungle and growth seemed luxuriant. Yet Keeling could produce only twenty-three species of plants. Darwin brought away specimens of all of them, excepting the coconut and the Cabbage tree which was not in flower. The Malays positively assured him that, apart from these two, his collection was complete. On this lonely island, thrown on their own resources, they knew every plant that grew.

With no cliffs to nest in, gannets, frigate birds and terns occupied the trees, and Darwin was charmed when a small snow-white tern made him its companion, smoothly hovering a yard away from his head, its large black eye scanning his face with a quiet curiosity. 'Little imagination,' he thought, 'is required to fancy that so light and delicate a body must be tenanted by some wandering fairy spirit.'

There were twenty-three islands and he explored most of them. Because of the coral nature of the soil, and the nature of the coral itself, he expected to find a purely littoral flora grown from seeds attached to the trunks of trees and other flotsam and jetsam washed up on the shores. Unlike the Galapagos the Keeling Islands had no true land birds, though there were one or two waders. The rest were birds of the sea, which could also bring seeds. Others could float here by themselves if they were able to resist the action of the sea water.

It was Henslow who classified and described the Keeling plants. He thought that all the introduced species were from the East Indian Archipelago or the neighbouring continent, though they had not all been noticed there. Two at least of Darwin's plants had not been described before, and one or two others were interesting from their rarity and the little that was known about them. All the rest had an extensive range throughout the intratropical regions.

When Joseph Hooker wrote his massive Linnean Society paper on the Galapagos plants he pointed out that an island like Keeling whose flora was wholly borrowed from other places, seldom had two species belonging to the same genus, saying that the more an island is indebted to a neighbouring continent for its vegetation, the more fragmentary is its flora, migration being of isolated individuals which are generally in no way related.

Darwin made careful notes. 'No sooner has a new reef become sufficiently elevated by the accumulation of sand upon its surface, but this plant is sure to be the first which takes possession of the soil,' he wrote of *Pemphis acidula* which had interesting seed capsules. 'A moderate sized tree, with small white flowers, very common' was *Tournefortia argentea* whose cymes bore flower and fruit at the same time. Three of the trees were evergreen: *Cordia orientalis, Guettardia speciosa*, and *Ochrosia parviflora* which made straight handsome trees with a smooth bark and bore bright green fruit like a walnut. Of the *Guettardia* he wrote: 'The flowers possess a delightful perfume.' They were white and fragrant of cloves. The tree grew to a height of up to thirty feet. Of the *Cordia*: 'The settlers have named this Keeling-teak, because it furnishes them with excellent timber. They have built themselves a vessel with it. A large tree, abounding in some of the islands, very leafy, with scarlet flowers; but only a few blossoms were expanded at the time, and they easily fell off.' A small tree also useful to the Keeling islanders was *Paritium tiliaceum* (now called *Hibiscus tiliaceus*) which was common on one of the islands. It was exceedingly useful throughout the Pacific, Darwin learned, in Otaheite particularly, the bark for making cordage, the light wood used by the fishermen for floats, while the natives readily procured fire from it by friction.

'From the Keeling Isles we came direct to this place,' Darwin wrote to his sister Catherine from Port Louis in Mauritius. 'All which we have yet seen is very pleasing. The scenery cannot boast of the charms of Tahiti & still less of the grand luxuriance of Brazil; but yet it is a complete & very beautiful picture. But there is no country which has now any attractions for us, without it is seen right astern, & the more distant & indistinct the better. We are all utterly home-sick.' There was a possibility of reaching England in eight weeks, if they escaped the heavy gales off the Cape of Good Hope. Their course beyond the Cape and St Helena was not certain, but he thought it would be the coast of Brazil, at Bahia. While at sea, Darwin spent the time rearranging his notes and sometimes rewriting them. He told Caroline: 'I am just now beginning to discover the difficulty of expressing one's ideas on paper. As long as it consists of description it is pretty easy; but where reasoning comes into play, to make a proper connection, a clearness & a moderate fluency, is to me as I have said, a difficulty of which I had no idea.'

Mauritius was also called the Isle of France, and Port Louis was French in character, the English people speaking French to their servants, and the place-names being French—evidence of its ninety-five years of French occupation from 1715 to 1810. The *Beagle* spent only five days here, giving Darwin time to do some exploring, but as most of the immediate country was cultivated and everybody was so hospitable, not much collecting was done. At the Cape, although it was not the flowering season, there were some very pretty mesembryanthemums, oxalis and heaths in bloom, and he had the luck to meet Sir John Herschel, dining at his comfortable country house and being shown his charming garden, full of Cape bulbs of his own collecting. On the 29th of June the *Beagle* crossed the Tropic of Capricorn for the sixth and last time, on the 8th of July arriving off St Helena.

Here the vegetation was decidedly British. The hills were crowned with plantations of Scots Pines, the sloping banks clad with thickets of gorse all in bright yellow flower. When Darwin learnt that the proportion of native plants was 52 to 424 species mainly from England, he could see the reason for this resemblance. Of them he wrote: 'These numerous species, which have been so recently introduced, can hardly have failed to have destroyed some of the native kind.' Many familiar English plants were flourishing here better than at home, and he thought it not improbable that even at the present day similar changes might be in progress. It was only on the highest and steepest mountain crests that the native plants were predominant, and this flora ran to the same pattern as in the Galapagos. Here, for instance, was a case of herbaceous plants becoming tree-like on an island where there was no competition with native or imported trees, for St Helena was destitute of a tree of any kind, except the pine. This arborescent development Darwin was to point out in the *Origin*.

At Ascension Island longed-for letters from home awaited them. There was one for Darwin from his sisters in which they told him that Sedgwick had called on their father and said that his son Charles would take a place among the leading scientific men. Darwin could not understand how Sedgwick could know anything of what he was doing. Later he heard that Henslow had read some of his letters before the Cambridge Philosophical Society and printed them for private distribution. At this moment he was so excited by the letter that 'I clambered over the mountains of Ascension with a bounding step and made the volcanic rocks resound under my geological hammer!'

Then it was Bahia in Brazil. In the four days they were there he was glad to find that his enjoyment of its tropical scenery was as great as it had been on his earlier visit. It was even intensified, especially his delight in the wild luxuriance of its vegetation, the bright red soil so strongly contrasting with the green, the numberless stately trees: all seeming to overpower the patches of cultivated ground with their houses, convents and chapels, and overpower even the cities. Learned naturalists had described these tropical scenes by naming a multitude of objects and mentioning some characteristic feature of each, and to a learned traveller this might possibly communicate some definite idea. 'But who else,' Darwin asked, 'from seeing a plant in an herbarium can imagine its appearance when growing in its native soil? Who, from seeing choice plants in a hot house, can multiply some into the dimensions of forest trees, or crowd others into an entangled mass?' It was at noon, when the sun had attained its greatest height, that one should visit this great wild, untidy, luxuriant hot-house of a land, when the dense splendid foliage of the mango hid the ground with its darkest shade, while its upper branches were made the more brilliant by the profusion of light.

In the last walk he took, he stopped again and again to gaze on such beauties, trying to fix for ever in his mind impressions which he knew must sooner or later fade away. The forms of the orange tree, the coconut, the palm, the mango, the banana, would remain clear and separate, but the thousand beauties which united them all into one perfect scene must, he knew, perish. 'Yet they will leave,' he wrote, 'like a tale heard in childhood, a picture full of indistinct but most beautiful figures.'

That afternoon they weighed anchor and stood out to sea.

On the 2nd of October 1836 the *Beagle* arrived at Falmouth. To Darwin's surprise and self-confessed shame the first sight of the shores of England inspired him with no warmer feelings than if it had been a miserable Portuguese settlement. That same night (and a dreadfully stormy one it was) he started by the Mail for Shrewsbury, and there, realising that at last he was home, dashing off a letter to his 'First Lord of the Admiralty', his Uncle Jos, he was able to say from the depths of his heart: 'I reached home late last night. My head is quite confused with so much delight, but I cannot allow my sisters to tell you first how happy I am to see all my friends again. . . . I am so very happy I hardly know what I am writing.'

5

Home is the Sailor

Home again, Darwin found the naturalists enthusiastic about the plants he had collected. 'I only wish,' he told Henslow, 'I had known the Botanists cared so much for specimens & the Zoologists so little; the proportional number of specimens should have had a very different appearance.'

He settled down to write the Journal of the voyage. He had thought of himself as a geologist, but it was the sublimity of the Brazilian forest that was brightest in his mind. To other intending young travellers he wrote: 'Be a botanist.'

The picture shows the herbarium sheet preserved at the Botany School, Cambridge, of Darwin's actual dried specimen of Scalesia darwinii. Note the small envelope bottom left. It contains seeds of the plant.

THOUGH HIS SISTERS were not allowed to be first in announcing his return from the voyage, Caroline begged for the blank space left unused to add a letter to their cousin Elizabeth. Charles, she wrote, 'is come home so little altered in looks from what he was five years ago and not a bit changed in his own dear self.' They had the very happiest morning, Charles so full of affection and delight at seeing their father looking so well and being with them all again. Charles himself was 'looking very thin but well—he was so much pleased by finding your and Charlotte's kind notes ready to receive him. . . . He feels so very grateful to Uncle Jos and you all, and has been asking about every one of you.'

She added: 'Now we have him really again at home I intend to begin to be glad he went this expedition, and now I can allow he has gained happiness and interest for the rest of his life.'

He spent eight days at Shrewsbury, waiting to hear from Captain FitzRoy as to when he must go to London to get his goods and chattels out of the *Beagle*, and meanwhile writing to Henslow that he was anxious for his advice, as he was 'in the clouds & neither know what to do, or where to go.' His chief puzzle was the geological specimens: who would have the charity to help him describe their mineralogical nature?

As no word came from FitzRoy he went off to Cambridge, spending a night in London on the way with 'good dear old brother Erasmus'.

Henslow was ready, as usual, with practical suggestions. From St Helena Darwin had written asking if he would propose him for the Geological Society. After the tour in North Wales, Sedgwick had offered to do this but had forgotten until Henslow jogged his elbow. On the 8th of September they had both signed Darwin's certificate. He was to be proposed on the 2nd of November. Henslow now urged him to meet his future fellow members, particularly Charles Lyell, the Society's president, and why not write to Sedgwick saying that he was home, and renew acquaintance with his old Edinburgh professor Robert Grant who was now at London University? He ought also to meet Richard Owen who had just been appointed Hunterian Professor at the Royal College of Surgeons. Write to them all and ask to see them, Henslow advised: between them they would know where his collections should go.

After five days of letter-writing Darwin returned to London and took up his quarters with Erasmus at 43 Great Marlborough Street.

The *Beagle* had followed a leisurely way from Falmouth, proceeding to Plymouth on the 4th of October and lying there till the 17th, on the following day sailing for the Thames, calling in at Portsmouth and Deal and getting up the river to Greenwich on the 28th before moving down to Woolwich to end, on the 17th of November 1836, an unusually long voyage of five years and one hundred and thirty-six days.

Darwin intercepted her at Greenwich and spent the day on board packing up his belongings. The Galapagos plants were his first concern. Henslow had begged him to bring them away with him (the other boxes could follow later), so that they could be examined immediately. Darwin wrote hoping they were 'in tolerable preservation'. This was on the 30th of October. An important day had intervened, the 29th, when at the house of Charles Lyell he met not only his hero of the *Principles of Geology* but Richard Owen.

It was an extraordinary meeting. Lyell, thirty-nine, was at the height of his career. Charles Darwin, twenty-seven, was only feeling his way. He had sought out Lyell as the head of British geology who might perhaps suggest where his geological specimens could go—usefully, for he knew enough to believe that they might be of prime importance. Also, he was proud to remember that the first place on the voyage where he had geologised—St Jago in the Cape Verde Islands—had convinced him of the infinite superiority of Lyell's views over those of any other geologist. By 1835 he was still 'a zealous disciple of Mr Lyell's views.' But, as he told Darwin Fox at that time: 'Geologising in South America I am tempted to carry parts to a greater extent even than he does.'

He was about to carry them now.

It was the older man who deferred to the younger. In Darwin's absence Henslow had also shown his letters to Sedgwick who had extracted from them remarks and information he thought would interest the Geological Society. On the 18th of November 1835 Sedgwick had read to the Society 'Geological Notes made during a survey of the East and West Coasts of South America, in the years 1832, 1833, 1834 and 1835, with an account of a transverse section of the Cordilleras of the Andes between Valparaiso and Mendoza.' An abstract of three pages was published in the *Proceedings*, but so little known was the author at this time that he was described as 'F. Darwin, Esq., of St John's College, Cambridge'!

Lyell had studied the paper with the greatest interest. He received his visitor as a friend, and was soon plying him with questions. Darwin found himself putting forward his ideas on the origin of coral reefs, a theory that had come to him in South America when he was studying the elevation of the land. He began tentatively, for he was in the presence of the greatest living geologist. And he was about to propound views directly opposed to those of his 'master'. But the questions Lyell shot at him were encouraging, and he went on. Emma Darwin, three years later, was to describe how Lyell never spoke above his breath. She complained that he was enough to flatten any party, because everybody kept lowering their tone to his. It was different when he was talking geology, and on this occasion 'Lyell, on receiving from the lips of its author a sketch of the new theory, was so overcome with delight that he danced about and threw himself into the wildest contortions as was his manner when excessively pleased.' In later years Darwin told Professor John Wesley Judd, who in his opinion was then the up-and-coming geologist, that until he discussed his coral-reef theory with Lyell he never fully realised its importance. To Lyell the meeting was a poignant one. Holding definite ideas of his own he relinquished them, writing to Sir John Herschel: 'I must give up my volcanic crater theory for ever, though it cost me a pang at first, for it accounted for so much.'

Richard Owen must have sat as a silent onlooker at the meeting while all this was being discussed. But when his turn came he was helpful. He suggested a letter to Sir Anthony Carlisle, president of the Royal College of Surgeons, who might welcome the fossil mammals for their museum. Alternatively there was the British Museum or, better still, the Paris Museum of Natural History where Georges Cuvier had laid the foundations of vertebrate palaeontology, culminating in his monumental work *Recherches sur les ossemens fossiles de quadrupèdes*. Owen had a partisan interest in the Paris Museum, having worked there with Cuvier five years before, examining bones and helping him classify them. Owen suggested, too, taking plaster casts. Not surprisingly when the *Zoology of the Voyage of the Beagle* came to be written it was Owen who contributed an article on Fossil Mammalia.

Darwin went to see Professor Grant who said he would be willing to examine 'some of the corallines'. A disappointing limitation, that 'some': Darwin had hoped that he would undertake the examination of the whole order. 'It is clear,' he wrote to Henslow, 'the collectors so much outnumber the real naturalists, that the latter have no time to spare.' He tried the museum of the Zoological Society, to learn that it was nearly full and that more than a thousand specimens remained unmounted. This information he gleaned at a meeting of the Society 'where the speakers were snarling at each other in a manner anything but that of gentlemen.' (The zoologists were to retrieve themselves when, at a meeting in January 1837, they exhibited 80 specimens of Darwin's mammals and 450 birds.) Twice he went to see William Yarrell, a founder member of the Zoological Society in 1826 and a vice-president of the Linnean Society, who carried on a family business of bookseller and newspaper agent, but Yarrell was so evidently oppressed with business worries that it was 'too selfish to plague him with my concerns.' Thomas Bell, Professor of Zoology at King's College, London, was so busy that there was 'no chance of his wishing for specimens of reptiles' (though he did eventually write them up for the *Zoology*). He had not yet seen Westwood, 'so about my insects I know nothing.' This was John Obadiah Westwood, secretary of the Entomological Society and later the first Hope Professor of Zoology at Oxford. He had seen William Lonsdale, the curator and librarian of the Geological Society, who had given him a most cordial reception, and if it was solely the geological specimens that were to be considered he was sure that Lyell's and Lonsdale's kindness would take care of that part of the collections.

As for his plants, this was a different matter. He had dined at the Linnean club and met David Don, Professor of Botany at King's College and librarian to the Linnean Society. He wrote to Henslow:

You have made me known among the botanists; but I felt very foolish, when Mr Don remarked on the beautiful appearance of some plant with an astoundingly long name, & asked me about its habitation.

Some-one else seemed quite surprised that I knew nothing about a carex from I do not know where. I was at last forced to plead most intire innocence,—that I knew no more about the plants which I had collected than the Man in the Moon.

From his various meetings with the naturalists the fact emerged that it was his plants that interested them more than anything else. 'I only wish,' he said to Henslow, 'I had known the Botanists cared so much for specimens & the Zoologists so little; the proportional number of specimens should have had a very different appearance.'

In the damp heights of James Island *Scalesia darwinii* grew tall and straight in groves 'deserving the title of a wood', Darwin wrote

He was out of patience with the zoologists, not because they were overworked 'but for their mean quarrelsome spirit'. The advice generally given him was that he must do all the work himself. Lyell had said the same thing, so, as he wrote to Henslow: 'Your plan will be not only the best, but the only one, namely to come down to Cambridge, arrange & group together the different families & then wait till people, who are already working in different branches may want specimens.'

He decided to spend a few months in Cambridge, and then, once he knew, with Henslow's help, what his collections amounted to, emigrate to London to complete his geology and try to push on the zoology. Having retrieved all his boxes from the *Beagle* he sent them by Marsh's Wagon to Cambridge. Henslow had repeatedly asked him to stay with them, and this Darwin did for two or three days, writing to Darwin Fox on the 15th of December that after a life of busy enjoyment he was looking forward to a period of quiet tranquillity. Henslow's new house at Cambridge was not large but it was comfortable, and he and Mrs Henslow were so very kind and affectionate that he quite felt to them as to his nearest relations. But these comfortable ways of life were not conducive to hard work, and tomorrow he was migrating to solitary lodgings in Fitzwilliam Street. In the mornings he was going to arrange his specimens into groups, and examine the geological fragments one by one, 'which will be a most tedious task', but Syms Covington would be helping him. In the evenings he would write, and as it was now the Christmas vacation he hoped to extract a good many solid hours out of each day.

This he did, paying one short visit to London to read a paper before the Geological Society, 'Observations of proof of recent elevation on the coast of Chile'; this on the 4th of January 1837 on which date he was formally admitted to the Society, and by Lyell himself.

On the 13th of March he moved to London, taking lodgings at 36 Great Marlborough Street, and after a round of welcoming dinner parties he was left in peace to go on with the writing of his *Journal*. This was to be the third volume of *The narrative of the voyages of H.M. Ships Adventure and Beagle*, FitzRoy 'making a plumpudding out of his own journal and that of Capt King's kept during the last voyage', as Darwin put it, to make the first two volumes.

His collections were gradually being classified. While at Cambridge his minerals and rocks were examined by William Hallowes Miller, Professor of Mineralogy, and Leonard Jenyns had undertaken to deal with his fish, including some from the Galapagos, and write them up for the *Zoology*. He was relying on Henslow to identify his plants, and meanwhile for the writing of the *Journal* had some botanical queries for him.

Will you keep a page in a memorandum book for queries for me? I will begin with two or three.—Name of plant (   )[1] from Fernando Noronha. The name of the Cardoon?—I found it out in Isabelle's voyages, but have forgotten it. He calls it also the Cardoon of Spain.—Do you know what is the giant thistle of the Pampas?—At some future time I shall want to know number [of] species of plants at Galapagos and Keeling, and at the latter whether seeds sent probably endure floating on salt water. I suppose after a little more examination you would be able to say what was the general character of the vegetation of the Galapagos? Pray take in hand as soon as your lectures are over, the potato from the Chonos Islds.

[1] Darwin left the bracket unfilled.

He wrote this on the 28th of March 1837.

By January of the previous year Henslow had got as far as numbering Darwin's specimens and sending William Hooker duplicates with a list of the more than 600 species he had separated, these being Darwin's South American collection up to June 1835—if indeed he had received and sorted the two boxes Darwin despatched from Lima in that month. He had offered to place the collection in Hooker's hands for identification. 'The public will have far greater confidence in your remarks & descriptions than in mine.' We do not know what Hooker's reply was, for no reverse letter can be traced. If it was a courteous reminder that he already had more work than he could conveniently deal with, this would be understandable, for with his lectures and the work of his botanic garden, the flood of new plants continually coming to him, which all had to be identified, the several journals he was editing, writing most of the material himself, and a never-ceasing correspondence, no busier botanist existed. What we do know is that he made a beginning by describing duplicates Henslow had sent him and publishing these in Volume III of the *Journal of Botany*, 1841.

Darwin's appeal to Henslow caught him at the time when he was moving his activities from Cambridge, having been presented to the living at Hitcham in Suffolk, and there he was going to live, returning to Cambridge only for his Easter-term lectures. He had given up his Friday evening receptions at the end of 1836, causing such a blank in the scientific life of Cambridge that the Ray Club was founded to take their place. One of the leading lights in the enterprise was his old pupil Charles Cardale Babington who was ultimately to succeed him in the botanical chair. In the difficulties of non-residence it was Babington who became his greatest helper, discharging many of his duties.

In the same March letter Darwin told Henslow that he had met Robert Brown who asked him 'in rather an ominous manner' what he meant to do with his plants. 'In the course of conversation Mr Broderip who was present remarked to him "you forget how long it is since Capt. King's expedition." He answered, "Indeed I have something, in the shape of Capt. King's undescribed plants to make me recollect it." '

The royal garden at Kew had not yet been acquired for the nation. There were still four years to go before Sir William Hooker would take it over as the Royal Botanic Gardens. But at the British Museum there was already a national herbarium, and Robert Brown was in charge of it.

The Captain King referred to was Philip Parker King, father of Midshipman King of the *Beagle*. In the earlier expedition to South America King had commanded the *Adventure*, and before then, from 1817 to 1822, had surveyed the coast of Australia, writing the *Narrative of the Survey of Australia* in collaboration with Allan Cunningham the botanist.

No wonder that Darwin, at mention of Captain King's undescribed plants asked Henslow indignantly: 'Could a better reason be given (if I had been asked) by me for not giving the plants to the Brit. Museum?'

But Brown was obviously angling for them. We remember that from Valparaiso Darwin had sent him specimens of the fossil trees. When they met again in May Brown told him that he was much pleased with them and had gone to the expense of having some of them cut and ground. He was now very curious about the fungi Darwin had found on the beech trees in Tierra del Fuego. He already had some specimens, he said, but would like to see Darwin's. 'But,' Darwin wrote doubtfully to Henslow, 'I do not know whether he wants to describe them: as your hands are so full, would you object to send them to me, & allow Mr Brown to do what he likes with them. If you particularly care about them, of course do not send them, but otherwise I should be glad to oblige Mr Brown.—I have introduced my imperfect account of them in my journal—so that I should for my own sake be glad of Mr Brown's inspection *soon*.'

Henslow sent some specimens of the fungi with a comment that brought the reply from Darwin: 'I fear by your letter you cared more about the edible fungi than I thought.—I took them to Mr Brown who said he had never seen anything of the sort before, & appeared interested on the subject, but whether he means to describe them, & for what he wants them,—I have not a guess,—at some future time, if I can summon courage, I will ask him, but I stand in great awe of Robertus Brown.' He added: 'I will copy out the list of Botanical questions on a separate piece of paper.'

In July after a visit to Shrewsbury he wrote to Henslow:

I am now hard at work, cramming up learning to ornament my journal with; you may guess the object of this letter is to beg a few hard names respecting my plants.—I believe I shall really begin printing in beginning of August, so that there is no time to lose.—Will you look over the list of questions, & try to answer me some of them.—For instance it will not take you long just to count the number of species in my collection from the Keeling Isds:—You can tell me something about the Galapagos plants, without any further examination:—You can tell me what genus of fungi the edible one from T. del Fuego comes nearest to; Mr Brown of course has not only never looked at it a second time, but cannot even lay his hand on the specimens.—I fear I must trouble you to send me one more *good* dried specimen, for I am thinking of having a wood cut.—To examine the potatoe from Chonos would not take you long; & it is probable you already know the name of some insignificant little plants, (the numbers of which are in the list of questions) which go to form the peat of that country.—Pray remember today is the 12th.—I know if possible you will answer the questions.

Again and again he appealed to Henslow. In August, with his *Journal* going off to the press in two days' time, he wrote to 'plague' Henslow about the Patagonian thistles. 'I verily believe no rider on the Pampas was ever more tormented by the living plants, than you have been by the dried ones.'

Henslow had written to say that Cambridge was in the midst of a Whig election. He took politics seriously, in 1835 when it was discovered that the Tories were using bribery in the borough election writing an open letter to the Vice-Chancellor denouncing the practice, when nobody else would come forward.

Darwin replied: 'I am very glad the election went off well. I am afraid amidst all the turmoil you would have hardly been able to have looked at my plants.—Pray write *soon* & tell me whether you can answer me any of the questions; so that I may know.'

It is interesting that Darwin particularly mentions the Keeling and Galapagos plants, for of all his collections these were the only two to be written up, except his fungi. But he knew them to be complete, as the Malays had assured him in the case of Keeling; while from the Galapagos he had brought away every plant he saw in flower, though he was anxious about these. 'Tell me whether you are disappointed with the Galapagos plants. I have some fears.' Again: 'Mr Brown also said that *you* must recollect that there are plants from the Galapagos Isds. at the Brit. Museum. It would be well to find out what they are.' And: 'I have forgotten to mention one bad bit of news, namely that Cuming was at the Galapagos.—Did he collect plants. I doubt it, because the far greater part of the plants only live near the summit of the mountains, some miles from the coast. I shall grieve if you lose your tiny botanical feast.'

The botanical feast of the Galapagos plants was not tiny to Henslow. It was a banquet whose riches were overloaded and indigestible. Some of the plants he could not place at all. They were unlike anything he had ever seen. He had no idea how to tackle the problem of classifying them nor how to sum them up in botanical philosophy. He looked more acutely at the Keeling plants: there were only twenty-one of them, a task not too formidable, and Darwin had written interesting notes about them. He sought the help of Sir William Hooker on points which completely puzzled him, and in the following year, 1838, in Volume I of the *Annals of Natural History*, appeared his 'Florula Keelingensis. An account of the Native Plants of the Keeling Islands.'

Another old pupil of Henslow's was the Rev. Miles Joseph Berkeley, a Christ's College man who had graduated in 1825. He was now curate of Apethorne and Wood Newton and was fast making his name as a fungologist. He was to become the founder of British mycology. To him Henslow sent Darwin's fungi collection. Berkeley wrote a notice of them which appeared in the *Annals of Natural History* in 1839. Several were new, of which *Daedalea erubescens* was one of the most beautiful of its race, as Berkeley wrote.

But why if the botanists cared so much about Darwin's botanical collections, was so little done to describe them? We can appreciate Darwin's nervousness at handing them over to the British Museum, especially when Robert Brown made no move to deal with a single interesting fungus. In Glasgow the active Sir William Hooker had described the botany of several voyages: the collections of Captain Edward Sabine during the *Griper* expedition of 1823; appendices to Captain Edward Parry's second and third voyages in search of a North-West passage, involving polar plants, and those from his fourth voyage; the botany of Captain Beechey's voyage to the Behring Sea and China, and Richard King's plants from the Arctic American land expedition of 1836 which was currently his concern. His son Joseph would eventually deal with the Galapagos plants, but in 1836 and 1837, when Darwin was vainly trying to find names for his plants, Joseph was still a medical student.

There was no one else, and Darwin's collections never were written up as a botany of the voyage. His specimens, still in a state of perfect preservation, are at the Botany School, Cambridge. The Galapagos plants have been extracted to keep them separate as a Darwin collection, and valuable work was recently done by Miss Mary McCallum Webster when she extracted the South American material. A few Keeling Islands plants have also been separated. But the rest of Darwin's plants are mingled with others in the main herbarium. Many of them are still unnamed.

So Darwin's *Journal* went to press without genus or species attached to his plants, only 'something about them' being written by Darwin himself, from his observations of them growing in the wild. That he deplored the want of these names is amply evident from his appeals to Henslow. That he deplored his botanical ignorance is evident from what he wrote in his diary on the last stage of the *Beagle*'s voyage, as she steered her course from the Azores to England. It was a summing up of what the voyage had meant to him, and what a similar voyage might mean to some other young man with a taste for adventure.

'Our voyage having come to an end, I will take a short retrospect of the advantages & disadvantages, the pains & pleasures, of our five years' wandering. If a person should ask my advice before undertaking a long voyage, my answer would depend upon his possessing a decided taste for some branch of knowledge which could by such means be acquired.

'No doubt it is a high satisfaction to behold various countries, & the many races of Mankind, but the pleasures gained at the time do not counterbalance the evils. It is necessary,' he pointed out, 'to look forward to a harvest however distant it may be, when some fruit will be reaped, some good effected. Many of the losses which must be experienced are obvious; such as that of the society of all old friends, & of the sight of those places, with which every dearest remembrance is so intimately connected. These losses, however, are at the time partly relieved by the exhaustless delight of anticipating the long-wished for day of return. If, as poets say, life is a dream, I am sure in a long voyage these are the visions which best pass away the long night.' Other losses were want of room, of seclusion, of rest; the jaded feeling of constant hurry; the privation of small luxuries, the comforts of civilisation, home life, and lastly even of music and the other pleasures of imagination. And if his intending voyager suffer much from seasickness, 'let him weigh it heavily in the balance,' Darwin cautioned, speaking from his experience of this haunting evil, 'and let him bear in mind how large a proportion of the time on a long voyage is spent on the water, as compared to the days in harbour.' No doubt there were some delightful scenes: a moonlight night, with the clear heavens, the dark glittering sea, the white sails filled by the soft air of a gently blowing trade wind; a dead calm, the heaving surface polished like a mirror, and all quite still excepting the occasional flapping of the sails. It was well once to behold a squall, with its rising arch and coming fury, though his imagination had painted something more grand, more terrible in the full grown storm. It was a finer sight on a canvas by Vandervelde, and infinitely finer when beheld on shore, where the waving trees, the wild flight of birds, the dark shadows and bright lights, the rushing torrents, all proclaimed the strife of the unloosed elements.

To him it was the days in harbour that were the brighter side of the five years, when the chain of the anchor rattled down to the sea-bottom and he could roam the countryside in each new land.

'The pleasures derived from beholding the scenery & general aspect of the various countries we have visited, has decidedly been the most constant & highest source of enjoyment.' Added to this, in comparing the character of the scenery in different countries, there was a pleasure distinct from mere enjoyment of their beauty. 'It more depends,' Darwin said, 'on an acquaintance with the individual parts of each view.' He believed that, 'as in Music, the person who understands every note, will, if he also has true taste, more thoroughly enjoy the whole; so he who examines each part of a fine view, may also thoroughly comprehend the full & combined effect.

'Hence,' Darwin advised, 'a traveller should be a botanist, for in all views plants form the chief embellishment. Group masses of naked rock, even in the wildest forms, for a time they may afford a sublime spectacle, but they will soon grow monotonous; paint them with bright & varied colours, they will become fantastick; clothe them with vegetation, they must form at least a decent, if not a beautiful picture.'

These were the words of Charles Darwin who had started the voyage as a geologist, and come home with the sublimity of the primeval forests brightest in his memory. 'Amongst the scenes which are deeply impressed on my mind, none exceed in sublimity the primeval forests, undefaced by the hand of man, whether those of Brazil, where the powers of life are predominant, or those of Tierra del Fuego, where death and decay prevail. . . . No one can stand unmoved in these solitudes, without feeling that there is more in man than the mere breath in his body.'

He gave his intending traveller no further advice—about arming himself with a knowledge of geology, zoology, or anything else. 'Be a botanist,' he said.

Was it not strange that Charles Darwin, who confessed himself 'unable to tell a daisy from a dandelion' (though in this he was rather poking fun at himself), should be the one not to follow the botanists but to lead them?

6

'These facts origin of all my views'

Simultaneously with the opening of his *Transmutation Notebook* Darwin turned his attention seriously to plants.

In the beautiful garden at Maer, home of Emma Wedgwood whom he married in 1839, and in his father's garden at Shrewsbury, he observed them with a botanical eye and made careful notes.

The note on Rhododendron azaloides *with the inset drawing of three stigmas was made at Maer in 1841. Done at his dictation it is in Emma's hand. There is an interesting footnote by Darwin.*

MARLBOROUGH STREET was hideous in Darwin's eyes. But it was convenient, and Syms Covington, still with him, an ideal clerk and amanuensis. ('He writes an excellent hand, and understands something of accounts.') He was also an artist, useful in doing drawings for him. Indeed, Syms Covington had been useful ever since he had become his personal servant on the *Beagle*. Darwin had taught him to shoot and skin birds, thus saving him much time and greatly augmenting his collections. At first Darwin had thought him 'an odd sort of person: I do not very much like him: but he is perhaps from his very oddity, very well adapted to all my purposes.' And so he had proved.

Marlborough Street had another virtue. 'It is very pleasant our being so near neighbours,' he wrote to William Darwin Fox, referring to Erasmus at Number 43. And next door were their cousins, the Hensleigh Wedgwoods. So there was quite a family colony. He told Fox that he had plenty of work for the next year or two, and until it was finished he would take no holidays. Cambridge had been too pleasant a place to work in, always with some agreeable party or other every evening. 'It is a sorrowful, but I fear too certain truth, that no place is at all equal, for aiding one in Natural History pursuits, to this odious, dirty smoky town, where one can never get a glimpse, at all, that is best worth seeing in nature.'

The publication of his *Journal* had been assured because it would be part of the official record of the *Beagle* voyage, but while he was still busy writing it he began to worry about the fate of his geological results. He thought Cambridge might provide an answer. Henslow had thought so, too, and Darwin wrote to remind him: 'Have you ever had an opportunity of sounding any of the great Cambridge Dons about the publication of my geology. I hope they will prove gracious for it would be a great bore to be half killed with seasickness, and then in reward half starved with poverty.'

And what of his zoology? The answer, in May, was that Henslow suggested that he should get the Duke of Somerset as president of the Linnean Society, Lord Derby, and William Whewell, president of the Geological Society whom Darwin had known at Cambridge, to sign their names to a statement of the value of his collections, and with this apply to the Government for a grant to engrave and publish the Zoology on some uniform plan. On the 16th of August Darwin was writing to thank Henslow 'for having so effectually managed my affair.' He had seen Thomas Spring Rice, the Chancellor of the Exchequer, that morning, and a Treasury grant of £1,000 would be forthcoming. The sumptuous work, *The Zoology of H.M.S. Beagle*, appeared in five parts between February 1838 and October 1843.

In May at the Geological Society he had read a paper on how coral islands had been formed, and one on the deposits containing extinct mammals in the neighbourhood of the Plata. His thoughts were very much on the origin of things. The giant fossils—Megatherium whose like did not exist today, armadilloes covered with armour like that on the present armadillo; and the diversity of the animals, birds and plants on the Galapagos Islands: 'The subject haunted me,' he wrote in his *Autobiography*. It was clear to him that such facts could be explained on the supposition that species gradually became modified—or were wiped out. It was equally clear that 'neither the action of the surrounding conditions, nor the will of the organisms (especially in the case of plants) could account for the innumerable cases in which organisms of every kind are beautifully adapted to their habits of life,—for instance, a woodpecker or tree-frog to climb trees, or a seed for dispersal by hooks or plumes. I had always been much struck by such adaptations, and until these could be explained it seemed to me almost useless to endeavour to prove by indirect evidence that species had been modified.'

He was seeing a lot of Lyell at this time. They discussed methods of producing evidence to substantiate theory. Lyell's way of working was to note down relevant facts as they came to him, and see how they added up. Darwin decided to follow his example, so that 'by collecting all facts which bore in any way on the variation of animals and plants under domestication and nature, some light might perhaps be thrown on the whole subject.' Accordingly, as he wrote in the 'little diary' in which he recorded his movements from place to place and the work on which he was engaged: 'In July opened first note Book on "transmutation of Species."—Had been greatly struck from about month of previous March on character of S. American fossils—& species on Galapagos Archipelago.' Later he added alongside: 'These facts origin (especially latter) of all my views.'

By 'transmutation' he meant the transmuting or changing of one species into another in the process of adapting to a changed environment.

In 1877 in a letter to Otto Zacharius he wrote: 'On my return home in the autumn of 1836, I immediately began to prepare my Journal for publication, and then saw how many facts indicated the common descent of species.'

He set himself to the task, combing agricultural and horticultural journals for information about variations in domestic stocks. He hobnobbed with horse-breeders and cattle-breeders, and joined the Columbarian and Philoperistera pigeon clubs. He read every book he could get hold of that might provide a clue, making a note of each fact or observation, together with its bibliographical reference. ('When I see the list of books of all kinds which I read and abstracted, including whole series of Journals and Transactions, I am surprised at my industry,' he wrote of this period.) Out of this mass of reading, and from his conversations with animal breeders and nurserymen, came a first conviction. 'I soon perceived that selection was the keystone of man's success in making useful races of animals and plants.' The question was, how could selection be applied to organisms living in a state of nature?

It was extraordinary how far at this stage he got along the path of his thinking. Lamarck who had believed in the mutability of species (because of the difficulty of distinguishing between species and varieties) had invented the symbol of a branching tree to replace the creationist static scale of living beings. An early note by Darwin reads: 'Organised beings represent a tree, *irregularly branched*; some branches far more branched,—hence genera.—As many terminal buds dying, as new ones generated.' Lamarck thought that animals and plants adapted themselves to their surrounding conditions by a 'slow willing'. Noted Darwin: 'Changes not result of will of animals, but law of adaptation as much as acid and alkali.' Besides, how could a plant 'will' itself to change? It was on this point that Darwin practically dismissed Lamarck.

In August 1837, after a last month of concentrated and uninterrupted work, he finished his *Journal*. Books were printed much more quickly in those days, and in September the printers were plaguing him for the corrected proofs. Less than a year had passed since his return from the voyage, and in that time he had gone from one anxiety to the next. His labours had been constant and unfamiliar. Apart from the brief visit to Shrewsbury in June he had been pent up in London. 'What a waste of a life,' he exclaimed in a letter to his cousin Elizabeth Wedgwood, 'to stop all summer in this ugly Marlborough Street and see nothing but the same odious house on the opposite side as often as one looks out.'

The strain began to tell, and on the 20th of September he wrote to Henslow that he was suffering from an uncomfortable palpitation of the heart, and that his doctors were urging him *strongly* to knock off all work and live in the country for some weeks. He thought he must do this but was in a puzzle how to go on correcting the proofs of his *Journal*. Henslow, who had been a second eye on the earlier proofs, now offered to receive them straight from the printers, leaving Darwin free to go to Shrewsbury; and there he spent about a month, going to the Wedgwoods for a few days before returning to London at the end of October.

It was on this visit to Maer that he made his first observations on earthworms. Thinking to interest him, Uncle Jos pointed out the large quantity of fine earth continually brought up to the surface by worms in the form of castings. Darwin concluded that all the vegetable mould over the earth's surface had passed many times through the intestinal canals of worms, and would again and again pass through them, in the process burying stones, paths, pavements, Roman villas and even cities, and ploughing the soil as they did so. On the 1st of November he read a paper on the subject at the Geological Society, in 1881 expanding it into a book. *The Formation of Vegetable Mould* sold 6,000 copies within a year and 13,000 before the end of the century.

Maer was responsible for more than that.

Now twenty-eight and with time on his hands to think of things other than work, his thoughts turned to the question of marriage. In July his sister Caroline had become engaged to their cousin Josiah Wedgwood III, a happy alliance, and no doubt this was what sparked off the balance-sheet Darwin drew up, heading it *This is the Question*.

MARRY	NOT MARRY
Children—(if it please God)—constant companion, (friend in old age) who will feel interested in one, object to be beloved and played with—better than a dog anyhow—Home, and someone to take care of house—Charms of music and female chit-chat. These things good for one's health. Forced to visit and receive relations *but terrible loss of time.* My God, it is intolerable to think of spending one's whole life, like a neuter bee, working, working and nothing after all.— No, no won't do.— Imagine living all one's day solitarily in smoky dirty London House.—Only picture to yourself a nice soft wife on a sofa with good fire, and books and music perhaps—compare this vision with the dingy reality of Grt Marlboro' St. Marry—Marry—Marry. Q.E.D.	No children, (no second life) no one to care for one in old age.—What is the use of working without sympathy from near and dear friends—who are near and dear friends to the old except relatives. Freedom to go where one liked—Choice of Society *and little of it.* Conversation of clever men at clubs. Not forced to visit relatives, and to bend in every trifle—to have the expense and anxiety of children—perhaps quarrelling. *Loss of time*—cannot read in the evenings—fatness and idleness—anxiety and responsibility—less money for books etc—if many children forced to gain one's bread.—(But then it is very bad for one's health to work too much) Perhaps my wife won't like London; then the sentence is banishment and degradation with indolent idle fool.

On the reverse side of the page he summed up his situation: 'It being proved necessary to marry—When? The Governor says soon for otherwise bad if one has children—one's character is more flexible—one's feelings more lively, and if one does not marry soon, one misses so much good pure happiness.' But if he married tomorrow the infinite trouble and expense in getting and furnishing a house—'fighting about no Society—morning calls—awkwardness—loss of time every day—without one's wife was an angel and made one keep industrious.' He ended: 'Never mind my boy—Cheer up—One cannot live this solitary life, with groggy old age, friendless and cold and childless staring one in one's face, already beginning to wrinkle. Never mind, trust to chance—keep a sharp look out.—There is many a happy slave—'

The balance-sheet was written on Maer notepaper, the watermark date of which was 1837, facts to which Dr Sydney Smith called my attention. Perhaps that far-seeing Uncle Jos had asked his thoughts on marriage, and Charles, taught by Paley to weigh pro and con, had promptly sat down to work out the equation. Was it done half in fun or wholly in earnest? In whatever vein, the decision to marry seemed to set his mind at rest. His *Beagle* writings went on smoothly. 'I set my shoulder to the work with a good heart,' he told Henslow. By the 25th of February he had finished his account of the geology of the Galapagos Archipelago, of Ascension Island, St Helena and the small islands in the Atlantic, on the 16th having been elected to the secretaryship of the Geological Society—unwillingly because it would consume precious time. But he did not give up this time-consuming post until 1841. Even writing up 'Birds for Zoology. much time thus lost.' Time now was for 'Existence of Species'.

Most of his first speculations had concerned animals, but now in his Transmutation notebook he began to link animals and plants together. 'All animals of same species are bound together just like buds of plants, which die at one time, though produced either sooner or later.—prove animal[s] like plants:—trace gradation between associated and non-associated animals—and the story will be complete.'

He sent out feelers in every direction. 'Fox tells me, that beyond all doubt seeds of Ribstone Pippins produce Ribstone Pippins, and Golden Pippins—goldens; hence *sub-varieties* and hence possibility of re-producing any variety, although many of the seeds will go back. Get instances of a *variety* of fruit tree or plant run wild in foreign country. Here we have avitism the ordinary event and succession the extraordinary.'

He struck a note which anticipated a powerful theme of his book on *The Effects of Cross- and Self-Fertilisation* when he asked himself: 'Do not plants, which have male and female organs together, yet receive influence from other plants—Does not Lyell give some argument about varieties being difficult to keep on account of pollen from other plants because this may be applied to show all plants do receive intermixture.'

A piece of the jigsaw came in a sentence culled from the *Edinburgh New Philosophical Journal*: 'Ancient Flora thought to [be] more uniform than existing.' Already he saw a relationship between plants and insects: 'Ask Entomologists whether they know of any case of *introduced* plant, which an insect has become attached to, that insect not being called omniphitophagous.'

It was to be the weight of his experiments with plants that was to prove his theories about them and convince science that he was right. Meanwhile he suggested: 'It really would be worth trying to isolate some plants under glass bells and see what offspring would come from them. Ask Henslow for some plant, whose seeds go back again, not a monstrous plant, but any marked variety.—Strawberry produced by seeds??—Universality of generation strongly shown by hybridity of ferns.—Hybridity showing connexion of two plants.'

There were doubts and difficulties. 'It is scarcely possible to get evidence of two races of plants run wild.—(For we know that such can take place without impregnating each other). For if they are different, then they will be called species, and mere producing fertile hybrids will not destroy that evidence, as so many plants produce hybrids, or else whole fabric will be overturned.—Hence extreme difficulty, argument in circle.'

Then, almost at the end of this first notebook, which he completed in
February 1838, came 'THE GRAND QUESTION: Are there races of plants run
wild or nearly so, which do not intermix—any cultivated plants produced
by seed.—Lychnis—Flax.'

The conclusion Darwin reached at this point was that transmutation
of species had taken place when populations were isolated (as in the
Galapagos) and no longer able to prevent the variation normally kept in
check by breeding throughout the population. So varieties became split
off from species and eventually became species themselves, while old
species became extinct.

He was now calling this line of thought 'my theory'.

In his second notebook he got as far as wondering how he could best
present it on paper. 'Argue opening case thus,' he instructed himself.

On the 23rd of September (1838) he went to Hackney to visit the
famous nursery of Conrad Loddiges. 'Saw in Loddiges garden 1279
varieties of roses!!! proof of capability of variation.' He went on to argue:
'Are not Loddiges 1279 roses kept in same soil, same atmosphere?—may
they not be transplanted?, & yet year after year successive roses & bud
are produced, like parent stock or if different deteriorating very slowly—
I presume most of these roses, without circumstances very unfavourable,
will continue of same variety as long as life lasts, yet they cannot
transmit through seeds these characters though transmitting them with
such facility to bud.—This must be owing to their unity in one stem.—A
bud may be transplanted & carry all these peculiarities.—not so a seed.' He
thought that 'Bud probably is like cutting off tail of Planaria, claw added
to crab, tail to lizard, healing of wound', and went on to reason that 'in the
separated part every element of the living body is present, in generation
something is added from one part of the body, (or of similar, body) to
another part of body.'

On the 13th of September he had written to Lyell of the 'delightful
number of new views which have been coming in thickly and steadily, . . .
bearing on the question of species. Note-book after note-book has been
filled with facts which begin to group themselves *clearly* under sub-laws.'

Then, on the 28th of September, he read—for amusement, as he
recorded, ' "Malthus on Population," and being well prepared to appreciate the struggle for existence which everywhere goes on from long-
continued observation of the habits of animals and plants, it at once
struck me that under these circumstances favourable variations would
tend to be preserved, and unfavourable ones to be destroyed. The result of
this would be the formation of new species. Here then I had at last got a
theory by which to work.'

The Rev. Thomas Robert Malthus, born at Dorking in 1766, was a
political economist who had human happiness much on his mind. Indeed,
this was part of his book's title, which was *An Essay on the Principle of
Population; or, a View of its Past and Present Effects on Human Happiness;
with an Inquiry into our Prospects respecting the Future Removal or
Mitigation of the Evils which it occasions*. It was published in 1798.

To Darwin it gave the clue he was looking for, the prodigality of nature being kept in check by the warring of one set of animals upon another, by the failure of a crop causing famine. The result would be that species must adapt to changes or perish. In the struggle for existence favourable variations would be the means of survival.

On page 175 of his third notebook Darwin wrote that it was absolutely necessary that 'some but not great difference . . . should be added to each individual before he can procreate. then change may be effect of difference of parents, or external circumstances during life', and that if the external circumstances which induced change were always of one nature, species would be formed. If not, 'the changes oscillate backwards & forwards & are individual differences. (hence every individual is different).' He added: 'All this agrees well with my view of those forms slightly favoured getting the upper hand and forming species.' This page is undated. But occasionally page numbers have no relation to dates. While page 152 is dated the 11th of September, 160 the 16th and 163 the 25th, the Malthus reference of the 28th of September is on page 134. It is possible therefore that his 'view of those forms slightly favoured getting the upper hand & forming species' was written before he read Malthus. At all events Darwin in his *Autobiography* credits Malthus with stimulating his own views.

In June 1838 there was a pleasant interruption to his weighty thoughts: a family reunion at the Hensleigh Wedgwoods when his sister Emma and Darwin's sister Catherine returned from Paris where the Sismondis and some of the other Wedgwoods had all met together. Emma wrote to her aunt Jessie Sismondi that her stay in London had been very enjoyable, Thomas Carlyle dining with them one night, and Hensleigh's father-in-law, Robert Mackintosh, very bright and pleasant, dining with them or coming in every evening, and Charles Darwin coming from next door. This she wrote on her return home to Maer.

A few weeks later, Charles also was at Maer. He went again in the autumn, though rather later than for the shooting. This time he had another quarry in view: he had made up his mind to marry his cousin Emma Wedgwood. 'He was far from hopeful, partly because of his looks, for he had the strange idea that his delightful face, so full of power and sweetness, was repellantly plain.' So wrote their daughter Etty in her book, *Emma Darwin: a Century of Family Letters*. But what happened on Sunday the 11th of November is told in Charles Darwin's own words, one brief ecstatic line: 'The day of days!' Emma had accepted him.

The exchange of letters between Uncle Jos and Robert Darwin is evidence of the joy their engagement brought to both families, Jos the practical in his letter telling Dr Darwin that he would do for Emma what he had done for her sister Charlotte and for three of his sons, give a bond for £5,000 and allow her £400 a year. Dr Darwin was already allowing Charles £2,000 a year. Charles wrote to Emma: 'I positively can do nothing, and have done nothing this whole week, but think of you and our future life.' Emma wrote to her Aunt Jessie Sismondi: 'When you asked me about Charles Darwin, I did not tell you half the good I thought of him for fear you should suspect something, and though I knew how much I loved him, I was not the least sure of his feelings, as he is so affectionate, and so fond of Maer and all of us, and demonstrative in his manners, that I did not think it meant anything.' His visit in August, when he was in very high spirits, and she very happy in his company, had given her 'the feeling that if he saw more of me, he would really like me.' His proposal was 'quite a surprise', as she had thought that very likely nothing would come of it after all. 'He is the most open, transparent man I ever saw, and every word expresses his real thoughts,' was Emma Wedgwood's summing up of her future husband.

Because of Charles's work they decided to make their home in London. After a few years they might feel that the pleasures of the country ('Gardens, walks, &c.') were preferable to society. But meanwhile they would reap the advantages of London life while they were compelled to live there. Charles and Erasmus on an initial scout, walking up one street and down another looking for the words *To Let*, found that 'houses are very scarce and the landlords are all gone mad, they ask such prices'. Hardly different from today. Charles saw that they would be obliged to pay a rent of at least £120. Emma joined in the search, and Hensleigh and his wife Fanny, and finally Charles succeeded in taking one they had both set their hearts on in Gower Street. It was a furnished house, and although everything was perfectly hideous—Macaw Cottage, they dubbed it from its odious yellow curtains—they loved every inch. An armchair was purchased for Charles, and Emma's father gave them a piano for Emma to play and Charles to listen to, a pleasure he enjoyed to the end of his days. At the beginning of January he moved in with the help of Syms Covington who was soon to emigrate to Australia. Friends and relatives found servants. At last they were equipped, and if it was still London at least there was a garden at the back. London had also meant buying clothes, principally Emma's wedding dress which was of a greenish-grey rich silk with a remarkably lovely white chip bonnet trimmed with blonde and flowers. On Tuesday the 29th of January 1839 Charles and Emma were married at Maer church. The wedding was quiet and they went at once to 12 Upper Gower Street where the servants gave them a welcome with blazing fires all over the house. Everything now looked very comfortable. They settled down to married life.

In the three years and eight months they spent in London, as Darwin wrote in his *Autobiography*, he did 'less scientific work' than at any other period of his life. This, he explained, was owing to frequently recurring unwellness and to one long and serious illness; though, he allowed, he worked as hard as he possibly could.

What was he doing?

Before his marriage he had begun *Coral Reefs*. This, though a small book, cost him twenty months of hard labour, as he had to read every work on the islands of the Pacific and consult many charts. He corrected the last proof on the 6th of May 1842. His theory of how barrier-reefs and atolls were formed—by the uplifting of an island by submarine volcanic

Emma Darwin in 1840, the year after she married

action, the colonising of its slopes by myriad coral polyps, and the gradual subsiding of the island into the sea, a process taking not less than a million years—is still the classic theory today.

He read two papers at the Geological Society, one on the erratic boulders of South America and the other on earthquakes.

But these were still the garnerings of the *Beagle* voyage. Now something else was stirring: a personal involvement with plants. This was something new. When the Transmutation theory struck him he had read every book and paper he could find, not only on animals but on

Charles Darwin in the same year, when he was thirty-one

plants. Now, having read what they had to say, he turned his attention to what the plants and the flowers themselves had to say.

So it is interesting that Darwin thought of himself as doing 'less scientific work' in these years, when it was in this period that he took up the study which was to become his major scientific work.

Preserved in the Darwin archives in the University Library, Cambridge, are notes he made from 1840 onwards, on variability and flower structures. They reveal that he had not forgotten the botany he had studied under Henslow. They reveal also that his mind was deeply probing into the meaning of flower structure and the relationship between structure and modification.

His studies of flowers in fact went back to '1838 or 1839' as he tells us. Page 129 of his fourth Transmutation notebook (October 1838 to 10 July 1839) refers to a belief by Henslow that 'Only red Lychnis grows in Wales & certainly only white in Cambridge', and that 'in some counties sometimes one and sometimes other.—there is some difference in habit between these varieties, so that they have been thought to be different species. Lychnis dioica generally dioecious [having the male and female flowers on different plants] yet parts only very slightly abortive & bed of female flowers will sometimes produce a few seeds.' This entry was dated April 3rd 1839. In 1841 he was investigating the Red Campion for himself. A field note reads: 'All Lychnis dioica in Wales red in Cambridgeshire all white. Suffolk and Staffordshire both colours.' He was in Staffordshire at Maer in July of that year when it would be in flower. (These two campions are now regarded as separate species of the genus *Silene*, *S. dioica* and *S. alba*.)

Darwin says in his *Autobiography* that 'During the summer of 1839, and I believe during the previous summer, I was led to attend to the cross-fertilisation of flowers by the aid of insects, from having come to the conclusion in my speculations on the origin of species, that crossing played an important part in keeping specific forms constant.' He was certainly interesting himself in the subject by October 1838, when he wrote in his fourth notebook on the 11th: 'Uncle John says he has no doubt bees fertilise enormous number of plants—it is scarcely possible to purchase seed of any cabbage where a great number will not return to all sorts of varieties, which he attributes to crossing.' This was John Hensleigh Allen of Cresselly in Pembrokeshire whose sister Elizabeth was Emma's mother. Darwin had probably seen him in the July when he was 'very idle at Shrewsbury'.

In the summer of 1839 he was studying dimorphism in the different forms of flax flowers. A field note accompanied a dried specimen: 'July 12th/41/Shrewsbury. Linum flavum.—an old plant has in *all flowers* stigma rather shrunken, & whole pistil *much* shorter than flower of one 2 years ago. Old cutting which has (normally) pistils standing above anthers in all its flowers . . .' We remember that he mentioned 'Flax' at the end of his first notebook, which he finished in February 1838.

He wrote in his fourth and last Transmutation notebook what might appear to be a very odd entry, when he asked himself: 'Is there any very sleepy mimosa, nearly allied to the Sensitive Plant.—' He went on: 'Shake some sleeping mimosa—do stamina of C. speciosus [*Clianthus formosus*, the Glory Pea] collapse at night, if so irritate them, as by an insect coming always at same time, see if by so doing can be made sensitive.' Darwin did not write of the sleep-movement of plants until 1880, but he was now investigating it for his Species work. The collapse of stamens at a touch he had already demonstrated on his Patagonian opuntia when he simulated a visiting insect.

Did a plant change when taken out of its natural place of growth? He

Two BRITISH ORCHIDS examined by Darwin:

(*above*) close-up and in situ, *Orchis morio*, the Green-winged Orchid

(*below*) in situ and close-up, the well-named Bee Orchid, *Ophrys apifera*

dug up a root of the knapweed *Centaurea nigra* and noted: 'kept habit when transplanted.' This was at Maer in June 1841. Emma sometimes helped him, as when she took down his notes on *Rhododendron azaloides* while he examined the flower: 'More delicate like Azalea, filament perfect, anthers *little* hard shrivelled triangle with folds, so that if filled would have form of anther of Rhod: therefore every organ perfect except pollen itself—aperture mere obscure slit. Scarcely rises higher than junction of filament. Sometimes turns up & is not parallel to filament as in 2. Carefully dissected under water not grain or rudiment of pollen. Stigma humid gummy & apparently perfect as in Rhod.' There was a *nota bene*. 'Is not this perfection of pistil general? Is it not analagous to stigma resisting conversion into petals longer than the stamens in double flowers.'

The useful conversion of one organ into another was an element of the success-story of plants which had modified themselves in order to survive, and this was one of the subjects Darwin was to discuss in the *Origin of Species*. It was related to the development of flowers during the early or embryonic stages of their lives, and he was to prove that embryos provided an important piece of evidence in relation to the origin of species.

In November 1841 Robert Brown (whom Humboldt called 'Facile princeps botanicorum') recommended him to read Christian Konrad Sprengel's *Das entdeckte Geheimnis der Natur*, a book Darwin found 'full of truth' but 'with some little nonsense'. But it had a bearing on one of the subjects he was studying, the fertilisation of flowers by the aid of insects. Sprengel had discovered that in many cases pollen must be carried from a flower to the stigma of another flower. What he did not understand was the advantage gained by the intercrossing of distinct plants, a weakness fatal to his book. Charles Darwin was to correct Sprengel's defective ideas.

It was in the well-stocked gardens at Maer and The Mount and the surrounding countrysides that Darwin was able to carry out his observations, watching, for instance, the effect on 'Thyme with abortive anthers [Maer June 8/42] moved from Pump to Garden door last year'. It was there that he amassed a collection of plants in spirits, his catalogue of these running into four figures. Away from London his health was better and he found he could work easily. In June 1842, dividing his time between both places, he 'wrote pencil sketch of my species theory.' It consisted of thirty-five pages, obviously written at great speed, the articles as in his notebooks often being omitted, words sometimes ending in mere scrawls. But at last he had it down on paper, and this was a satisfaction.

On the 18th of the following month he recorded in his diary that he was 'employed about Down'. It was the happy ending of a hunt for a country house. The Darwins, who now had two children, had escaped as often as they could to their family homes, but since September of the previous year, as Darwin told his cousin Fox, they had been 'taking steps to leave

London, and live about twenty miles from it on some railway.'

The village of Downe in Kent was miles from a railway station, eight and a half to be precise, and Down House (without the 'e') about a quarter of a mile from the village, which had only about forty houses. There was no main thoroughfare. Lanes converged on its centre where stood some old walnut trees and an old flint church. There was an infants' school, the villagers touched their hats as in Wales and sat at their open doors in the evening. The little pot-house was also a grocer's shop and its landlord the village carpenter. Thus Darwin described it to his younger sister Catherine. As for Down House itself, it stood close to a tiny lane and was ugly, looking neither old nor new and with rather small windows. But it was in good repair and had a capital study, eighteen feet square. There were three storeys and plenty of bedrooms. 'We could hold the Hensleighs, and you, and Susan and Erasmus all together.' There were two bath-rooms. The water came from a deep well.

With the house went a fifteen-acre field looking down into flat-bottomed valley on both sides. Outside the drawing-room there were some old but very productive cherry trees, walnut trees and an old mulberry, which made rather a pretty group. A purple magnolia flowered against the house. The kitchen garden was 'a detestable slip' and the soil looked wretched from the quantity of chalk and flints, but they were told it was productive. But to Darwin 'the charm of the place is that almost every field is intersected (as alas is ours) by one or more foot-paths. I never saw so many walks in any other country. The country is extraordinarily rural and quiet with narrow lanes and high hedges and hardly any ruts. It is really surprising to think London is only 16 miles off.'

To Emma the whole thing was initially a disappointment: the house and the country around the house. She first saw it on the 22nd of July, a day that was cold and gloomy with a north-east wind. And she had toothache and a headache. They stayed that night at the little pot-house and went back the following day. This time she liked the field and house even more than her husband did, and in driving away from it, as the countryside rolled by, up one steep hill and down the next, she turned to him and told him that they would be very happy there. So, both approving, Charles was able to write home a full description of the place, to satisfy his father who was advancing the purchase price of £2,020. In September 1842 the Darwins moved their home from Upper Gower Street. On the 23rd, nine days after her arrival at Down House, Emma gave birth to a daughter, Mary Eleanor, who lived only three weeks. It was a sad beginning, but already there were two others: William Erasmus, born on the 27th of December 1839, and his fifteen-months-younger sister Anne Elizabeth.

What a house of children it was to be! What a house of genius, flowing down to them! Meanwhile, just before Christmas, Charles Darwin set an experiment going in his field near the house, spreading part of it with broken chalk so that he could observe 'at some future period to what depth it would become buried.' This was for his earthworm book. For his book on volcanic islands, to be published in 1844, he was rearranging the notes Syms Covington had made for him. He was busy with plans to improve his new home and extend the garden. His days, as he said, were very full.

It was marvellous how, like the man in the circus, he managed to keep so many plates so skilfully turning.

7

'At last gleams of light have come'

When Joseph Hooker was preparing a paper
for the Linnean Society on plants collected over
the years in the Galapagos Islands, Darwin was
most anxious to know if his own outnumbered
those collected by other visiting botanists. To
this, Hooker was able to write a resounding Yes.

Hooker, the most honoured botanist of all time,
gave Darwin the palm for his masterly sketch of
the unique zoology and unequal dispersion of
plants over the various islets, 'and to whose
comprehensive view of the natural history of the
Galapagos this essay can be considered as
supplementary only.'

I N 1839 ANOTHER young man had set out on a voyage of discovery. He was Joseph Dalton Hooker, assigned to the expedition Captain James Clark Ross was taking to the Antarctic for the purpose of locating the south magnetic pole. He was twenty-two years old, the same age as Charles Darwin at the start of the *Beagle* voyage, and by the time *Erebus* and *Terror* triumphantly returned from their mission, in September 1843, his father, Sir William, had begun the improvements and extensions which were to make the Royal Botanic Gardens, Kew, the greatest in the world.

Sir William was active in every direction, and sadly overworked. It was his dream to see his son established as his assistant and, one day, successor. Meanwhile, Joseph's first task was to write up the botany of the voyage, and he was anxious to see the plants Darwin had collected. Having devoured his *Journal* he knew already that those of the Galapagos Archipelago were of a novel character, and Darwin had wondered 'to what district or "centre of creation"... they must be attached.' To Joseph this was a speculation of much interest.

His father was a great arranger. He wrote at once to Henslow who replied belatedly on the 13th of November 1843 that he would be delighted to let Joseph make use of the Darwin plants. The trouble was the sending of them from Cambridge where he was unlikely to be until April for his Easter-term lectures, though if Joseph were anxious to have them sooner he would contrive to get over for a day or two and pack them off. Henslow told Darwin of all this and said he had sent Joseph Hooker the plants, including those from South America. On the 21st of November Darwin wrote to young Hooker congratulating him on his safe return from his long and glorious voyage. He had hoped before now to have had the pleasure of meeting him. He was glad that Henslow had sent him 'my small collection'—Darwin was ever modest. 'You cannot think how much pleased I am,' he wrote, 'as I feared they wd have been all lost & few as they are, they cost me a good deal of trouble.' He explained that there were several notes, which he believed were with Henslow, describing the habitats of some of the more remarkable plants, and that he had paid particular attention to the alpine flowers of Tierra del Fuego and was sure he had collected everything in flower in Patagones.

He added:

I have long thought that some general sketch of the Flora of the point of land, stretching so far into the southern seas, would be very curious.—Do make comparative remarks on the species allied to the European species, for the advantage of Botanical Ignoramuses like myself. It has always struck me as a curious point to find out, whether there are many European genera in T. del Fuego, which are not found along the ridge of the Cordillera; the separation in such cases wd be so enormous.—Do point out in any sketch you draw up, what genera are American & what European & how great the differences of the species are, when the genera are European, for the sake of the Ignoramuses. I hope Henslow will send you my Galapagos Plants (about which Humboldt even expressed considerable curiosity)—I took much pains in collecting all I could.—A Flora of this archipelago would, I suspect, offer a nearly parallel case to those of St Helena, which has long excited interest.

Alexander Humboldt would have been pleased to know that Joseph Hooker was going to examine Darwin's plants. In a long letter of praise for the *Journal*, there was one thing he deplored. 'How much I regret that Mr Henslow could not finish his examination of your curious collection, or at least the keying of the families containing some known species. The vegetation of a country exhibits a fundamental character. By tracing the main features, one gets an image which will remain in one's mind, something like a stereotype . . .'

Joseph replied to Darwin on the 28th of November from West Park, Kew.

My dear Sir,

Many thanks for your kind letter of congratulations & also for your offer of assistance in examining the plants you collected, of which I shall most thankfully avail myself. It is very liberal of you to place them as at my disposal & I do hope that I shall shew myself not to be altogether unworthy of the trust.

Prof[r.] Henslow has promised me your plants, but they have not arrived yet, he having hardly got home after his late visit to Town. Amongst them I hope to find much that is curious & something new too, as it is the alpine plants of the Fuegian Islands which I want more than anything else. In the month of October I could get few in flower, even on the low grounds of Hermite Island, the mountain plants are of course in a much more unsatisfactory state, but as I collected every scrap for the sake of illustrating the Geog. distrib. of the species whenever better specimens should turn up to identify them, they may yet prove of interest.

The cryptogamic plants being much more widely distributed & being in a tolerable state in all seasons, I was enabled to form a pretty good collection of them, including 60 species of Mosses alone from that little Isld: as also a new species of your & Mr Berkeley's genus Cyttaria from the duodecimus leaved Beech (a much smaller plant with only 4 cells) & a tolerable collection of Lichens. . . .

I am exceedingly glad to think you attach so much importance to the comparison of the Arctic plants with the Antarctic as it was my aim throughout to establish an analogy between the two hemi-spheres, & to draw up tables upon several plans, shewing for instance the proportion of plants in each of the predominant Nat. Ord[s.] common to both, as also how that proportion diminishes in leaving the lower forms & ascending to the higher.

Joseph wondered what he was to take as the northern limit of the flora of the west coast of South America. He was anxious to take in the glacier-bound Gulf of Peñas and the Tres Montes peninsula, 'which would be very tolerable geographical features', and, he added, 'if I knew any Botanical ones it would be much better—Can you tell me what the northern limit of the Deciduous or Evergreen beech may be at the level of the sea, you mention one as common in the Chonos Archipelago, but not so abundant in proportion to other trees, as it is to the Southwd.'

Darwin had taken careful note of the beeches and their leaves, but he thought that the Fuegian differed from the Chonos beech. In fact he thought there were two separate ones in the Archipelago. Joseph Hooker in his *Flora Antarctica*, when discussing *Fagus antarctica* and *Fagus forsteri* (the generic name *Nothofagus* is now used for the Southern Beeches), wrote the following credit: 'Trees allied to these seem to have charac-terised the ancient or fossil flora of Fuegia, for I owe to Mr Darwin's kindness impressions of three apparently distinct species of deciduous Beech.' And he was able to fix the Chonos Archipelago as the limit of his antarctic flora.

This was the start of a correspondence that was to go on unin-terruptedly for thirty-nine years. It was also the beginning of a great scientific friendship that was to result in the bombshell of the *Origin*.

The two were not entirely strangers. Shortly before the *Erebus* sailed Joseph was walking through Trafalgar Square with a fellow officer who had served in the *Beagle*. He, recognising Charles Darwin coming towards them, hailed him and introduced Joseph. The meeting was brief but young Hooker bore away the memory of 'a rather tall and rather broad-shouldered man, with a slight stoop, an agreeable and animated expression when talking, beetle brows, and a hollow but mellow voice; and that his greeting of his old acquaintance was sailor-like—that is, delightfully frank and cordial.' To Joseph, Charles Darwin was already a hero, for he had read some of the proofs of his then unpublished *Journal*. These Darwin had sent to Charles Lyell who had sent them to his father, Charles Lyell of Kinnordy in Forfarshire, a botanist and old friend of Sir William Hooker, who with a kind interest in Joseph's projected career as a naturalist, had passed them on to him—at the moment when Joseph was hurrying on his studies so as to take his medical degree before the expedition sailed. 'So pressed for time was I,' Joseph later told Darwin's son Francis, 'that I used to sleep with the sheets of the "Journal" under my pillow, that I might read them between waking and rising. They impressed me profoundly, I might say despairingly, with the variety of acquirements, mental and physical, required in a naturalist who should follow in Darwin's footsteps, whilst they stimulated me to enthusiasm in the desire to travel and observe.'

Knowing how Darwin began his career had even helped Joseph to begin his own. His appointment was 'Assistant Surgeon and Botanist to the *Erebus*'. The surgeon, Dr Robert Maccormick (he who had sailed in the *Beagle*), was to be Zoologist, an arrangement that angered Joseph because it would 'completely interfere with all my duties'. He asked Ross whether he would take a 'Naturalist' if the Government appointed one. Ross's reply was that if he did it would have to be 'such a person as Mr Darwin', to which Joseph retorted: 'What was Mr Darwin before he went out?' Ross, amused, but admiring the young man's sense of responsibility, had compromised by appointing him Assistant Surgeon to the *Erebus* and Botanist to the Expedition, a hair-splitting distinction which, however, satisfied Joseph's principles and gave him the status he needed. He sailed with a copy of Darwin's now published *Journal*, a gift from Charles Lyell.

Darwin and Hooker had therefore much in common. Both, at the same age, had sailed in ships of discovery. Both had viewed the world through a naturalist's eyes. There was a difference. Joseph, even at twelve an able assistant in his father's herbarium, had sailed as a botanist. Darwin, though an all-round naturalist, had regarded himself first as a geologist and become converted to botany. ('rocks . . . will soon grow monotonous', 'in all views plants form the chief embellishment'.) Both had come home stirred by the mystery of how plants had wandered over the face of the earth. To Joseph the subject was a map whose uncharted regions had to be filled with botanical facts. To Darwin it had a deeper significance.

For this reason he was disappointed that Joseph did not expect to find the Galapagos plants very interesting. He hastened to point out in his next letter, dated the 12th of December 1843: 'Pray be careful, to observe, if ever I mark the individual Is^d of the Galapagos islands, for the reason you will see in my Journal.'

It was the individual peculiarities of these island plants which captured Joseph's attention when at last, in the middle of December, he received them from Henslow, with the result that Darwin was acknowledged as laying the foundations of a new flora. Previously Joseph had received the South American plants, and these he used for his *Flora Antarctica*, describing them and giving the discoverer's name. Many times 'C. Darwin, Esq.' was thus credited. This first Flora of Joseph Hooker's three great works on the botany of the expedition was published in two parts between 1844 and 1847. In his Introduction, in mentioning St Paul's Rocks, he drew the reader's attention to page seven of 'Mr Darwin's Journal', for his 'admirable description of these remarkable rocks, distant 350 miles from the nearest land (the island of Fernando Noronha).'

Darwin had long been very anxious to have a summing-up of how the plants of the southern regions were distributed and how they were related to the plants of other regions. He had drawn Joseph's attention to the importance of correlating the Fuegian flora with that of the Cordillera and of Europe. 'This,' as Hooker recalled to Francis Darwin, 'led to me sending him an outline of the conclusions I had formed regarding the distribution of plants in the southern regions, and the necessity of assuming the destruction of considerable areas of land to account for the relation of the flora of the so-called Antarctic Islands. I do not suppose that any of these ideas were new to him but they led to an animated and lengthy correspondence full of instruction.'

As for the alpine plants from Fuegia, which Joseph was particularly anxious to see, Darwin regretted that his first and best set had been lost (this in the gale that stove in one of the boats and practically swamped the ship).

So at last Darwin had a botanical confidant, someone with whom he could exchange views and with knowledge enough to be the touchstone of facts at which he had been hammering so long and so hard. Eagerly he put it to Joseph: 'Is not the similarity of plants of Kerguelen Land and S.S. America very curious,' and asked: 'Is there any instance in the Northern Hemisphere of plants being similar at such great distances.'

His next letter (the third) shows how quickly a trusted friendship sprang up between them, for in it Darwin divulged a conclusion he had come to, which was, to say the least of it, portentous in its implications, and because it was so serious a thing he made light of it, almost a joke of it. The letter was dated 'Jan 11, 1844'.

He began by discussing the botany of South Patagonia ('I collected *every* plant in flower at the season when there'), saying that it would be worth comparing with the North Patagonian collection by d'Orbigny, and went on to tell Joseph that 'My cryptogamic collection was sent to Berkeley; it was not large; I do not believe he has yet published an account, but he wrote to me some years ago that he had described & mislaid all his descriptions. Wd it not be well for you to put yourself in communication with him; as otherwise some things will perhaps be twice laboured over.—My best (though poor) collection of the cryptogam. was from the Chonos Islands.' Next he asked Joseph to observe whether any species of plant *peculiar* to any island such as the Galapagos, St Helena or New Zealand, where there were no longer quadrupeds, had hooked seeds. 'Such hooks as if observed here would be thought with justness to be adapted to catch into wool of animals.'

And then he wrote:

Besides a general interest about the southern lands, I have been now ever since my return engaged in a very presumptuous work & which I know no one individual who wd not say a very foolish one.—I was so struck with distribution of Galapagos organisms &c &c & with the character of the American fossil mammifera, &c &c, that I determined to collect blindly every sort of fact, which cd bear any way on what are species.—I have read heaps of agricultural & horticultural books, & have never ceased collecting facts—at last gleams of light have come, & I am almost convinced (quite contrary to opinion I started with) that species are not (it is like confessing a murder) immutable. Heaven forfend me from Lamarck nonsense of a 'tendency to progression' 'adaptations from the slow willing of animals' &c.—but the conclusions I am led to are not widely different from his—though the means of change are wholly so—I think I have found out (here's presumption!) the simple way by which species become exquisitely adapted to various ends.—You will now groan, & think to yourself 'on what a man have I been wasting my time on writing to.'—I sh^d, five years ago, have thought so:—I fear you will also groan at the length of this letter—excuse me, I did not begin with malice prepense.

When Darwin next wrote, on the 23rd of February 1844, he began his letter 'Dear Hooker'. The 'My dear Sir' was dropped for ever, and soon the formal 'Yours truly' was ousted by variations on 'Ever yours' and, as their friendship ripened, 'Yours affectionately'. He hoped that Joseph would excuse the freedom of his address, but felt that as 'co-circum-wanderers and as fellow labourers (though myself a very weak one)' they might throw aside some of the old-world formality. Almost from the first he had dropped into an easy style of letter-writing. Indeed, he called it 'talking to Hooker'.

In 1844 two particular pieces of work were occupying Darwin. He was 'very busy scientifically and unscientifically in planting.'

The Down House garden was unsheltered. One of the first things he did was lower the lane by about two feet and use the excavated earth to make banks and mounds round the lawn. These he planted with evergreens. At the same time he built a high flint wall along the lane to give privacy. A bowed front was added to the south-west face of the house, long sash windows replacing the ones they had found rather small. The house was then stuccoed, and creepers were now beginning to cover its walls. Plants came from Maer, including a root of the Double Apple Rose (*Rosa villosa* var. *duplex*) which is still growing at Down. George Darwin, born in July

Down House as it was when Darwin lived there, from 1842 till his death in 1882

1845, retained a deep love of Down all his life, remembering the lawn with its bright strip of flowers, the row of big lime trees that bordered it, the two yews between which was the swing their father put up for them, and the Sand Walk which became known as his 'thinking walk', with its acre of copse planted in 1846. There were a thousand other delights for children, a hollow ash and an enormous beech Darwin had remarked on when he viewed the house. This was always called the Elephant Tree because of something like the head of a monstrous beast growing out of its trunk where a branch had been cut off. It grew by the Sand Walk and is still there.

The Sand Walk, Darwin's 'Thinking Walk', a narrow strip of land 1½ acres in extent, around which he took a certain number of turns every day, counting them by means of a heap of flints, one of which he kicked on to the path each time he passed. A plaque on the tree now reads:
'Elephant Tree Preserved through the generosity of Lois and Lewis Darling, Connecticut, U.S.A.'

'Very busy scientifically' referred to a new version of the Sketch he had written two years before. In January after sixteen months of work he had finished his book on *Volcanic Islands* which formed the second part of the *Geology of the voyage of the Beagle*. This was published in November 1844. After catching up with current scientific reading (including the *Vestiges of the Natural History of Creation*, published anonymously in 1844 but later known to have been written by Robert Chambers, founder of *Chambers's Journal*, which, though it created a great deal of talk, Darwin dismissed as strange and unphilosophical, its geology bad and zoology far worse) he was ready to get back to his Species theory in earnest.

The manuscript Sketch of 1844 is preserved at the University Library, Cambridge, as is the fair copy which is written on 231 folio pages in the hand of Mr Fletcher, the village schoolmaster whom Darwin employed for this task. Blank pages alternate for revisions and corrections. In the margins Darwin has pencilled his own criticisms and remarks such as 'Hooker says goodish', for Joseph went through it in detail and wrote many comments. Like the brief earlier Sketch it is divided into two parts: I.'On the variation of Organic Beings under Domestication and in their Natural State.' II. 'On the Evidence favourable and opposed to the view that Species are naturally formed races descended from common stocks.'

The first part embodied the main argument he was to put forward in the *Origin*. He started off:

> The most favourable conditions for variation seem to be when organic beings are bred for many generations under domestication: one may infer this from the simple fact of the vast number of races and breeds of almost every plant and animal, which has long been domesticated. Under certain conditions organic beings even during their individual lives become slightly altered from their usual form, size, or other characters: and many of the peculiarities thus acquired are transmitted to their offspring.

Horticulturists even then could breed different varieties of tulip, carnation and other flowers. But these had a tendency to 'go back'. There were disastrous failures: thus, twenty thousand seeds of the Weeping Ash had been sown and not one came up true, though out of seventeen of the Weeping Yew all came up true. Darwin followed this with discussions on variation under nature, and the struggle for life, which was Natural Selection. In Chapter III, which concluded the first part, he wrote of the variations which occur in the instincts and habits of animals. This formed a complement to the chapters dealing with variation in structure and seems to have been placed early in the Sketch to prevent the hasty rejection of the whole theory by any reader to whom the idea of natural selection acting on instincts might seem impossible. The second part of the Sketch was 'devoted to the general consideration of how far the general economy of nature justifies or opposes the belief that related species and genera are descended from common stocks.'

Now Darwin felt that even if the worst happened to him before he could write a word more, he would have stated the case for Natural Selection as the means by which species had descended, slowly and gradually becoming adapted to various conditions of life by an infinite number of small modifications. On the 5th of July he sat down to write a formal letter to his wife:

> I have just finished my sketch of my species theory. If, as I believe my theory in time be accepted even by one competent judge, it will be a considerable step in science.
>
> I therefore write this in case of my sudden death, as my most solemn and last request, which I am sure you will consider the same as if legally entered in my will, that you will devote £400 to its publication, and further will yourself or through Hensleigh, take trouble in promoting it.

He wished that the Sketch be given to some competent person, with this sum to induce him to take trouble in its improvement and enlargement; and he was to have all his books on natural history which were either scored or had references at the end to the pages. There were many instructions about 'scraps roughly divided into eight or ten brown paper portfolios'; these with copied quotations from various works might aid his editor. 'I also request that you, or some amanuensis, will aid in deciphering any of the scraps which the editor may think possibly of use.' Darwin's writing at its worst needed careful understanding.

> With respect to editors, Mr Lyell would be the best if he would undertake it; I believe he would find the work pleasant and he would learn some facts new to him. As the editor must be a geologist as well as a naturalist the next best editor would be Professor Forbes of London. The next best (and quite best in many respects) would be Professor Henslow. Dr Hooker would be *very* good.

Later he added: 'Lyell, especially with the aid of Hooker (and of any good zoological aid), would be best of all.' In August 1854 he was to write on the back of the letter: 'Hooker by far best man to edit my species volume.'

Personal friendship began in 1844 when Darwin invited Joseph to join him at breakfast at his brother's house, now in Park Street: he was up in town for one of his rare visits. The Hookers, father and son, reciprocated with an invitation to Kew. It would have given Darwin real pleasure to accept. 'But I assure you a morning's work in London totally unfits me for everything, even the quietest conversation in the evening.' He had been compelled to relinquish the Geological Society's evening meetings and now only went to those of its Council. Would Joseph come to Down? He would send the phaeton to the station: he had lent it to someone but would try to borrow it back. Sydenham or Croydon would be the station.

Joseph managed to go in December, the first of many visits. 'And delightful they were,' he was to recall. 'A more hospitable and more attractive home under every point of view could not be imagined.' Sometimes brother naturalists were there, most often Dr Hugh Falconer the palaeontologist and botanist (for whom Joseph was to name a

rhododendron), Professor Thomas Bell who had written up Darwin's reptiles, and George Robert Waterhouse the living mammals. Professor Edward Forbes who shared Darwin's interest in geographical distribution was another visitor.

(left) Charles Darwin when he was thirty-three with his eldest child, William

(above) George, Darwin's second son

There were long walks, romps with the children on hands and knees, and 'music that haunts me still,' Joseph told Francis Darwin.

He had been toiling at Darwin's Galapagos plants. He read three papers on them at the Linnean Society—in March, May and December of 1845, and two others in December of the following year. After his first lecture Darwin, remembering from Robert Brown that Hugh Cuming had collected in the Archipelago, was anxious to know: 'Was my collection your main materials? *Roughly* [twice underlined] what percentage in my collection to that of others?' Joseph Hooker was able to give him resounding reassurance. Although on the eve of departure for Edinburgh to deputise for poor Professor Robert Graham who was dying, he wrote out the full details, an Enumeration in miniature, in his elegant hand. He promised to send a copy of the Linnean paper when it was printed.

This was a noble piece of work contained in ninety-nine large pages and divided into two sections: first his *Enumeration of the Plants of the Archipelago* which he read on the three dates in 1845, followed by his discourse *On the Vegetation of the Galapagos as compared with that of some other Tropical Islands and of the Continent of America.*

It was Charles Darwin, as he acknowledged in his Introduction, who 'drew my attention to the striking peculiarities which mark the Flora of the Galapagos group, and to the fact that the plants composing it not only differ from those of any other country, but that each of these islands has some particular productions of its own, often representatives of the species which are found in the others of the group.'

His first attempt to explain these peculiarities in the vegetation was frustrated, he confessed, by the novelty of the species themselves, forbidding any direct comparison of the flora with that of adjacent countries. Thus, classification of the plants themselves was a first requisite. He had given himself to the task of naming them, describing what proved to be new, and bringing to notice the ranges of the known plants. These, with some others previously collected by various voyagers, had enabled him to make some general remark on the botany of these islands and its relation to that of other countries.

Then followed the Enumeration. There were 239 Galapagos plants, which number did not include varieties of the same species. Of any collector in the Islands Darwin was solely responsible for by far the most plants and by far the most new species; in all 180. The remaining 59 plants were credited in two ways. David Douglas is solely credited with 1 plant, John Scouler 6, Hugh Cuming 1, Thomas Edmonstone 2, James Macrae 21, Goodridge 1, and 3 to Admiral Abel du Petit-Thouars, a total of 35. Then, sometimes a plant was collected by more than one man. The remaining 24 plants were in this category, and of these Darwin collected 19, so that of this group there were only 5 he did not collect. His total out of 239 plants was 199.

The peculiar or new species were mainly allied to plants of the cooler parts of America or the uplands of the tropical latitudes. The non-peculiar were the same as abounded chiefly in the hot and damper regions such as the West Indian islands and the shores of the Gulf of Mexico. Less than half the plants had come from the American continent.

As for the individual floras of the islands, it was a truly amazing fact that in James Island, of its 38 Galapagos plants or those found in no other part of the world, 30 were exclusively confined to this one island; and in Albemarle Island, of its 26 aboriginal plants only four grew in the other islands of the Archipelago.

In discussing the geographical distribution of the plants Joseph pointed out that as a field of observation this group of islands possessed the rare advantage of being one whose vegetation had never been interfered with by any aborigines of the human race, and that it was only very recently that man, or the animals he had introduced into the islands, had disturbed the indigenous flora, and this only to a very limited extent. It possessed the further singularity of containing a flora differing by upwards of one-half its species from that of the rest of the world. There were 123 new species, of which only three had previously been described.

The most remarkable feature of the flora was the number of its composites or daisy-flowered plants, including the curious tree-like ones Darwin had seen growing on the lush mountains of the islands. There were eight species in this group and they had no near allies in any other part of the globe, though all were closely related to one another.

To Darwin, Joseph Hooker gave the palm for his masterly sketch of the unique zoology and of the unequal dispersion of the plants over the several islets, and 'to whose comprehensive view of the natural history of the Galapagos this essay can be considered as supplementary only.'

8

'Hurrah for my Species work!'

In 1854 Darwin began sorting his notes for his Species theory, the climax of which was to be a book that shook the thinking world.

To find out what reception his theory was likely to have he wrote articles for the **Gardeners' Chronicle**. Thus his 'Nectar-secreting Organs of Plants' was an open challenge to the Natural Theologians who believed that Nature provided flowers with nectar for the purpose of attracting insects so that pollination can occur.

The pea flower has nectar but is self-pollinating. Of what use is the bee? In alighting on the flower it disturbs the pollen which then drops on to the stigma. But this, Darwin pointed out, was merely coincidental—because he found that many bees nibbled holes in the nectaries to obtain honey, without ever entering the flower—and these peas became fully fertile.

I N OCTOBER 1846 Darwin started work on—at first sight—a most unlikely subject, the study of barnacles. His interest in them began on the coast of Chile when he found a most curious kind that burrowed into the shell of a gastropod called the Concholepas. He had to form a new sub-order for its sole reception. In order to understand the structure of the creature he examined and dissected many of the common ones, and this led him on to study the whole group, including great rarities amongst parasitic forms.

Such was Darwin's industry that he was ultimately to publish two thick volumes on the Cirripedia, describing all the known living species (Syms Covington sent him many specimens from Pambula, New South Wales) and two thin quartos on species that were extinct.

The work took him eight years, during which time the young Darwins were so accustomed to seeing their father daily poring over them at his dissecting table, let into a window of his study, that they accepted barnacles as being the preoccupation of all fathers. The story is told of a visit to their neighbour Sir John Lubbock, who was vice-president of the Royal Society. They were taken to see over the house, High Elms, near Downe, and came back looking puzzled. 'Then where does he do his barnacles?'

In after years Darwin wondered if it had not all been a waste of time. On the other hand he realised that his work on the cirripedes, a highly varying and difficult group of species to classify, was of considerable use to him when he came to discuss the principles of the natural classification in the *Origin of Species*. The five years of the *Beagle* had given him sound practical experience in physical geography, in geology proper, in geographical distribution, and in palaeontology; but, as T. H. Huxley was to point out, what he still lacked was the relation to taxonomy of anatomy and development. As Darwin wrote to Joseph Hooker in October 1846: 'Are you a good hand at inventing names? I have a quite new and curious genus of Barnacle, which I want to name, and how to invent a name completely puzzles me.' He applied himself to taxonomy and came to the *Origin* equipped to discuss the natural classification that had been Linnaeus's goal and had since eluded every systematist.

Much of his work was done with his simple dissecting microscope, and in a letter to FitzRoy, recently returned from New Zealand where he had been Governor, he spoke of being 'for the last half-month daily hard at work in dissecting a little animal about the size of a pin's head, from the Chonos archipelago, and I could spend another month and daily see more beautiful structure.' Finding it a strain and requiring higher magnification he bought a compound microscope which, as he said to Joseph Hooker, was a 'splendid plaything'.

This was in one of the last letters he wrote to Joseph before losing him to India where Hooker was going as a plant collector for Kew. He sailed in November 1847, sharing his cabin with Hugh Falconer who was going out to take charge of the Calcutta Botanic Garden. The magnificent Sikkim rhododendrons Joseph sent home were to mark a new era in gardening. His *Himalayan Journals* were dedicated 'To Charles Darwin by his affectionate friend, Joseph Dalton Hooker.' The four years of his absence were a sore loss to Darwin, who wrote: 'It will be a noble voyage and journey, but I wish it was over, I shall miss you selfishly and all ways to a dreadful extent ...' Before leaving, Joseph had become engaged to Henslow's daughter Frances.

Two of the eight cirripedian years were lost by illness, as he reckoned, though this was by Darwin standards, for scientific papers still came from his pen and the barnacles marched on. Certainly the work took him longer than it would have done but for visits to Malvern for hydropathic treatment and periods of 'much sickness & failure of power'. But he 'worked on all well days'. On the 13th of November 1848 his father died in his eighty-third year. He was unable to attend his funeral, which added to his misery of losing him. His sister Susan was with their father till the end and wrote to her brother: 'God comfort you, my dearest Charles, you were so beloved by him.' Charles had been at Shrewsbury for a fortnight the month before.

The nature of Darwin's perpetual illness remains a mystery, though various theories have been put forward. One, that he had Chaga's disease: at Luxan on his last ride through the Andes he was bitten by the Benchuca insect, but Professor A. W. Woodruff has pointed out that it is not the mere bite of this insect which causes the disease but contamination of the wound by its excreta, and that in any case people who contract the disease have been exposed to the infection for several years. A strong case has been put forward by John Winslow that he was suffering from the 'Victorian Malady', chronic arsenical poisoning, and we know that at Cambridge his father prescribed arsenic for eczema on his hands, and that he asked his father about taking it with him on the voyage. Arsenic, still considered useful in treating chronic and atopic eczema, was regarded in Victorian days as a panacea and was frequently given as a tonic. Another explanation, suggested by Dr Douglas Hubble, is that Charles Darwin's illness arose from the suppression and non-recognition of fear, guilt, or hate felt towards his father who had unjustly condemned him for being idle during his youth. This we can dismiss. Darwin deferred to his father but purely out of love and respect. Visits to Shrewsbury were frequent and longed-for, and he became the loving father of his own children. Dr Hubble further suggested that Darwin's illness was a psychoneurosis to protect him from social intercourse, a theory recently developed by Sir George Pickering. In the profit and loss account Darwin drew up when contemplating marriage we see how much he feared sociability as an interruption to his work. Near the end of his *Autobiography* he wrote that 'Even ill-health, though it has annihilated several years of my life, has saved me from the distraction of society and amusement.' He got out of engagements when he could: an invitation to some official dinner produced stomach trouble; when his presence was requested at some gathering entailing a wearisome journey he was seized with giddiness. 'To have cured Charles Darwin's illness,' wrote Douglas Hubble, 'would have both lessened his ambition and destroyed his way of achievement.' Perhaps there is more to be written on the subject: but so far as Charles Darwin knew, 'anything that flurries me completely knocks me up afterwards.' He believed himself to be suffering from dyspepsia.

He corrected the final revise of the Cirripedia work on the 18th of July 1854, and a fortnight later was 'sending ten thousand Barnacles out of the house all over the world.' With the second volume published he wrote to Thomas Henry Huxley, the young zoologist who had first made contact with him on his return from the *Rattlesnake* voyage, sending him a report on the specimens he had collected. Darwin now asked him for the names of continental naturalists to whom he should present copies. Theirs was to be another scientific and personal friendship.

Then it was 'Hurrah for my Species work!'

In the interval much had been happening at Down House. His family had been increasing. By July 1850 there were four boys and three girls: William, aged ten, George five, Francis not quite two, Lenny six months; Annie nine, Etty six, and Elizabeth three. In that summer Annie's health began to break down and in March of the following year Darwin took her to Malvern to try the effect of the water-cure, which had greatly benefited himself, brought to England by Dr James Gully. With them went Etty and their old nurse Brodie. They were joined by the children's governess, Miss Thorley. Emma could not go with them, as she was about to have another baby. Very shortly afterwards Annie fell ill with a fever. She died on the 23rd of April. Darwin was with her for the last few days, and Etty never forgot how he flung himself on the sofa in an agony of grief. Annie was his favourite child, as he confessed to William Fox. 'Her cordiality, openness, buoyant joyousness and strong affection made her most lovable. Poor dear little soul.' He went home to comfort Emma and there write a portrait of Annie, recalling how she used to come running downstairs with a stolen pinch of snuff for him (he kept it in the attic to be out of temptation), her whole form radiating with the pleasure of giving pleasure, and how when she accompanied him round the Sand Walk, although he walked fast yet she often went before him, 'pirouetting in the most elegant way, her dear face bright all the time with the sweetest of smiles', and how she never once complained when she was ill, never once became fretful, and was always considerate of others.

Less than a month afterwards Horace was born, the Darwins' fifth son. He grew up like the others to know his father as a delightful playfellow. One of the boys when about four years old tried to bribe him with a sixpence to come out and play in working hours. They all knew the sacredness of working-time, but that anyone could resist a sixpence was beyond comprehension. He had unbounded patience with them, suffering them to make raids into his study when they had an 'absolute need' of sticking plaster, string, pins, scissors, stamps, foot-rule or hammer. Only once did he remonstrate, and that in the gentlest tone. 'Don't you think you could not come in again, I have been interrupted very often.' They loved it when he told them stories about the *Beagle*, and any child who was unwell was tucked up on the study sofa to be near him. He was passionately attached to his children, caring for all their pursuits and interests, putting his whole mind into answering their questions, making them feel that their opinions and thoughts were valuable to him, so that Etty wrote: 'Whatever there was best in us came out in the sunshine of his presence.' He was absolute truth and law to them, but both he and their mother respected their liberty and never asked what they were doing or thinking unless they wished to tell. 'After lessons we were always free to go where we would,' Etty wrote, 'and that was chiefly in the drawing-room and about the garden, so that we were very much with both my father and mother.' They were completely a happy family. To Charles Darwin, Emma was the 'angel' he had sought for his life partner. She was his shield against annoyances, his comfort in illness. Whatever concerned him concerned her, as she wrote to him once, so that 'I should be most unhappy if I thought we did not belong to each other for ever.' By this she simply meant in a life hereafter. His lack of religious faith was the one thing that distressed her. But having promised never to speak about it, she never did.

Apart from his scientific reading he never opened a book, preferring that Emma read aloud to him, some light novel which must always have a happy ending. For years they played two games of backgammon every evening, a serious contest. He won most games, but she won most gammons. They kept a tally, and it was a great day (in 1875) when Emma had won only 2,490 games 'whilst I have won, hurrah, hurrah, 2795 games!'

The servants knew him as a most considerate master. He never gave an order but always asked 'Would you be so good . . .' when he wanted anything. In 1845 he rebuilt the kitchen quarters to make them more comfortable, with a bigger pantry for Joseph Parslow, the butler they had brought from London. They had complained to him what a nuisance it was to have a passage for everything only through the kitchen. Darwin wrote to his sister Susan: 'It seemed so selfish to make the house so luxurious for ourselves and not so comfortable for our servants that I was determined if possible to effect their wishes.'

On the 8th of September 1854 began an investigation which resulted in a curious discovery. George, who was then nine, came to tell him that he had seen some bumble-bees (Darwin called them humble-bees) entering a hole at the foot of a tall ash tree. Was it the entrance to a bee's nest? Darwin went to see, but no nest could he find. While he was examining the hole another bee entered, flew off, returned almost immediately and then, flying upwards for about a yard, flew away through a fork between two large branches of the ash tree. Darwin now removed the grass and other plants growing around the hole but still could not find an entrance. After a minute or two another bumble-bee appeared. It buzzed over the area and then flew up and passed, like the previous one, through the same fork. This pattern was repeated every few minutes by other bees. Darwin then followed a bee from the ash to a bare spot at the side of a ditch where again he watched them repeating the buzzing process. Several yards farther on was an ivy leaf. This, too, the bees were making a buzzing place, and from the ivy leaf they went into a dry ditch covered over by a thick hedge and flew slowly along the ground between the dense branches of hawthorn. The tunnel was too small for Darwin to make his way along it. He called two of his other children, Franky who was six and Etty ten, and by stationing a child at the buzzing places as each was discovered, calling out 'Here is a bee!', they were able to log a flight path following a wide circle. They kept up these observations for several years, and Darwin found that all the bees were males of *Bombus hortorum*. He was astonished at his discovery but could not understand how bees born in different years could learn exactly the same habits, following the same flight path and buzzing at the same places. He tried to fool them by sprinkling flour over one spot and pulling up the grass and plants from another at the foot of an oak, but he never succeeded. The ritual had nothing to do with a search for a queen. Today the mystery of these flight paths is still unsolved.

On the day following the discovery of the bee-hole Darwin began sorting his notes for the Species theory, and he was soon in the thick of experiments. Important as a foundation was the subject of geographical distribution, and in order to test how long seeds would remain viable in sea water he began immersing different kinds in salted water in a large tank in the cellar. The winter of 1854–55 was a very hard one, and this work was interrupted by a month's stay in London so that Emma could enjoy a little society. They rented a house in Upper Baker Street, at 27 York Place. The visit was not a success. Neither Emma nor Charles was well. They went to concerts, and it was not the music-loving Emma who enjoyed them as much as her husband. They returned to Down to find the snow so deep on the lawn that it was level with the top of the iron railings. But the snow was the answer to the problem of keeping the seeds in water that was sea-cold. The Darwin children had fun piling snow into the tank. Thereafter their father was able to keep the water at a temperature of 32°–33°F, writing in triumph about this to Hooker who at first had been inclined to scoff at his friend's experiments but who by the spring of 1855 was a co-experimenter and competitor in the game of seeing which of them could keep the seeds living longest, a game in which the Darwin children joined eagerly, demanding at each stage: 'Are you going to beat Dr Hooker?'

Some of the seeds were put in small bottles out of doors and exposed to a variation of night and day, sun and shade temperatures. These were seeds of cress, radish, cabbages, lettuce, carrots, celery and onion—representatives of four great families. At Kew Joseph was faithfully following the same method but declaring that the cress would be killed in a week. The two scored off each other in 'little triumphs'. Darwin was able to report that all his had germinated in exactly one week, 'which I did not in the least expect (and thought how you would sneer at me); for the water of nearly all, and of the cress especially, smelt very badly.' But they all grew splendidly, the germination of all the

A Darwin note on sea-
water germination.
The sea-water was
'made artificially with
salt procured from Mr
Bolton, 146 Holborn
Bars' and tested by
numerous sea animals
and algae having lived
in it for more than a
year

seeds—especially the cress and lettuce—being accelerated, except the
cabbages which came up very irregularly and of which a good many died.

Darwin went on to see what would happen after 14 days' immersion,
reckoning that on the average rate of the Atlantic current being 33 miles
a day, some seeds could be transported 300 miles or even more. The main
Equatorial current ran at the rate of 60 miles, the Cape Stream at 80 miles.
On the 14th of April he had 'a nice little triumph', for the cress and lettuce
vegetated well after being immersed for 21 days.

Berkeley (of the fungi) was also enlisted in the experiments and nobly
undertook to test 53 different kinds. He sent his seeds to Ramsgate tied up
in little bags, to be immersed in real sea water renewed every day. After
three weeks they were sent back to him, partially dried but still damp,
but unfortunately were not unpacked for four days, so that the immersion
period had to be reckoned as equivalent to more than a month. Their
results sometimes differed, Darwin finding that tomato seeds would

germinate after 22 days' immersion but most being killed by 36 and 50 days, Berkeley finding them still viable after a month. The survival record was won by fresh seed of the wild cabbage sent from Tenby, which germinated excellently after 50 days, very well after 100 days, while two seeds out of some hundreds germinated after 133 days' immersion.

Dr Hooker was not doing so well: many of his seeds sank, and there seemed no way out of the problem of what could 'arrest their everlasting descent into the deepest depths of the ocean.' Darwin had also been experiencing this difficulty and called the seeds ungrateful rascals. But he had a thought. He soaked some seeds and took them to the fish at the Zoo. In imagination they were swallowed by a fish which in turn was swallowed by a heron which flew a hundred miles and voided them on the banks of another lake, and lo and behold! the seeds germinated splendidly. In fact the fish at the Zoo swallowed the seed and then to his dismay ejected them vehemently. Darwin told Hooker that he would have to rely on floods, slips and earthquakes to deal with these. It was not impossible. As far back as 1843 Darwin had germinated seeds found by 'a Mr Kemp (almost a working man)' in a layer at the bottom of a deep sand pit near Melrose. The man might have been suspect had he not shown himself to be a most careful and serious observer. The pit, he thought, was the site of an ancient lake. Darwin sent some of the seeds to Henslow and the Horticultural Society where Lindley also germinated them. They turned out to be those of a common *Rumex* and a species of *Atriplex* which neither Lindley nor Henslow had ever seen, and it was certainly not a British plant. Poor Mr Kemp! Babington doubted that the seeds were ages old. He sowed some seeds of *Atriplex angustifolia* and they came up the same as the supposedly new *Atriplex* of the sand pit. But the exercise was not altogether in vain, as later experiments were to prove.

Henslow by now was completely involved in village affairs. He had written a Flora of Hitcham and was teaching botany to the children of the village school. Darwin wrote to ask if *Geum rivale* or *Epilobium tetragonum* grew in his neighbourhood: neither were to be found around Down because of the dryness. To this, Henslow responded by saying he had organised a group of little girls to collect seeds for him. Darwin sent him a list of what he wanted and promised the little girls '(if not too big and grandiose)' sixpence each and threepence for each packet. The list was of all the European plants found in the Azores, as a central oceanic archipelago. He wanted to do a salt-water test to see if originally they could have travelled thus far by sea.

Similarly, remembering how he had met British weeds in America, he was interested in finding out the number and species of European plants that had wandered across the Atlantic. In April 1855 he had begun corresponding with Asa Gray. The two had met at Kew, Gray remembering Darwin as 'the heavy-browed young explorer'. The great American botanist was about to publish a new edition of his *Manual of the Botany of the Northern American States*, and Darwin suggested his appending '(EU)' against any plant of European origin, a plan that was adopted. He asked for information on the alpine plants of America, explaining that for several years he had been collecting facts on 'variation', and that 'when I find that any general remark seems to hold good amongst animals, I try to test it in plants.'

June 1855 saw Darwin with a botanical companion in Miss Thorley, the children's governess, who helped him to make a collection of all the plants growing in a field which had been allowed to run waste for fifteen years but previously had been cultivated from time immemorial. They also collected all the plants in an adjoining but cultivated field—'just for the fun of seeing what plants had arrived or died out'—, as he wrote to Hooker. How dreadfully difficult it was to name plants, he groaned. But 'I have just made out my first grass, hurrah! hurrah! I must confess that fortune favours the bold, for, as good luck would have it, it was the easy *Anthoxanthum odoratum*: nevertheless it is a great discovery; I never expected to make out a grass in all my life, so hurrah!' He added: 'It has done my stomach surprising good . . .' By the end of the month he had identified 28 species.

In the following month buried seeds (shades of Mr Kemp) again came under discussion. A letter to Hooker asked how many years he thought Charlock seed might live in the ground. The copse in the Sand Walk had been planted on a piece of ground laid down as pasture in 1840. Before the trees were planted it was deeply ploughed, and Darwin remembered that Charlock had sprung up plentifully the following summer. In the spring of 1855 he had some thorn bushes pulled up, and to his surprise Charlock plants sprang up and duly flowered in July. This inspired him to have three separate plots of ground—each two feet square and in different and open parts of the wood—cleared of thick grass and weeds and dug one spit deep. By the beginning of August many seedlings had come up. Most of them proved to be Charlocks, though there was no Charlock growing anywhere near the beds, the wood being surrounded by grassland. Darwin wrote an informative note for the *Gardeners' Chronicle*, pointing out that the power in seeds of retaining their vitality when buried in damp soil for many years might well be an element in preserving the species.

He wrote to Hooker begging a plant of *Hedysarum*, this on the 5th of June, the very day that Joseph was appointed Assistant Director to his father at Kew. 'I do hope it is not very precious, for . . . it is for probably a *most* foolish purpose. I read somewhere that no plant closes its leaves so promptly in darkness, and I want to cover it up daily for half an hour and see if I can teach it to close by itself, or more easily than at first in darkness.' The Telegraph Plant, *Desmodium gyrans*, to use its modern synonym, behaved in exactly the way he had hoped. The closing of the leaves, he was to discover, was a survival mechanism.

During all his experiments, carried on throughout the rest of his life, he sent out what he called 'printed inquiries'. These took the form of notes reporting on experiments he was doing, and asking if anyone knew of similar work being done. He also wrote longer articles that inevitably called forth replies because of the very original (and therefore controversial) line of his thought. His medium was the *Gardeners' Chronicle*, still the journal of leading gardeners and nurserymen, to which he started contributing in 1841, the first year of its publication, between April and December 1855 writing four articles for it on seeds in salt water.

The articles sometimes had another motive. With his growing confidence in his Transmutation theory he used them as probes to find out what reception his theory was likely to have. Thus his 'Nectar-secreting Organs of Plants' which appeared in the issue of 21 July 1855 was an open challenge to the Natural Theologians and their explanation of natural adaptations. He pointed out the presence of nectar-secreting glands in vetches and beans. Bees and other insects were attracted to the nectar— but these were self-fertilising plants. The secretion of the nectar within the flowers was therefore not done deliberately by Nature for the purpose of attracting insects to pollinate them: it was merely a coincidental adaptation. Coincidental because the bees in alighting on the flower disturbed the pollen which then dropped on the stigma.

Again in 1855, discussing in 'Seedling Fruit Trees' varieties that bred true, and suggesting that the appearance of new types might be attributed to accidental crosses, he showed how clearly he had grasped the significance of heredity, both in supplying variability and in maintaining uniformity in species.

Other articles dealt with cross-breeding, double flowers, and the conversion of plant organs from one form to another—significant in tracing the development of his Natural Selection theory; variegated leaves, and a brief report on the results of an experiment in which he crossed different varieties of carnations and pinks. He was also writing articles in more scientific vein for *Nature* and for the *Annals and Magazine of Natural History*. All were on topics embraced by his theory and links in that chain: the struggle for existence, competition, the means of survival.

Living at Ternate, a small and remote island in the Malay Archipelago, was the naturalist Alfred Russel Wallace. He made his living by collecting tropical birds, animals, and moths and butterflies, which he despatched to London for sale. His life was a lonely one. Apart from his Malay and Chinese helpers he had no companion, though he and his naturalist friend H. W. Bates, half a world away on the Amazon, managed to keep up a correspondence. He had plenty of time to think, and he thought much and deeply about the beautiful creatures whose exotic world he was invading. A King Bird-of-Paradise brought to him one day, which repaid him for months of delay and waiting, moved him to reflect on the long ages of the past during which successive generations of such beautiful little creatures had run their course, living and dying among the dark and secret trees of the tropical forest, 'with no intelligent eye to gaze on their loveliness'. He wrote exquisitely about them: his description of the courtship dance of the Great Bird-of-Paradise, with its two silky cascades of yellow plumes, is said to be still unsurpassed. But while he thought it wrong that these living jewels should be unseen, he also thought it would be better that civilised man should never come to these virgin forests. For civilised man was destructive. 'He will so disturb the nicely balanced relations of organic and inorganic nature as to cause the disappearance, and finally the extinction, of these very beings whose wonderful structure and beauty he alone is fitted to appreciate and enjoy.'

In February 1855, while at Sarawak, he wrote a paper 'On the Law that has regulated the Introduction of New Species' which was published in the *Annals and Magazine of Natural History*. Sir Charles Lyell was one who read it. He noted Wallace's remarks on organic changes, his proposal that 'a like gradation and natural sequence from one geological epoch to another' had taken place. Lyell himself had suggested the slow and gradual extinction and creation of species, but had stopped short at descent. Wallace's 'Law' supported a hypothesis that might explain the past and present distribution of life upon the earth, which, he said, had occurred to him about ten years earlier. There were thirteen pages of argument that took in such points as the part played by geographical isolation in the origin of peculiar forms of life. Wallace had read the *Journal of the Beagle* and instanced the Galapagos. There was also some original thinking.

Lyell, who by this time knew the lines on which Darwin was working,

A study of Charles Darwin, Charles Lyell and Joseph Dalton Hooker painted by a Russian artist for the centenary of the *Origin*. It hangs in the Darwin Museum, Moscow

was thoroughly alarmed. For Wallace's hypotheses were parallel. He wrote at once, drawing Darwin's attention to the paper and suggesting that he should not delay in publishing a sketch of his views. To this Darwin replied on the 3rd of May.

I hardly know what to think, but will reflect on it, but it goes against my prejudices. To give a fair sketch would be absolutely impossible, for every proposition requires such an array of facts. If I were to do anything, it could only refer to the main agency of change—selection—and perhaps point out a very few of the leading features, which countenance such a view, and some few of the main difficulties. But I do not know what to think; I rather hate the idea of writing for priority, yet I certainly should be vexed if any one were to publish my doctrines before me. Anyhow, I thank you heartily for your sympathy. I shall be in London next week, and I will call on you on Thursday morning for one hour precisely, so as not to lose much of your time and my own; but will you let me this time come as early as 9 o'clock, for I have much which I must do in the morning in my strongest time? Farewell, my dear old patron,
                    Yours,
                        C. Darwin

In a postscript he asked: 'If I did publish a short sketch, where on earth should I publish it?'

He was going up to London to read a paper on 'Seeds in Salt Water' before the Linnean Society, of which he had been elected a Fellow the year before. On the 9th, three days later, he wrote to Hooker.

I very much want advice and *truthful* consolation if you can give it. I had a good talk with Lyell about my species work, and he urges me strongly to publish something. I am fixed against any periodical or Journal, as I positively will *not* expose myself to an Editor or a Council, allowing a publication for which they might be abused. If I publish anything it must be a *very thin* and little volume, giving a sketch of my views and difficulties; but it is really dreadfully unphilosophical to give a *resumé*, without exact references, of an unpublished work. But Lyell seemed to think I might do this, at the suggestion of friends, and on the ground, which I might state, that I had been at work for eighteen years, and yet could not publish for several years, and especially as I could point out difficulties which seemed to me to require especial investigation. Now what think you? I should be really grateful for advice. I thought of giving up a couple of months and writing such a sketch, and trying to keep my judgement open whether or not to publish it when completed. It will be simply impossible for me to give exact references; anything important I should state on the authority of the author generally; and instead of giving all the facts on which I ground my opinion, I could give by memory only one or two. In the Preface I would state that the work would not be considered strictly scientific, but a mere sketch or outline of a future work in which full references &c., should be given. Eheu, eheu, I believe I should sneer at anyone else doing this, and my only comfort is, that I *truly* never dreamed of it, till Lyell suggested it, and seems deliberately to think it advisable.

I am in a peck of trouble, and do forgive me for troubling you.

Yours affectionately,

C. Darwin

Hooker joined with Charles Lyell in urging him to publish, agreeing that a separate 'Preliminary Essay' would be the thing. 'But,' Darwin wrote back, 'I cannot bear the idea of *begging* some Editor and Council to publish, and then perhaps have to *apologise* humbly for having led them into a scrape.' He was torn. 'It yet strikes me as quite unphilosophical to publish results without the full details which have led to such results. . . . I confess I lean more and more to at least making the attempt and drawing up a sketch and trying to keep my judgement, whether to publish, open. But I always return to my fixed idea that it is dreadfully unphilosophical . . .' And finally: 'I certainly think my future work in full would profit by hearing what my friends or critics (if reviewed) thought of the outline.'

Darwin's contemporaries:

(*below*) Asa Gray, professor of natural history at Harvard University

(*right*) Thomas Henry Huxley who was 'Darwin's Bulldog'

(*far right*) Alfred Russel Wallace

He sat down obediently on the 14th of May 1856 to write out a sketch. But he found it unsatisfying. Fox had protested against Lyell's advice of a brief essay and this, coinciding with his own aversion, decided Darwin. In November he confessed to Lyell: 'I am working very steadily at my big book; I have found it quite impossible to publish any preliminary essay; but am doing my work as completely as my present materials allow without waiting to perfect them. And this much acceleration I owe to you.'

He worked almost uninterruptedly. His diary records several visits in the next two years to Dr Lane's watercure establishment at Moor Park, near Farnham, but apart from that he was either writing or out in his garden or fields attending to experiments in connection with his work. He had said to Lyell that he would use only 'present materials'—but the experiments went on. One was on the struggle for existence, which made him 'see a little clearer how the fight goes on,' as he wrote to Hooker. 'Out of sixteen kinds of seed sown on my meadow, fifteen have germinated, but now they are perishing at such a rate that I doubt whether more than one will flower. Here we have choking which has taken place likewise on a great scale, with plants not seedlings, in a bit of my lawn allowed to grow up. On the other hand, in a bit of ground, 2 by 3 feet, I have daily marked each seedling weed as it has appeared during March, April and May, and 357 have come up, and of these 277 have *already* been killed, chiefly by slugs.' He told him that at Moor Park he had seen 'rather a pretty case of the effect of animals on vegetation: there are enormous commons with clumps of old Scotch firs on the hills, and about eight or ten years ago some of these commons were enclosed, and all round the clumps nice young trees are springing up by the million, looking exactly as if planted, so many of the same age.' In other parts of the common, not yet enclosed,

he had looked for miles and could not see a single young tree. He then looked closely down into the heather, and there were tens of thousands of young Scots Pines (thirty in one square yard) with their tops nibbled off by the few cattle which occasionally roamed over the heaths. One little tree, only three inches high and with a stem about as thick as a stick of sealing-wax, was twenty-six years old: he had counted the rings. 'What a wondrous problem it is,' he marvelled, 'what a play of forces, determining the kind and proportion of each plant in a square yard of turf! It is to my mind truly wonderful. And yet we are pleased to wonder when some animal or plant becomes extinct.'

From Moor Park he wrote to Hooker about observations he was doing on embryonic seedlings, sending him specimens. 'Look at the enclosed seedling gorses, especially one with the top knocked off. The leaves succeeding the cotyledons being almost clover-like in shape, seems to me feebly analogous to embryonic resemblances in young animals, as, for instance, the young lion being striped. I shall ask you whether this is so.'

The likeness of many embryonic seedlings to each other and their dissimilarity to their parents was another link in Darwin's evolutionary theory.

This was in 1857. In the same year he had two letters from Wallace, to which he replied encouragingly, agreeing to the truth of 'almost every word' of his paper in the *Annals*, adding that "This summer will make the 20th year (!) since I opened my first note-book, on the question how and in what way do species and varieties differ from each other. I am now preparing my work for publication but I find the subjects so very large, that though I have written many chapters I do not propose I shall go to press for two years.'

Wallace had been disappointed that no notice had been taken of his paper. Darwin told him that while he agreed with his conclusions, 'I believe I go much further than you; but it is too long a subject to enter on my speculative notions.'

He was determined not to say too much to Wallace about his own work, which was going ahead smoothly: by June 1858 he had written ten chapters, about one half of the projected book. Struggles, yes, glorious tussles with Hooker ('fighting a battle with you always clears my mind wonderfully'), letters to Lyell about geology, to Asa Gray on botanical geography and 'close species', rousing him to ask: 'What are you up to?' Finally Darwin had told him, first in a letter of the 20th of July 1856; then in one dated the 5th of September 1857 writing a full outline of his theory.

Out in Ternate, in February 1858, Wallace was lying ill with malaria. His thoughts were jumbled. Something (he never said what) brought to mind a book he had not read for twelve years, Malthus's *Principle of Population*. Wallace, too, had been working on a theory, as he had confided to Bates; and now, recalling what Malthus had said about checked and unchecked populations, a 'flash of light' came to him. He seized a pen and paper.

On the morning of the 18th of June a packet arrived at Down House. Opening it and scanning the sheets of thin foreign paper, Darwin was appalled. It was 'a bolt from the blue'.

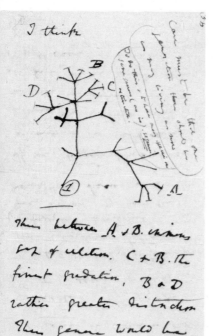

9

The Quiet War

Darwin's evolutionary tree: an early attempt to explain how living matter had developed from a single root. The drawing is a facsimile taken from page 36 of his first Transmutation Notebook. It shows how some branches have thrown out new branches, while others have come to a stop.

Later, Darwin developed the symbol into a widely branching tree of which, as he wrote, few of the original twigs remained, but some had grown into great branches. Of these, some had decayed and dropped off: they represented extinct families, genera and species known to us only as fossils; and some had survived to bear other branches, branchlets and twigs: they represented modified descendants.

'As buds give rise by growth to fresh buds,' said Darwin, 'and these, if vigorous, branch out and overtop on all sides many a feebler branch, so by generation I believe it has been with the great Tree of Life, which fills with its dead and broken branches the crust of the earth, and covers the surface with its everbranching and beautiful ramifications.'

(For Darwin's genealogical tree, see page 174.)

T HAT SAME DAY, CHARLES Darwin wrote to Lyell, summing up the prelude that had led to the disaster which had befallen him, and putting himself completely at the mercy of his friend's reproaches and of his own conscience.

He did so in a spirit of humility and generosity.

Down, 18th [June 1858]

My dear Lyell,

Some year or so ago you recommended me to read a paper by Wallace in the 'Annals', which had interested you, and as I was writing to him, I knew this would please him much, so I told him. He has to-day sent me the enclosed, and asked me to forward it to you. It seems to me well worth reading. Your words have come true with a vengeance—that I should be forestalled. You said this, when I explained to you here very briefly my views of 'Natural Selection' depending on the struggle for existence. I never saw a more striking coincidence; if Wallace had my MS sketch written out in 1842, he could not have made a better short abstract! Even his terms now stand as heads of my chapters. Please return me the MS., which he does not say he wishes me to publish, but I shall, of course, at once write and offer to send to any journal. So all my originality, whatever it may amount to, will be smashed, though my book, if it will ever have any value, will not be deteriorated; as all the labour consists in the application of the theory.

I hope you will approve of Wallace's sketch, that I may tell him what you say.

My dear Lyell, yours most truly,

C. Darwin

Thoughts crowded upon thoughts. Was there a way out of the dilemma? A few days later he again wrote to Lyell. 'There is nothing in Wallace's sketch which is not written out much fuller in my sketch, copied out in 1844, and read by Hooker some dozen years ago.' There was another fortuitous piece of evidence. 'About a year ago I sent a short sketch, of which I have a copy, of my views . . . to Asa Gray, so that I could most truly say and prove that I take nothing from Wallace. I should be extremely glad now to publish a sketch of my general views in about a dozen pages or so.' This would wrap up the case completely—except for one thing. He went on: 'But I cannot persuade myself that I can do so honourably.' As he had not intended to publish any sketch, could he do so now because Wallace had sent him an outline of his doctrine? 'I would far rather burn my own book,' he declared, 'than that he or any other man should think that I had behaved in a paltry spirit.' He ended his letter by asking if Lyell would object to sending this letter and his answer to

Hooker to be forwarded to him with Hooker's answer. 'For then I shall have the opinion of my two best and kindest friends.'

The blow could hardly have come at a more cruel time. Illness had plagued the Darwin household for a year: Etty ill, Lenny with a very intermittent pulse, recovering, poorly again, and Etty worsening; and there were weary months of suffering for Emma after the birth of Charles Waring Darwin on the 6th of December 1856. Then, in the same week that the fateful letter arrived from Wallace, scarlet fever appeared in the village, plunging the house into grief and panic. Darwin recorded in his diary on the 28th: 'Poor dear Baby died.'

In the midst of it all, the repercussions of the Wallace bombshell were echoing back and forth in letters between Darwin, Lyell, and Hooker, with Darwin urging them 'to make the case as strong as possible against myself', arguing that 'Wallace might say: "You did not intend publishing an abstract of your views till you received my communication. Is it fair to take advantage of my having freely, though unasked, communicated to you my ideas, and thus prevent me forestalling you?" the advantage which I should take being that I am induced to publish from privately knowing that Wallace is in the field. It seems hard on me that I should be thus compelled to lose my priority of many years' standing, but I cannot feel at all sure that this alters the justice of the case.'

Hooker and Lyell, after much heart-searching, weighed the case fairly to both. With Darwin's consent they arranged for Wallace's paper to be read before the Linnean Society jointly with 'Extracts from an unpublished Work on Species' by Charles Darwin, and an abstract of the letter to Asa Gray dated Down, September 5th, 1857. They had hoped to present Darwin's 'general views' in the dozen pages or so he had finally consented to write. Because of the state of things at Down House he was unable to do so. No wonder he told Hooker: 'I am quite prostrated and can do nothing.' They used instead a portion of a chapter from his 'big book', 'On the Variations of Organic Beings in a state of Nature; on the Natural Means of Selection; on the Comparison of Domestic Races and the True Species'.

The papers were read on the 1st of July (replacing one by George Bentham on the fixety of species!), and in a letter Sir Joseph Hooker wrote many years later to Francis Darwin he recorded that the interest excited was intense but the subject too novel and too ominous for the Old School to enter the lists before armouring. It was talked over after the meeting with 'bated breath'. Hooker thought that Lyell's approval 'and perhaps in a small way mine, as his lieutenant in the affair, rather overawed the Fellows, who would otherwise have flown against the doctrine.' The occasion was a quiet scientific one. But it was the quiet before the storm broke like a thunderclap on the publication of the *Origin of Species*. How ironic that Thomas Bell, the president, should record for the Society's *Transactions* that in 1858 no particularly important papers had been read!

Meanwhile, things had gone prosperously, as Darwin was glad to learn. He was grateful to his two friends for all they had done. 'You must know that I look at it, as very important, for the reception of the view of species not being immutable, the fact of the greatest Geologist and Botanist in England taking *any sort of interest* in the subject: I am sure it will do much to break down prejudices.'

At Down House, too, things were on the mend. There was solace even in the death of eighteen-month-old Charles who had been born subnormal and never learnt to walk or talk. After the first sharp grief they could only feel thankful. To Hooker Darwin wrote on the 5th of July: 'We are become more happy and less panic-struck, now that we have sent out of the house every child, and shall remove H., as soon as she can move. The first nurse became ill with ulcerated throat and quinsy, and the second is now ill with scarlet fever, but, thank God, is recovering. You may imagine how frightened we have been. It has been a most miserable fortnight.'

They all went on holiday, first to The Ridge, the home of Elizabeth Wedgwood, Emma's eldest sister, on the fringe of Ashdown Forest near Hartfield in Sussex.

It had now been arranged that Darwin would write an abstract for the Linnean Society *Journal*, an idea he accepted, though still doubting how it could be made scientific without giving factual evidence. However, directly after returning home he would 'begin and cut my cloth to my measure'. In fact he began writing it before then. His diary records: 'July 20th to Aug 12th at Sandown, began abstract of Species book.' His task was to condense it into thirty pages—which Darwin knew from the start would be impossible. He offered to help pay for the printing if it proved longer, and finally Hooker after consulting the editor of the *Journal* was able to assure him that the Society was willing to publish it in parts. Accordingly, 'Sept 16th Recommenced Abstract' was the entry in his diary for that day.

It was never published by the Linnean Society, for by October it was clear that it would 'in bulk make a small volume'.

The small volume became a fat one of 502 pages, but he still called it his Abstract: the real book was his 'big book', never to be finished.

But at least and at last he had begun his great work, *The Origin of Species by means of Natural Selection, or the Preservation of Favoured Races in the Struggle for Life.*

There was one 'little triumph' for Darwin connected with this vexacious period. Indeed, it was something of magnitude to boast of—the conversion of Joseph Hooker to the idea of Natural Selection. He who had supported Darwin (and was to continue to do so in the greater trials ahead) did not until after the Linnean meeting unreservedly accept his Theory. From Elizabeth Wedgwood's on the 13th of July Darwin wrote to him: 'You cannot imagine how pleased I am that the notion of Natural Selection has acted as a purgative on your bowels of immutability.' (Hugh Falconer had attacked Darwin the year before 'most vigorously, but quite kindly' over this. 'You have already *corrupted* and half-spoiled Hooker!')

So here was the 'one competent judge' Darwin wanted. Lyell did not come over to Darwin's side on the question until November 1859.

Before we travel in the realms of the *Origin* to discover what Darwin said in 1859, it would be well to ask: '*Why* the *Origin*?'

It was not that all thinking men believed implicitly that the world had been created in six days. There had been a succession of evolutionists before Darwin, among them his own grandfather, all searching inquisitively for a scientific explanation of how the world and its living creatures had come into being. Isaac Newton, Robert Boyle and John Ray saw nature as an orderly system of matter in motion whose beauty and usefulness was created by God for the benefit of intelligent Man. But matter in motion implied change, and already René Descartes had opened the way to doubt. Georges Louis Leclerc, Comte de Buffon, Intendant of the Royal Garden and Keeper of the Royal Cabinet of Natural History, begged the naturalist to leave the interpretation of the Scriptures to theologians and confine himself to working out probable hypotheses based on accurate observations of nature. James Hutton's investigations of the earth's crust assumed a history of millions of years: not the 6,000 of the theologians. There had been a succession of worlds, he declared, forming and reforming. For every mountain worn down by the ravages of wind and rain (in order to provide soil for the growth of plants), there was a new land-mass consolidating under the sea, eventually to be elevated above the waters. William ('Stratum') Smith, surveyor and engineer, developed the technique of identifying and tracing geological strata by their fossil contents. In Paris, Cuvier and his associate Alexandre Brongniart, examining fossil shells and the bones of extinct animals, were able to assign them to eleven distinct geological formations. Cuvier explained these as creatures destroyed by some revolutions of our globe; beings whose place those which exist today have filled, perhaps themselves to be destroyed and replaced some day by others. Many of the great fossil quadrupeds he examined had been dug up in America. James G. Graham was a physician who witnessed the digging up of huge bones on a farm at Shawangunk, New York. He described them, adding: 'And why Providence should have destroyed an animal or species it once thought proper to create, is a matter of curious inquiry and difficult solution.' At Punta Alta in Patagonia on the 22nd of September 1832 twenty-three-year-old Charles Darwin uncovered the fossil shell of a huge armadillo. Jean Baptiste Pierre Antoine de Monet, Chevalier de Lamarck, had been taught that the earth's surface had undergone constant slow change. Reflection on the nature of the system of matter in motion convinced him that heavenly bodies had also imperceptibly changed. How unlikely, then, that living bodies had remained unchanged. If living species were compared with extinct species, it would be seen that they had descended from lost ancestors, in the course of time having undergone modification. The world had changed: they had changed with it, developing new organs to meet new needs. Each 'felt need' had stirred their inner consciousness (*sentiment intérieur*), and little by little the organ was produced and

The old study at Down House, as it was in Darwin's lifetime. Here he wrote the *Origin of Species* and most of his other books

developed by reason of its constant vigorous use. Thus had the giraffe, in need of reaching to higher branches in order to browse upon them, developed its long neck. Charles Darwin agreed that the giraffe could certainly have its neck elongated for this purpose, but he disagreed entirely with Lamarck's line of reasoning. We remember his remark to Hooker in 1844: 'Heaven forfend me from Lamarck nonsense of "adaptations from the slow willing of animals"...' As Lyell wrote to him, referring to Lamarck and after reading the *Origin*: 'You may say that in regard to animals you substitute natural selection for volition to a certain considerable extent, but in his theory of the changes of plants he could not introduce volition.'

No, agreed Darwin, therefore we will find out how plants have changed, and we will have the complete story.

His theory was based on the belief that each new variety and ultimately each new species was produced and maintained by having some advantage over those with which it came into competition. Accordingly he arranged his book under three great divisions: Variation, the Struggle for Existence, and the Survival of the Fittest, with two chapters on Geographical Distribution.

A corner of Darwin's
study showing an
experiment in progress

First was the fact that plants had varied under cultivation. Man the nurseryman and florist had seen to that.

There was the steadily increasing size of the common gooseberry and the astonishing improvement in many florists' flowers. Compare, said Darwin, the flowers of the present day with drawings made only twenty or thirty years ago. When a race of plants was established, the raisers did not pick out the best plants but merely went over their seedbeds and pulled up the 'rogues'. Nature provided variations: 'Man adds them up in certain directions useful to him.' It was easy to see the accumulated effects of selection by comparing the plants in the flower garden with those in the kitchen garden: in the first, the different flowers of the many varieties of the same species; in the second the variety of leaves, pods or tubers—whatever part was valued. See, he said, how different are the leaves of the cabbage family, and how alike the flowers; how unlike the flowers of the pansy and how alike the leaves. The continued selection of slight variations, either in the leaves, the flower or the fruit, produced races differing from each other in these characters.

So, said Darwin, 'the key is man's power of accumulative selection.'

But this could be called Unconscious Selection. It was a long and gradual process of improvement, through the occasional preservation of the best individuals, whether or not sufficiently distinct to be recognized at their first appearance as distinct varieties and whether or not two or more species or races had become blended together by crossing. What was plainly to be seen now was the increased size and beauty of the pansy, rose, pelargonium, dahlia and other plants when compared with the older varieties or with their parent stocks. 'No one,' said Darwin, 'would ever expect to get a first-rate heartsease or dahlia from the seed of a wild plant. No one would expect to raise a first-rate melting pear from the seed of a wild pear, though he might succeed from a poor seedling growing wild if it had come from a garden stock. The pear though cultivated in classical times appears from Pliny's description to have been a fruit of a very inferior quality. I have seen great surprise expressed in horticultural works at the wonderful skill of gardeners, in having produced such splendid results from such poor materials. But the art has been simple, and, as far as the final result is concerned, has been followed almost unconsciously. It has consisted in always cultivating the best-known variety, sowing its seeds, and, when a slightly better variety chanced to appear, selecting it, and so onwards.'

There were, Darwin pointed out, circumstances favourable to man's power of selection. A high degree of variability was obviously favourable, as freely giving the materials on which selection could work. 'But as variations manifestly useful or pleasing to man appear only occasionally, the chance of their appearance will be much increased by a large number of individuals being kept. Hence, number is of the highest importance for success.' Nurserymen, from keeping large stocks of the same plant, were generally far more successful than amateurs in raising new and valuable varieties. But what was probably the most important element in selection was that the closest attention be paid to even the slightest deviations. Otherwise, improvements could not be made. 'I have seen it gravely remarked,' said Darwin, 'that it was most fortunate that the strawberry began to vary just when gardeners began to attend to this plant. No doubt the strawberry had always varied since it was cultivated, but the slight varieties had been neglected. As soon, however, as gardeners picked out individual plants with slightly larger, earlier, or better fruits, and raised seedlings from them, then (with some aid by crossing distinct species) those many admirable varieties of the strawberry were raised which have appeared during the last half-century.'

And what of plants in the wild? Are these subject to any variation, asked Darwin. In the Galapagos he had seen how species varied from island to island and how they were all related to South American plants. He had seen in the mountains of Tierra del Fuego dwarfed plants fitted to the wind-swept heights by their habit of growing close to the ground.

This brought him to the question that has tormented botanists: What is a species? 'No one definition has satisfied all naturalists; yet every naturalist knows vaguely what he means when he speaks of a species.' Generally the term included the unknown element of 'a distinct act of creation'. The term 'variety' was almost equally difficult to define; but here community of descent was almost universally implied, though it could rarely be proved.

To Darwin the question of what was a species and what a sub-species was important.[1] 'No one supposes that all the individuals of the same species are cast in the same actual mould. These individual differences are of the highest importance to us, for they are often inherited, as must be familiar to everyone; and they thus afford materials for natural selection to act on and accumulate, in the same manner as man accumulates in any given direction individual differences in his domestic productions.' The trouble was that no two botanists looked at these individual differences in exactly the same way. Classification was therefore arbitrary. 'Compare,' said Darwin, 'the several floras of Great Britain, of France, or of the United States, drawn up by different botanists, and see what a surprising number of forms have been ranked by one botanist as good species, and by another as mere varieties.' He had asked three leading botanists to mark in the *London Catalogue of British Plants* those generally considered to be varieties but which had all been ranked by botanists as species. Hewett Cottrell Watson (editor of the *Catalogue*) marked 182, Charles Babington 251 and George Bentham only 112—a difference of 139 doubtful forms!

These doubtful or intermediate forms interested Darwin because they conveyed the idea of a passing from one to the other. 'Hence,' he said, 'I look at individual differences, though of small interest to the systematist, as of the highest importance for us, as being the first steps towards such slight varieties as are barely thought worth recording in works on natural history. And I look at varieties which are in any degree more distinct and permanent, as steps towards more strongly-marked and permanent varieties; and at the latter, as leading to sub-species, and then to species.'

He thought that some interesting results might be obtained by tabulating all the varieties in the floras of twelve different countries, to see which species varied most. The great French–Swiss botanist Alphonse de Candolle had shown that plants which have very wide ranges generally produce varieties. This was to be expected because they are exposed to different conditions and are competing with different plants. Darwin's tables showed this, and further showed that the species which are the most common within their country are the ones to produce varieties sufficiently distinct to be recorded by the botanist. He proved, too, by his tables, that species of the larger genera vary more than the species of the smaller genera. Where many trees grow, we expect to find many saplings, was his simile. He now turned an eye to Genesis. 'If we look at each species as a special act of creation, there is no apparent reason why more varieties should occur in a group having many species, than in one having few.'

But, he went on, the mere existence of individual variability does not help us much to understand how species come into being. 'How,' he asked, 'have all those exquisite adaptations of one part of the organisation to

[1] Here Darwin is using the terms 'variety' and 'sub-species' as interchangeable. In modern usage the hierarchy, descending down, is genus, species, sub-species, variety, sub-variety, form.

another part, and to the conditions of life, and of one organic being to another being, been perfected? We see these beautiful coadaptations most plainly in the woodpecker and the mistletoe: the woodpecker with its feet, tail, beak, and tongue, so admirably adapted to catch insects under the bark of trees; the mistletoe which draws its nourishment from certain trees, which has seeds that must be transported by certain birds, and which has flowers with separate sexes absolutely requiring the agency of certain insects to bring pollen from one flower to the other.' There were these beautiful adaptations everywhere, in every part of the living world.

And, he asked, how is it that varieties ultimately become converted into good and distinct species? How do these groups of species, called genera, arise? The answer, Darwin suggested, is that they follow as a result of the struggle for life. Variations however slight, if in any way beneficial to the individual in its complex relationship with other beings and with its environment, will tend to its preservation and will generally be inherited by its offspring. This was what he meant by the term Natural Selection, as different from man's selection which was unnatural. Man could produce great results, but Natural Selection was an infinitely greater power and one that was always ready to go into action. Two animals in a time of dearth fight with each other to get what food there is and so live. But a plant on the edge of a desert has many struggles for life. Dependent on moisture, it struggles against drought. If annually it produces a thousand seeds, of which on average one comes to maturity, it has to struggle against plants of its own and other kinds already in possession of the ground. The Mistletoe is dependent on the apple and a few other trees, and if too many of these parasites grow on the same tree it languishes and dies. Several seedling Mistletoes growing close together on the same branch struggle with each other. The existence of its race is dependent on birds, but it has to struggle with other fruit-bearing plants in tempting the birds to devour and thus disperse its seeds.

The struggle for existence was constantly being waged against nature's own prodigality. Linnaeus had calculated that if an annual plant produced only two seeds—and there is no plant so unproductive as this—and their seedlings next year produced two, and so on, then in twenty years there would be a million plants. It was known that plants introduced from another country could become common in less than ten years. The Cardoon and the tall thistle he had seen covering square miles of the wild plains of La Plata had been introduced from Europe. Hugh Falconer told him of plants introduced into India from America, which now ranged from Cape Comorin to the Himalayas. The explanation was that the conditions of life were highly favourable to them, their native predators were absent, and consequently there was little destruction of the plants and their seedlings. Their geometrical rate of increase explained their rapid and wide diffusion in their new homes. Then, to maintain the population of a tree that lived for a thousand years, it would be necessary for the tree to produce only a single seed in its lifetime, assuming that the seed was not destroyed and could germinate and grow to maturity. But in that time it would produce thousands upon thousands of seeds. This was where Malthus came in with his 'checks' to populations. In plant life there is a vast destruction of seeds, and Darwin had seen that seedlings suffered most from germinating in ground already thickly stocked with other plants. They were also destroyed in huge numbers by various enemies. In his 'weed garden', three feet long and two feet wide, dug and cleared, he had marked all the seedlings as they came up, and out of 357 no fewer than 295 were destroyed, chiefly by slugs and insects. He had left part of his lawn unmowed: the more vigorous plants gradually killed the less vigorous. Out of 20 species growing in a patch three feet by four, 9 perished in their unequal struggle with other species allowed to grow up freely. Browsing animals had accounted for the destruction of the myriad little Scots Pines he had discovered at Moor Park. Climate was another slayer; particularly seasons of extreme cold or drought. In the Arctic regions, on snow-capped summits or in deserts, the struggle for life was almost entirely with the elements. A large stock of the same species was necessary for its preservation. We can easily grow plenty of corn in our fields because the seeds vastly outnumber the birds that feed on them; nor can the birds, even with a superabundance of food at this one season, greatly increase in number, because the cold and dearth of food in winter provides a check. 'But anyone who has tried,' said Darwin, 'knows how troublesome it is to get seed from a few wheat or any other such plants in a garden: I have in this case lost every single seed.'

Not all introduced plants succeed in a new country. One was *Lobelia fulgens*, a native of Mexico. Darwin grew it in his garden and had noticed that it was never visited by insects: consequently it never set a seed. He had studied the wild orchids that grew near Down. Nearly all of them required insects to carry their pollen from one flower to another and so fertilise them. He had found from experiments that bumble-bees were the pollinators of the Heartsease (*Viola tricolor*), other bees never visiting this flower. He had also found that the visits of bees are necessary for the fertilisation of some kinds of clover, and that only bumble-bees visit Red Clover. So if all the bumble-bees became extinct, the Heartsease and Red Clover might completely disappear! Now, said Darwin, the number of bumble-bees in any district depends on the number of field mice, which destroy their combs and nests. It was believed that two-thirds of them were destroyed in this way all over England. And the number of mice was regulated by the number of cats. The hymenopterist Colonel H. W. Newman, writing on the subject in the 1851 *Proceedings* of the Entomological Society, reported that 'Near villages and small towns I have found the nests of humble-bees more numerous than elsewhere, which I attribute to the number of cats that destroy the mice.' Darwin thought it quite credible that the presence of many cats in a district might determine the numbers of Red Clovers and Heartsease!

What a struggle, he exclaimed, must have gone on during long centuries between the several kinds of trees, each annually scattering its seeds by the thousand. What war between insect and insect—between

insects, snails, and other animals with birds and beasts of prey—all
striving to increase, all feeding on each other, or on the trees, their seeds
and seedlings, or on the other plants which already clothed the ground
and thus checked the growth of the trees!

As far back as March 1839 he had confided his horror of this slaughter
to a page of his Transmutation notebook. 'It is difficult to believe in the
dreadful but quiet war of organic beings going on in the peaceful woods
and smiling fields.'

Despite the slaughter enough remained to keep the races constant.
He set out the means by which they survived. The down on fruit and the
colour of the flesh were considered by botanists as characters of the most
trifling importance, but the famous American pomologist Charles
Downing wrote that smooth-skinned fruits suffered far more from the
beetle Curculio than those with down, and that purple plums suffered
more from a certain disease than yellow plums, while another disease
affected yellow-fleshed peaches more than those with flesh of another
colour. If, said Darwin, with all the aids available to the fruit-grower,
these slight differences make a great difference in cultivating the
varieties, assuredly in the wild—where the trees would have to struggle
with other trees and with a host of enemies—such differences would
effectually settle which variety, the smooth or the downy, the yellow or
purple fleshed, would succeed.

This was in Chapter IV of the *Origin* where Darwin asked if the
principles of selection, so potent in the hands of man, could apply under
nature. He reminded his readers that variability was not produced by
man, who could neither originate nor prevent its occurrence but could
only preserve and accumulate the variations that occurred. Now he set
out to prove that the same applied in nature, that Natural Selection
implies only the preservation of variations as they arise and if they are
beneficial to the plant or animal.

Vigour and fertility were necessary to a plant, as to an animal, for its
survival, and Darwin had found in a series of experiments that in order to
keep their vigour and fertility even hermaphrodite flowers, though self-
fertilising, must occasionally be crossed with other flowers. In his garden
and greenhouse he had studied the many beautiful and curious adaptations by which nature ensured the visit of a particular insect to a
particular kind of flower. He had seen from his tables that variability was
an element of success, which was another way of saying that natural
selection was fitting the plant to adapt to its environment. But, he wrote,
'Though nature grants long periods of time for the work of natural
selection, she does not grant an indefinite period; for as all organic beings
are striving to seize on each place in the economy of nature, if any one
species does not become modified and improve in a corresponding degree
with its competitors, it will be exterminated.' This was the principle of
the Survival of the Fittest, a term invented by Herbert Spencer, which
Darwin first used in 1869, in the fifth edition of the *Origin*. So as the
favoured forms—those which through natural selection had acquired

some advantage in a beneficial modification—had increased in number, so had the less favoured decreased and become rare, and rarity was the precursor to extinction.

After writing his 1844 Sketch Darwin was aware that he had overlooked one important problem. He had dealt with the branching and sub-branching of the evolutionary tree, the branching or splitting of one species into two. But, as these became modified, was there a tendency for their descendants to diverge in character? He recalled in his *Autobiography* 'the very spot in the road, whilst in my carriage, when to my joy the solution occurred to me: and this was long after I had come to Down.' It may have been in 1852. He wrote to George Bentham: 'It is to me really laughable, when I think of the years which elapsed before I saw what I believe to be the explanation of some parts of the case; I believe it was fifteen years after I began before I saw the meaning and cause of the divergence of any one pair.' The explanation was an ecological one: 'that the modified offspring of all dominant and increasing forms tend to become adapted to the many and highly diversified places in the economy of nature.'

Coke of Holkham, Thomas William, first Earl of Leicester, 'the greatest farmer in the world', was famed for inventing the system of the rotation of crops. Darwin carried out an experiment to prove that nature follows what may be called a simultaneous rotation. Using two similar plots of ground he sowed one with one species of grass and the other with several distinct kinds of grasses. The second yielded a greater number and a greater weight of dry herbage. Farmers had tried this with wheat. This proved that the greatest amount of life can be supported by diversity. It could be thought that plants introduced into foreign countries would succeed only if they were closely related to the native plants, these being looked at as specially created and adapted to their native land. In fact they were more successful if they were new genera than if they were merely new species.

How, Darwin now asked, had nature taken care that there should be this diversity? Plants were either 'high' or 'low' in organisation, and some botanists ranked those plants as highest which had every organ—sepals, petals, stamens, and pistils—fully developed in each flower; other botanists regarding as highest those which had their organs modified and reduced in number. Darwin favoured the latter view, arguing that as natural selection led to specialisation, organs might therefore become superfluous or useless. A multitude of the lowest forms existed throughout the world; and if it was no advantage to them to be improved, natural selection would leave them as they were, for in nature there was room for lowly organised forms: they were not in close competition with higher forms.

The affinities of all the beings of the same class could be represented by a great tree.

The green and budding twigs may represent existing species; and those produced during former years may represent the long selection of extinct species. At each period of growth all the growing twigs have tried to branch out on all sides, and to overtop and kill the surrounding twigs and branches, in the same manner as species and groups of species have at all times overmastered other species in the great battle for life. The limbs, divided into great branches and these into lesser and lesser branches, were themselves once, when the tree was young, budding twigs; and this connection of the former and present buds by ramifying branches may well represent the classification of all extinct and living species in groups subordinate to groups. Of the many twigs which flourished when the tree was a mere bush, only two or three, now grown into great branches, yet survive and bear other branches; so with the species which lived during long-past geological periods, very few have left living and modified descendants. From the first growth of the tree, many a limb and branch has decayed and dropped off; and these fallen branches of various sizes may represent those whole orders, families, and genera which have now no living representatives, and which are known to us only in a fossil state. As we here and there see a thin straggling branch springing from a fork low down in a tree, and which by some chance

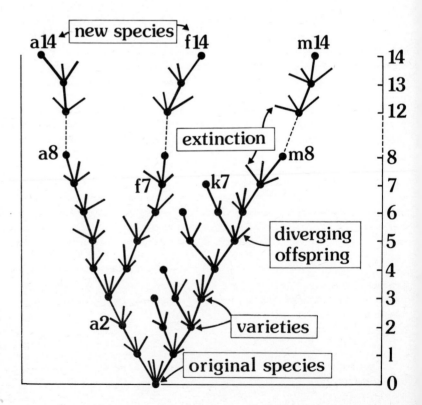

Darwin's genealogical tree. The *original species* of a large genus has variable offspring. The most *divergent offspring* tend to survive, giving rise eventually to new *varieties*. The vertical scale at the side represents the passage of generations, each step being 1,000 generations. As a result of *divergence of character* and *extinction* of varieties (*k*), the descendants of a common ancestor eventually become new species (*a*, *f* and *m*), in this case after 14,000 generations

has been favoured and is still alive on its summit, so we occasionally see an animal like the Ornithorhynchus or Lepidosiren, which in some small degree connects by its affinities two large branches of life, and which has apparently been saved from fatal competition by having inhabited a protected station. As buds give rise by growth to fresh buds, and these, if vigorous, branch out and overtop on all sides many a feebler branch, so by generation I believe it has been with the great Tree of Life, which fills with its dead and broken branches the crust of the earth, and covers the surface with its ever-branching and beautiful ramifications.

Thus Darwin in choosing a many-branching tree was able to give us a model for the system of natural classification covering not only the whole world of plants and animals living today and those which are extinct but, in providing the link between them, explain how natural selection has worked from the beginning of life itself.

10

The Great Migrations

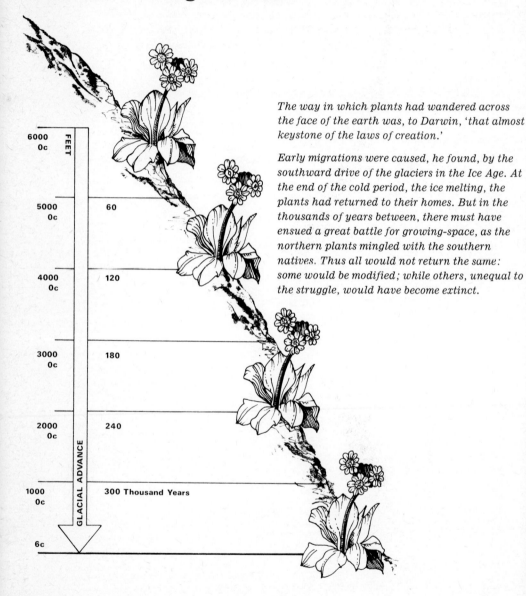

FEET

6000
0c

5000
0c   60

4000
0c   120

3000
0c   180

2000
0c   240

1000
0c   **300 Thousand Years**

6c

GLACIAL ADVANCE

12c   **Glacier now retreats Plants return**

The way in which plants had wandered across the face of the earth was, to Darwin, 'that almost keystone of the laws of creation.'

Early migrations were caused, he found, by the southward drive of the glaciers in the Ice Age. At the end of the cold period, the ice melting, the plants had returned to their homes. But in the thousands of years between, there must have ensued a great battle for growing-space, as the northern plants mingled with the southern natives. Thus all would not return the same: some would be modified; while others, unequal to the struggle, would have become extinct.

THE THREE QUARTO volumes Joseph Hooker wrote on the botany of the voyage of *Erebus* and *Terror* were each prefaced by a masterly essay. The Introduction to his *Flora Antarctica,* which was Volume I of the trilogy, contained the first pronouncement of his theory of the geographical distribution of plants, a subject which he believed would be 'the key which will unlock the mystery of species.'

In a letter to Hooker of the 10th of February 1845 Darwin capped this. 'I know,' he said, 'I shall live to see you the first authority in Europe on that grand subject, that almost keystone of the laws of creation, Geographical Distribution.'

The two were agreed on the importance of the subject.

Hooker had added to his botanical explorations of the southern hemisphere four years of travel in India. He came home in 1850, already a master of plant geography: thoroughly acquainted from boyhood with his father's world-wide herbarium, he had now completed his personal experience of how plants had wandered from continent to continent. On the same Indian mountain he had climbed from the seething heat of the plains through a temperate English spring to Arctic winter—finding on its summit the same lichen, *Lecanora miniata,* that crimsoned the rocks of Crozier Island where grew the most southerly vestiges of vegetation. He had seen how the islands south of New Zealand, and the Falkland Islands, South Georgia, Tristan da Cunha, and Kerguelen Island five thousand miles away, had all borrowed plants from Fuegia. And in Fuegia he had found a host of plants that had wandered from England: the Thrift, the Dandelion, Starwort and Wild Celery, with a primrose so like our own that the two were scarcely distinguishable. Dozens of the common genera and nineteen genera of grasses were the same, while almost every Fuegian lichen was a prevalent species in Britain. Recognising them as familiar plants of home his mind was drawn to 'that interesting subject— the diffusion of species over the surface of our earth.'

The same interesting facts had struck Darwin. On Joseph's return from the Antarctic voyage, when the two started corresponding, it was the subject of geographical distribution that was paramount in their letters.

First, Darwin drew the younger man's attention to the importance of correlating the Fuegian flora with that of the Cordillera and of Europe; and, as we know, Henslow had sent him the collections Darwin made in the Galapagos Islands, Patagonia, Fuegia, and in fact those of the entire voyage. Darwin was delighted and astonished at the results of Hooker's examinations of the Galapagos specimens. How wonderfully they supported his assertion of the difference in the species of the different islands, 'about which I have always been fearful,' he wrote. He plied Hooker with questions, apologising for their number and frequency, which brought the reply: 'If you knew how grateful the turning from the drudgery of my "professional Botany" to your "philosophical Botany" was, you would not fear bothering me with questions. The truth in its primitive nakedness is, that I really look for and count upon such questions, as the best means of keeping alive a due interest in these subjects. I indulge vague hopes of treating them some day, but days and years fly over my head and all I do is done in correspondence with you, but for which I should soon lose sight of the whole matter.'

Darwin's questionings were sometimes of direct and practical help to Hooker, as a letter to Asa Gray reveals:

> I was led by one of my wild speculations to conclude (though it has nothing to do with your statistics) that trees would have a strong tendency to have flowers with dioecious, monoecious or polygamous structure. Seeing that this seemed so in Persoon[1] I took one little British Flora, and discriminating trees from bushes according to Loudon, I have found that the result was in species, genera and families, as I anticipated. So I sent my notions to Hooker to ask him to tabulate the New Zealand Flora for this end, and he thought my result sufficiently curious, to do so; and the accordance with Britain is very striking, and the more so, as he made three classes of trees, bushes, and herbaceous plants. (He says further he shall work the Tasmanian Flora on the same principle.) The bushes hold an intermediate position between the other two classes. It seems to me a curious relation in itself, and is very much so, if my theory and explanation are correct.

[1] Christian Hendrick Persoon (1761–1836), the South African botanist.

Hooker and Darwin approached the subject from completely different angles. As a taxonomist Hooker was interested in 'species problems' from the phytogeographical standpoint, the relationships of plants in one country with those in another. Darwin painted on a wider canvas, with his knowledge able to trace plants back to their first wanderings, and so understand why they had become modified in the process or why they had stayed the same.

Apart from the voluminous correspondence that developed between them (which would make a book in itself) there were frequent sessions at Down. Of these visits Hooker told Francis Darwin: 'It was an established rule that he every day pumped me, as he called it, for half and hour or so after breakfast in his study, when he first brought out a heap of slips with questions botanical, geographical, &c., for me to answer, and concluded by telling me of the progress he had made in his own work, asking my opinion on various points.'

But however much Hooker contributed, so profound was Darwin's knowledge on almost any subject they discussed, that he 'always left with the feeling that I had imparted nothing and carried away more than I could stagger under.'

He always brought work of his own to do, and after the morning session he saw no more of Darwin till about noon 'when I heard his mellow ringing voice calling my name under my window—this was to join him in his daily forenoon walk round the sand-walk. On joining him I found him in a rough grey shooting-coat in summer, and thick cape over his shoulders in winter, and a stout staff in his hand; away we trudged through the garden, where there was always some experiment to visit, and on to the sand-walk, round which a fixed number of turns were taken, during which our conversation usually ran on foreign lands and seas, old friends, old books, and things far off to both mind and eye.

'In the afternoon there was another such walk, after which he again retired till dinner if well enough to join the family; if not, he generally managed to appear in the drawing-room, where seated in his high chair, with his feet in enormous carpet shoes, supported on a high stool—he enjoyed the music or conversation of his family.'

So nearly did they work together, so easily did Hooker assimilate his arguments, while advancing powerful arguments of his own, that Darwin began to fear that in discussing his ideas so freely and fully with him he had checked Hooker's own original lines of thought. On the 14th of November 1858 he wrote:

> I have for some time thought that I had done you an ill-service, in return for the immense good which I have reaped from you, in discussing all my notions with you; and now there is no doubt of it, as you would have arrived at the figorific mixture independently.

It was inevitable that their minds should converge. In his Introductory Essay to the New Zealand Flora (1864), Hooker, while putting forward some general opinions about the origin of species, still held to the accepted doctrine that species had been created as such and were immutable. Then on his return from India he had plunged into comparative studies of the vast flora of that continent and of Australia, and related them to those of the neighbouring countries. Affinities were his clue. This brought him to variation, and in his Introductory Essay to the Tasmanian Flora he advanced his own theory of the origin of species. But it was nowhere to be compared with the scope of Darwin's and the weight of evidence with which Darwin backed it up. It was this weight of fact ('the application of the theory') that converted and convinced the scientific world, and Hooker was aware how much he owed to his botanico-philosopher. In accepting that species were 'derivative and mutable' he wrote that 'whatever opinions a naturalist may have adopted with regard to the origin and variation of species, every candid mind must admit that the facts and arguments upon which he has grounded his convictions require revision since the recent publication by the Linnean Society of the ingenious and original reasonings and theories of Mr Darwin and Mr Wallace.'

In the physical approach to geographical distribution they were, however, partners from the start.

To Darwin the point at issue hung on the question: had the same species been simultaneously created in different parts of the world—or had they sprung from common parents?

There was difficulty in understanding how the same species could possibly have migrated from some one place to the distant and isolated places where they were now found. 'Nevertheless,' Darwin wrote, 'the simplicity of the view that each species was first produced within a single region captivates the mind. He who rejects it, rejects the *vera causa* of ordinary generation with subsequent migration, and calls in the agency of a miracle.' He set out to prove how migrations could have taken place.

He looked first at the great barriers of oceans. His salt-water experiments had proved that on average the seeds of 14 out of 100 plants belonging to one country might be floated across 924 miles of sea to another country, and when stranded, if blown inland by a gale to a favourable spot, would germinate. Drift timber was thrown up on most islands, even in the midst of the widest oceans. On the coral islands of the Pacific he had found that when irregularly shaped stones were embedded in the roots of trees, as they frequently were, particles of earth were often enclosed in their interstices and behind them. Out of one small portion of earth thus enclosed he had germinated three dicotyledonous plants. The carcases of birds floating on the sea were another source: many kinds of seeds in their crops long retained their vitality. In his tank of artificial sea-water he had floated a dead pigeon for thirty days. Seeds of peas and vetches he took out of its crop nearly all germinated. In his garden he had picked twelve kinds of seed out of the excrement of small birds; some of them germinated. The feet of birds often had little cakes of earth attached to them. Alfred Newton, Professor of Zoology at Cambridge, had sent him the leg of a partridge with a hard ball of earth weighing six and a half ounces adhering to it. The earth had been kept for three years, but when Darwin broke it up, watered it and placed it under a bell glass, no fewer than 82 plants grew from it, consisting of 12 monocotyledons including the common oat and a grass, and 70 dicotyledons of at least three distinct species. 'With such facts before us, can we doubt,' asked Darwin, 'that the many birds which are annually blown by gales across great spaces of ocean, and which annually migrate—for instance, the millions of quails across the Mediterranean—must occasionally transport a few seeds embedded in dirt adhering to their feet or beaks?'

Icebergs were sometimes loaded with earth and stones, and were known to have carried brushwood, bones, and the nest of a land bird. They also must occasionally have transported seeds from one part to another of the arctic and antarctic regions, and during the Glacial Age from one part of the now temperate regions to another. Darwin suspected that the Azores had been partly stocked by ice-borne seeds, judging by the large number of plants common to Europe in comparison with the species on the other Atlantic islands nearer to the mainland. But were erratic boulders to be found in the Azores? Lyell sought the answer from one of his geological correspondents who replied that he had found large fragments of granite and other rocks not native to the archipelago. 'Hence,' said Darwin, 'we may safely infer that icebergs formerly landed their rocky burthens on the shores of these mid-ocean islands, and it is at least possible that they may have brought thither some few seeds of northern plants.'

Geographical
distribution by air,
water, and land.
Darwin once picked
twelve kinds of seeds
from the excrement of
small birds. Some of
them germinated.
Fish, he found, also
swallowed seeds.
Burred seeds were
easily caught in the
fur of animals, to be
shaken off at a
different place

The Glacial Age was responsible for another type of migration. Darwin instanced the identity of many plants on mountain summits separated from each other by hundreds of miles of lowlands where alpine species could not possibly exist. How had they been able to cross this barrier? He had asked Asa Gray about the alpine plants in America, and back came the remarkable fact that all the plants on the White Mountains were the same as those of Labrador, and nearly the same as those on the loftiest mountains of Europe.

It was known that within a very recent geological period central Europe and North America had an arctic climate. Darwin on geological expeditions in Wales and Scotland had seen evidence of this in the mountains with their scored flanks, polished surfaces and perched boulders. In northern Italy vines and maize grew on gigantic moraines left by old glaciers. Throughout a large part of the United States erratic boulders and scored rocks told the same tale of a former cold period. Darwin explained what had happened to the plants. As the cold came on, driving downwards and chilling the southerly regions, these places would become fitted for the plants of the north which would then spread southward, and thus they would go on spreading unless stopped by barriers, when they would perish. By the time the cold had reached its maximum an arctic flora would have covered the central parts of Europe as far south as the Alps and Pyrenees and even stretching into Spain. North America would likewise be covered by arctic plants and these would be nearly the same as those of Europe, Darwin pointing out that the present circumpolar species 'which we suppose to have everywhere travelled southward' are remarkable uniform round the world. Then as the warmth returned, the arctic forms would return northward, closely followed up by plants of the more temperate regions. And as the snow melted from the bases of the mountains, the arctic plants would seize on the cleared and thawed ground, always ascending, as the warmth increased and the snow still farther disappeared, higher and higher, while their brethren were pursuing their northern journey. So, when the warmth had fully returned, the same species which had lately lived together on the European and North American lowlands would again be found in the arctic regions of the Old and New Worlds and on many isolated mountain summits far distant from each other.

There was another interesting aspect of the Glacial Age. In their journey to the south and back to the north the arctic forms, travelling in unison with the cold climate, would not have been much changed. But not all the present alpine plants on the one mountain were the same as on every other. There were some varieties, some doubtful forms, and some distinct though closely allied species. Darwin assumed that in the returning warmth modification had taken place of ancient pre-Glacial alpines which had not travelled so far south, while others were the result of mingling with northward-bound plants returning to their old homes. There would also be isolated pockets bypassed by the flow of the glaciers. Again, the southern hemisphere in its turn had been subjected to a Glacial period, with the northern hemisphere becoming warmer. The trek then had been from the south to the north, and there must have ensued a great battle for growing-space, with plants unequal to the struggle becoming extinct, others becoming modified. Darwin likened these treks to living waters which flowed from the north during one period and during another from the south, both reaching the Equator. 'As the tide leaves its drift in horizontal lines, rising higher on the shores where the tide rises highest, so have the living waters left their living drifts on our mountain summits, in a line gently rising from the Arctic lowlands to a great altitude under the equator. The various beings thus left stranded may be compared with savage races of man, driven up and surviving in the mountain fastnesses of almost every land, which serves as a record full of interest to us, of the former inhabitants of the surrounding lowlands.'

He considered how fresh-water plants could have migrated. Lakes and rivers were separated from each other by barriers of land, and it might be thought that their inhabitants could not have ranged widely within the same country, and with the sea a still more formidable barrier, that they could never have extended to distant countries. It was exactly the reverse. 'When first collecting in the fresh waters of Brazil,' Darwin recalled, 'I well remember feeling much surprise at the similarity of the fresh-water insects, shells, &c., and at the dissimilarity of the surrounding terrestrial beings, compared with those of Britain.' In most cases short and frequent migrations from pond to pond and from stream to stream in the same country was the explanation. Whirlwinds accounted for wider dispersal: it was not unknown for live fishes to be swept up and dropped at distant places. Floods joining one river to another were a certain proof; and wading birds, which frequent the muddy edges of ponds, had the widest flying range. They were to be found on the most remote and barren islands of the open ocean. Gaining the land, the dirt on their feet would be washed off in new muddy haunts. 'I do not believe,' wrote Darwin, 'that botanists are aware how charged the mud of ponds is with seeds. I have tried several little experiments.' He instanced one of them. 'I took in February three table-spoonfuls of mud from three different points, beneath water, on the edge of a little pond: this mud when dried weighed only 6¾ ounces; I kept it covered up in my study for six months, pulling up and counting each plant as it grew; the plants were of many kinds, and were altogether 537 in number; and yet the viscid mud was all contained in a breakfast cup! Considering these facts, I think it would be an inexplicable circumstance if water-birds did not transport the seeds of fresh-water plants to unstocked ponds and streams, situated at very distant points.'

But the barriers of ocean certainly limited the numbers of plants to be found on oceanic islands. The whole of New Zealand together with the outlying islands of Auckland, Campbell, and Chatham, had only 960 kinds of flowering plants between them compared with the 847 in the county of Cambridge and the 764 of the little island of Anglesey, though in both these numbers a few ferns and a few introduced plants were included. The barren island of Ascension had less than half a dozen flowering plants that were native (as Joseph Hooker knew from his visit there), yet many species had now become naturalised on it as they had in New Zealand and on every other oceanic island. In St Helena the naturalised plants had exterminated nearly all the native kinds. So, said Darwin, he who admits the doctrine of the creation of each separate species, will have to admit that a sufficient number of the best adapted plants and animals were not created for oceanic islands; for man had unintentionally (and sometimes intentionally) stocked them far more fully and perfectly than had nature.

It was Darwin's practice to put himself in the position of being his own severest critic. This served as a challenge which often helped him to work out the completeness of his statements. It also gave him an inkling of how these statements would be received by the public. In stating his belief 'that the innumerable species, genera and families, with which this world is peopled, are all descended, each within its own class or group, from common parents', and taking this 'one step farther, namely, to the belief that all animals and plants are descended from some one prototype', he foresaw that his readers might well ask for proof. All very well to talk of variation, which any gardener or naturalist might accept, but what of the intermediary forms showing this transition? Were the links to be found in fossils?

Unfortunately there was no such unbroken chain. Only a small portion of the earth's surface had been geologically explored, and few parts thoroughly. One of the most striking instances was that of the Flysch formation, which consists of shale and sandstone several thousand, and here and there six thousand, feet in thickness, extending for at least 300 miles from Vienna to Switzerland. Although this great mass had been most carefully searched only a few fossils—of plants—had been found.

Darwin was convinced that nearly all the ancient geological formations rich in fossils had been formed during subsidence, and that fossils were to be found only where the supply of sediment was sufficient to embalm the remains before they had time to decay. Where the sea was shallow these beds of sediment when upraised and subjected to the pounding of the coastal waves would be destroyed. Sediment could also be formed in thick and extensive accumulations in the profound depths of the sea, but at such depths there were not so many or such varied forms of life. So when this mass was upraised there would only be an imperfect record of the life existing in the neighbourhood during the period of its accumulation. Wherever sediment did not accumulate, or where it did not accumulate rapidly enough to protect organic bodies from decay, no remains could be preserved.

There had been a long succession of land upheavals. Formation had piled upon formation but in no neat layers whereby it was possible to cut a section and see how a plant had modified and perfected itself through a series of gradations. The land had been folded, crumpled, fractured, and piled up in a confusion of movements, and between these movements were immense lapses of time. In the case of Darwin's silicified trees, he had found them standing upright as they grew. But this was the result of an uncomplicated upheaval recent in geological time, and concerned only a single species. What of the cases of lower beds which had been upraised, denuded, submerged, and re-covered by the upper beds of the same formation? 'It should not be forgotten,' Darwin wrote, 'that at the present day, with perfect specimens for examination, two forms can seldom be connected by intermediate varieties, and thus proved to be the same species, until many specimens are collected from many places; and with fossil species this can rarely be done.' Therefore, he added, 'we have no right to expect to find in our geological formations, an infinite number of those fine transitional forms which, on our theory, have connected all the past and present species of the same group into one long and branching chain of life. We ought only to look for a few links, and such assuredly we do find—some more distantly, some more closely, related to each other.' However wide were the breaks in the succession of fossil relics, enough remained to show that the thread of life itself had never been broken. As with the plants that had wandered far across the earth and become changed yet preserved affinities with their stay-at-home relatives, so there was the same deep organic bond preserved through space and time; and that bond was simply inheritance.

His look into the past was far and deep. He saw that continents may have existed where oceans are now spread out; and clear and open oceans existed where continents now stand. We could not assume, he said, even if the bed of the Pacific Ocean were now converted into a continent, that we should find sedimentary formations in a recognisable condition older than the Cambrian strata, supposing such to have been deposited. Imagine the enormous press of water and the metamorphic action they must have undergone. There were immense areas in some parts of the world, for instance in South America, of naked metamorphic rocks which must have been heated under great pressure. Was it surprising that though we find in geological formations many links between a species now existing and one which formerly existed, we do not find infinitely numerous transitional forms closely joining them all together?

'For my part,' wrote Darwin, 'I look at the geological record as a history of the world imperfectly kept, and written in a changing dialect. Of this history we possess the last volume alone, relating only to two or three countries. Of this volume, only here and there a short chapter has been preserved; and of each page, only here and there a few lines. Each word of the slowly-changing language, more or less different in the succeeding chapters, may represent the forms of life which are entombed in our consecutive formations, and which falsely appear to have been abruptly introduced. On this view, the difficulties above discussed are greatly diminished or even disappear.'

Supposing, though, his readers were still unconvinced? There was an answer. 'Nature,' wrote Darwin, 'may be said to have taken pains to reveal her scheme of modification, by means of rudimentary organs, of embryological and homologous structures, but we are too blind to understand her meaning.'

We could discover the lines of descent by the permanent characters.

With all organic beings, excepting perhaps some of the very lowest, sexual reproduction was essentially similar. With all, the germinal vesicle was the same. So, he said, all organisms start from a common origin. 'If we look even to the two main divisions—mainly to the animal and vegetable kingdoms—certain low forms are so intermediate in character that naturalists have disputed to which kingdom they should be referred.' Asa Gray had remarked that 'the spores and other reproductive bodies of many of the lower algae may claim to have first a characteristically animal and then an unequivocally vegetable existence.'

Therefore, said Darwin, 'on the principle of natural selection with divergence of character, it does not seem incredible that, from some such low and intermediate form, both animals and plants may have been developed; and if we admit this, we must likewise admit that all the organic beings that have ever lived on this earth may be descended from some one primordial form.'

That a flower could dispense with some of its parts was a discovery that helped Darwin trace its ancestry. The diagram shows the basic components of an archetypal flower. In the transverse section (TS) the five whorls of components can be seen, and these are also shown in the longitudinal section (LS). Any one of the components of one or more whorls can be modified or become rudimentary

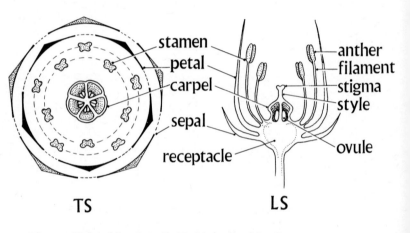

The one living kingdom divided into two kingdoms.

Animal and plant organs in a rudimentary condition plainly showed that an early common progenitor had the organ in a fully developed condition. The petals of a flower are sometimes rudimentary and sometimes well-developed in the individuals of the same species. In certain plants having separate sexes the botanist Gottlieb Kölreuter had found that by crossing a species whose male flowers had the rudiment of a pistil, with a hermaphrodite species with its well-developed pistil, the rudiment in the hybrid offspring was much increased in size. This clearly showed that the rudimentary and perfect pistil were essentially alike in nature. An organ acting for two purposes may become rudimentary or completely aborted for one purpose, even the more important, and remain perfectly efficient for the other. The use of the pistil is to allow the pollen-tubes to reach the ovules within the ovarium. The pistil consists of a stigma supported by a style below which is the ovarium. But in some Compositae (daisy-like flowers) the male florets, which of course cannot be fertilised, have a rudimentary pistil, for it is not crowned with a stigma. Yet the style remains well developed and is clothed in the usual way with hairs which serve to brush the pollen out of the surrounding anthers. Rudimentary organs may be completely aborted. In most of the Scrophulariaceae the fifth stamen is lacking. Yet we could conclude that a fifth stamen once existed, for a rudiment of it is found in many species of the family, and it occasionally becomes perfectly developed, as may sometimes be seen in the common Snapdragon. In tracing the homologies of any part in different members of the same class, nothing is more common, or, in order fully to understand the relations of the parts, more useful, said Darwin, than the discovery of rudiments. Disuse, the main agent in rendering organs rudimentary, might be compared with the letters in a word, still retained in the spelling but become useless in the pronunciation, but which serve as a clue to its derivation. 'On the view of descent with modification,' Darwin said, 'we may conclude that the existence of organs in a rudimentary, imperfect, and useless condition, or quite aborted, far from presenting a strange difficulty, as they assuredly do on the old doctrine of creation, might even have been anticipated in accordance with the views here explained.'

The embryos of animals and plants also helped us to see the long-

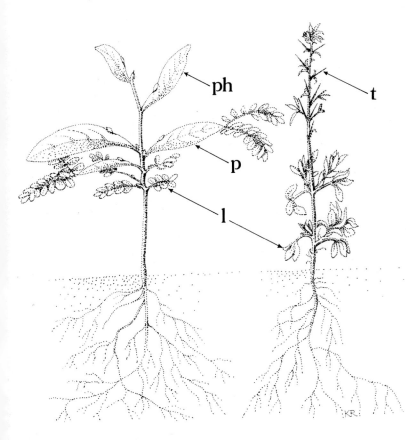

The similarity of some seedlings, as those of the gorse and mimosa, gave Darwin another clue to common ancestry. Shown on the left is a young *Acacia pycnantha*, the Australian mimosa, and on the right is a young gorse seedling, both to the same scale. The young embryonic leaves are in both cases small, smooth and soft (*l*); in the acacia the petiole eventually becomes flattened vertically and (*ph*) forms a phyllate; whereas in the gorse the leaves eventually become modified to thorns (*t*)

continued process of modification. Darwin reminded his readers that the main divisions of flowering plants are founded on differences in the embryo—on the number and position of the primary leaves and the way in which the radicle, or first root, and plumule (the minute bud between the primary leaves) develop. Embryonic forms of plants, as well as animals, often resembled each other more closely than they did as adults. Thus, said Darwin, the first leaves of the gorse and the first leaves of the mimosa were both pinnate, or divided, like the mature leaves of the pea. Yet how different in their adult stage were the thorny leaves of the gorse and the branched leaves of the mimosa.

Hermaphrodite flowers were another stage of development, able to fertilise themselves without the aid of insects. A vast number of plants are hermaphrodites, among them peas, on which Darwin did many interesting experiments. He did not treat the subject at length in the *Origin*, 'though I have the materials prepared for an ample discussion'. This was for a future book. Long and detailed as it was, the *Origin* was still the Abstract of his 'big book', part of which was to go into *The Variation of Animals and Plants*.

It took some 200 million years for vegetation to develop nectar-bearing flowers, a climax of time culminating also in the development of flower-visiting insects. This we now know, but Darwin could point to the fact that in a flower the sepals, petals, stamens, and pistils, are arranged in a spire and that they are really metamorphosed leaves. So this was looking back to the time when plants had no flowers.

Darwin called his *Origin of Species* 'one long argument'. It may seem incredible to us that, little over a century ago, he had to convince the scientists of his day that, for instance, rudimentary organs had not been created 'for the sake of symmetry' or in order 'to complete the scheme of nature'.

He well knew that there must be a battle ahead before his theory of Natural Selection won acceptance, but he was confident of the outcome. 'When the view advanced by me in this volume, and by Mr Wallace, or when analogous views on the Origin of Species are generally admitted, we can dimly foresee that there will be a considerable revolution in natural history.' (This was a prime understatement.)

He looked to the younger men to carry the banner. 'In the future I see open fields for far more important researches.' And again: 'A grand and almost untrodden field of enquiry will be opened, on the causes and laws of variation, on correlation, on the effects of use and disuse, on the direct action of external conditions, and so forth.'

How many fields of inquiry he opened—in genetics, in cytology, and in other branches of biology. How many are yet to be trodden.

Meanwhile, he wrote, concluding the book that was to cause a revolution in man's thinking:

It is interesting to contemplate a tangled bank, clothed with many plants of many kinds, with birds singing on the bushes, with various insects flitting about, and with worms crawling through the damp earth, and to reflect that these elaborately constructed forms, so different from each other, and dependent on each other in so complex a manner, have all been produced by laws acting around us. These laws, taken in the largest sense, being Growth with Reproduction; Inheritance which is almost implied by reproduction; Variability from the indirect and direct action of the external conditions of life, and from use and disuse; a Ratio of Increase so high as to lead to a Struggle for Life, and as a consequence to Natural Selection, entailing Divergence of Character and the Extinction of less-improved forms. Thus, from the war of nature, from famine and death, the most exalted object which we are capable of conceiving, namely, the production of the higher animals, directly follows. There is grandeur in this view of life, with its several powers, having been originally breathed into a few forms or into one; and that, whilst this planet has gone cycling on according to the fixed law of gravity, from so simple a beginning endless forms most beautiful and most wonderful have been, and are being evolved.

The *Origin of Species* was published by John Murray on the 26th of November 1859—and the storm broke. Darwin was reviled from the pulpit for his blasphemous views. Why? The hypothesis that living things could share a common descent was not new: Darwin's own grandfather had groped as far as Selection. What was so shocking was the use of the word 'Natural', which disposed at once of *super*natural intervention. It was not Bible teaching.

In the following year at the British Association meeting at Oxford two pitched battles took place. On the Thursday, the 28th of June, a paper was read to Section D by Dr Charles Giles Bridle Daubeny 'On the final causes of the sexuality of plants, with particular reference to Mr Darwin's work on the "Origin of Species".'

Huxley was called on by the president of the section but tried to avoid a discussion, on the grounds that 'a general audience, in which sentiment would unduly interfere with intellect was not the public before which such a discussion should be carried on.' This did not suit Richard Owen, who declared that there were facts 'by which the public could come to some conclusion with regard to the probabilities of the truth of Mr Darwin's theory.' Huxley challenged him (and later crushingly defeated him). Battle had been joined.

The encounter was renewed on the Saturday when Dr Draper of New York read a paper on 'The Intellectual Development of Europe considered with reference to the Views of Mr Darwin'. Joseph Hooker described the tense atmosphere, the tremendous excitement. The Museum lecture room was not large enough to hold the crowd that came pouring in. They moved to the long West Room, and even it became crowded to suffocation. In the chair was Henslow. For more than an hour 'Dr Draper droned out his paper'. Then discussion began. The first three speakers were shouted down as irrelevant. Hooker was there, Huxley and Sir John Lubbock. In the opposite camp were Richard Owen, Captain FitzRoy and Dr Samuel Wilberforce, the Bishop of Oxford ('out to "smash" Darwin'). When the Bishop's turn came it was obvious that he had been well coached by Owen and knew nothing at first hand. He ridiculed Darwin badly and Huxley savagely, all in dulcet tones. He then asked whether Huxley was related by his grandfather's or grandmother's side to an ape. To this Huxley replied that if he did have an ancestor whom he felt shame in recalling, it would be a *man* who plunged into scientific questions with which he had no real acquaintance, only to obscure them by an aimless rhetoric.

But Huxley could not command the audience, and by now Hooker's blood was boiling. He handed his name up to the president and mounted the platform 'beside saponaceous Sam', and, as he gleefully related to Darwin:

> There and then I smashed him amid rounds of applause. I hit him in the wind at the first shot in ten words taken from his own ugly mouth; and then proceeded to demonstrate in as few more: (1) That he could never have read your book and (2) that he was absolutely ignorant of the rudiments of Bot. Science. I said a few more on the subject of my own experience and conversion, and wound up with a very few observations on the relative positions of the old and new hypotheses, and with some words of caution to the audience. Sam was shut up—had not one word to say in reply, and the meeting *was dissolved forthwith*, leaving you master of the field after 4 hours' battle.

11

'A millionaire in odd and curious little facts'

The flower mechanisms of Orchis mascula, showing how a pollen-mass is attached by a stalk to a saddle. The saddle, which is sticky, glues itself to the proboscis of a visiting bee, in an upright position. Darwin reckoned that it took the bee 30 seconds to leave the first flower and find and fly to a second flower. During the 30 seconds the pollen-mass bends on its stalk through an angle of 45°, so that it is in the forward position for ramming the stigma of the second flower.

How orchids are pollinated in their partnership with bees and other insects, Darwin found 'most wonderful, most beautiful'.

A T MURRAY'S AUTUMN sale on the 22nd of November 1859 the *Origin* had more than sold out. The printing was for 1,250 copies, of which 1,192 were for sale. The bids rose to 1,500. 'I am infinitely pleased and proud at the appearance of my child,' Darwin told his publisher. Murray, too, was delighted and immediately decided to bring out a second edition of 3,000 copies. Darwin began the work of revision at Ilkley Wells in Yorkshire where he had been taking the water-cure since October, having left home the minute the book was off his hands. For despite Emma's willing help with the proofs, he had felt terribly overdone and ill.

They were back at Down on the 10th of December. A multitude of letters had followed him to Ilkley. More awaited him, and a pile of papers with reviews that kept pouring in. There was a magnificent essay of three and a half columns in *The Times*. It was unsigned, but Darwin thought there was 'only one man in England' who could have written it—Huxley. He wrote to him, 'Who can the author be? I am intensely curious. It included an eulogium of me which quite touched me, though I am not vain enough to think it at all deserved.' The reviewer was obviously a profound naturalist. He had refrained from dogmatism, and claimed a respectful hearing for the *Origin*. 'Well, whoever the man is, he has done great service to the cause,' Darwin wrote, 'far more than by a dozen reviews in common periodicals.' Eventually the secret leaked out: the reviewer was indeed Huxley, henceforward to be nicknamed 'Darwin's Bulldog'.

Sedgwick's review in the *Spectator* was abusive. 'And what a misrepresentation of my notions!' Darwin was forgiving. 'But my dear old friend Sedgwick, with his noble heart, is old, and is rabid with indignation.'

There was one 'prodigy of a review' by François Jules Pictet the palaeontologist, 'namely, an *opposed* one' but which Darwin thought '*perfectly* fair and just'. 'Of all the opposed reviews, I think this is the only quite fair one, and I never expected to receive one.'

In America Asa Gray defended him, and Louis Agassiz, a guardian of the divine fixation of species, attacked him. ('He growls over it like a well-cudgelled dog,' wrote Gray.)

Edinburgh opinion also was in opposition, led by John Hutton Balfour, the professor of botany. Richard Owen in the *Edinburgh Review* added gall, wrote Hooker, 'to the Balfourians' bitterness of spirit, for not content with snubbing me and spitefully entreating Darwin and Huxley, the cool fish hedges for a transmutation theory of his own!' Nevertheless, in a conversation with Henslow, Owen admitted that the *Origin* was the 'Book of the Day'.

There was another claimant for priority. In its issue of the 7th of April the *Gardeners' Chronicle* gave an extract from a book on 'Naval Timber and Arboriculture' published in 1831 by Patrick Matthew, in which he 'briefly but completely anticipates the theory of Natural Selection', as Darwin at once admitted. He bought the book and read it. Matthew's theory was in an appendix and was 'an incomplete and not developed anticipation' in less than a page. Darwin in a reply to the *Gardeners' Chronicle*, while hoping he might be excused for not having discovered it in a work on naval timber, offered to call attention to it in any future edition of the *Origin*, and this he faithfully did.

In March the *Gardeners' Chronicle* had reprinted word for word Huxley's *Times* review. Previously, on the 31st of December 1859 it had printed its own thoughts on Darwin, as 'an author of first-rate standing in science, of great popularity, and a frequent contributor to our columns.' It said:

> We have risen from the perusal of Mr Darwin's book much impressed with its importance and have moreover found it to be so dependent on the phenomena of horticultural operations, for its facts and results, and so full of experiments that may be repeated and discussed by intelligent gardeners and of ideas that may sooner fructify in their minds than in those of any other class of naturalists, that we shall be doing them (and we hope also science) a service by dwelling in some detail upon its contents.

A month later the paper applied the principles of Natural Selection and the struggle for life to a discussion on how economic plants such as wheat, cotton and sugar might be improved in the colonies.

Darwin was reaching his public.

His intention now was to complete his 'big book'. But on the way home from Ilkley he had stopped for two days in London, staying as usual at Queen Anne Street with Erasmus, and called on John Murray, pleased to tell him that he was making progress with the second edition of the *Origin* and what he had in mind for the future. No, said Murray, and proceeded to give him sound advice, coinciding with what Hooker, Lyell and Huxley were urging him to do, as a letter to Huxley tells us:

> You have hit on the exact plan, which, on the advice of Lyell, Murray, &c, I mean to follow—viz. bring out separate volumes in detail—and I shall begin with domestic productions.

He began looking over his old half-finished manuscript early in January 1860, this time for the *Variation of Animals and Plants under Domestication.*

There were many interruptions, but towards the end of March he was writing the Introduction to the book, explaining that in the *Origin* space had been limited: he could not give references or always fully explain his statements and readers therefore had to take many of these on trust. *Variation* was to remedy this by presenting the facts he had collected or observed on the changes animals and plants had undergone while under man's dominion.

Work went on until May when the cowslips were in bloom at Down. On the 7th he wrote to Hooker: 'I have this morning been looking at my experimental cowslips, and I find that some plants have all flowers with long stamens and short pistils, which I will call "male plants", others with short stamens and long pistils, which I will call "female plants".' This oddity had been pointed out to him by Henslow, as he told Hooker. 'But,' he went on, 'I find (after looking at my two sets of plants) that the stigmas of the male and female are of slightly different shape, and certainly different degree of roughness, and what has astonished me, the pollen of the so-called female plant, though very abundant, is more transparent, and each granule is exactly only $\frac{2}{3}$ of the size of the pollen of the so-called male plants. Has this been observed?'

It had not been observed. He was on the threshold of a discovery, for which he had to invent new botanical terms. A whole string of Linnean Society papers was to result. Finally he was to publish a book on the subject, one of the most fascinating of all his studies of flowers.

In May his daughter Etty fell ill again. She was sufficiently recovered by July to be moved to her Aunt Elizabeth's home on the fringe of Ashdown Forest, 'the kindly hospital for all who are sick or sorry', as the family called it. Her parents went with her, and it was here that a new interruption occurred in the writing of *Variation*. In his *Autobiography* Darwin tells us of his new discovery.

> In the summer of 1860 I was idling and resting near Hartfield, where two species of Drosera abound; and I noticed that numerous insects had been entrapped by the leaves. I carried home some plants, and on giving them insects saw the movements of the tentacles, and this made me think it probable that the insects were caught for some special purpose. Fortunately a crucial test occurred to me, that of placing a large number of leaves in various nitrogenous and non-nitrogenous fluids of equal density; and as soon as I found that the former alone excited energetic movements, it was obvious that here was a fine new field for investigation.

In August Emma wrote to Lady Lyell that 'He is treating Drosera (the sun-dew plant) just like a living creature, and I suppose he hopes to end in proving it to be an animal.'

He did not exactly hope that, but he certainly put his *Drosera* plants through exhaustive tests, telling Hooker: 'I have been working like a madman at Drosera. Here is a fact for you which as certain as you stand where you are, though you won't believe it, that a bit of hair $\frac{1}{78000}$ of one grain in weight placed on gland, will cause *one* of the gland-bearing hairs of Drosera to curve inwards, and will alter the condition of the contents of every cell in the foot-stalk of the gland.'

This was in November. *Drosera* had become a passion. 'I care more about Drosera than the origin of all the species in the world,' he declared to Lyell.

Again, they were to be the subject of another book, *Insectivorous Plants*, which he was to publish in 1875.

There was yet another interruption to *Variation*. It had begun long

Bessy. Henrietta Horace Emma Lenny Frank Jos Down.
Darwin School friend called Spitta

The Darwin family, outside the old drawing-room at Down House

ago. A quarter of a mile from Down House, above the quiet Cudham Valley, was what he and Emma called Orchis Bank. The Fly Orchid and green-flowered Musk Orchid were common there, growing among the junipers, and the helleborines *Cephalanthera* and *Epipactis* under the beech trees. It was one of Darwin's favourite walks. In the summer of 1838 or 1839, having come to the conclusion that crossing played an important part in keeping species constant, he had begun studying the ways in which flowers were cross-fertilised by insects. The finding of Orchis Bank led him to study these flowers particularly, for in no other flower were pistils, stamens and petals so wonderfully formed to secure cross-pollination by the aid of one particular insect or another.

It was in the summer of 1860 while at Hartfield that he began making his important discoveries about them. To Hooker he wrote excitedly: 'I have been examining *Orchis pyramidalis*, and it almost equals, perhaps even beats your Listera case.' He marvelled at the movements by which pollen was deposited. 'I never saw anything so beautiful.' And to Lyell: 'Talk of adaptation in woodpeckers, some of the orchids beat it.' He told him that he was doing some interesting work on them.

So here he was with *Variation* on his hands, and primulas, and sundews, and now orchids. As he confessed to Hooker of this period: 'I am a complete millionaire in odd and curious little facts.' The trouble was, on which should he concentrate first? He wrote a note on the Bee Orchid for the *Gardeners' Chronicle*, and this touched off so much interesting discussion that he found himself becoming deeply involved. Indeed, orchids were such a delightful study that he felt he must be wasting time on them which ought to be given to *Variation under Domestication*, declaring that 'There is to me incomparably more interest in observing than in writing; but I feel quite guilty in trespassing on these subjects, and not sticking to varieties of the confounded cocks, hens and ducks. I hear that Lyell is savage at me.'

He could not leave orchids alone. But really he could not hold off *Variation* any longer. Neither would the orchids wait! They were still in flower! He divided his time between the two, and even with experiments on his *Drosera* going on was able to note in his diary the progress he had made when on 'August 11th Began Ch. III'. It was slow going. Not till March of the following year was the chapter finished and Chapter IV begun.

There was a wonderful holiday at Torquay in the summer of 1861 which lasted '8 weeks and a day', when Etty at last began to get well. The boys were full of enjoyment. They were growing up: George, now sixteen, and Frank, twelve, were both at Dr Pritchard's school at Clapham; Lenny was eleven and Horace ten; William had graduated at Cambridge (where he occupied his father's old rooms) and was now a fledgling lawyer at Lincoln's Inn. Their father managed to achieve a walk of four miles, which was a grand feat. He spent much of his time studying a box of exotic orchids Joseph Hooker sent him from Kew. He wrote a paper on orchids. And then at last came the entry in his diary: 'All rest of year Orchid Book.'

Darwin was anxious for his readers to share his excitement over orchid flowers. 'The contrivances by which orchids are fertilised are as varied and almost as perfect as the most beautiful adaptations in the animal kingdom,' he told them. More than that, he wanted his readers to enjoy for themselves the adventures of discovering them. 'As orchids are universally acknowledged to rank amongst the most singular and most modified forms in the vegetable kingdom, I have thought that the facts to be given might lead some observers to look more curiously into the habits of our several native species.' He was confident that they would not be disappointed. 'An examination of their many beautiful contrivances will exalt the whole vegetable kingdom in most persons' estimation.'

In his Introduction to the book—its full title was *The Various Contrivances by which Orchids are Fertilised by Insects*—he explained the special terms used in describing these contrivances. For each kind of orchid has its special kind of mechanism, complex and fascinating, though the general design is the same. The diagram shows the parts of the Early Purple Orchid (*Orchis mascula*), which was the first Darwin described. This is a British species blooming in April and May. It has a spike of red-purple flowers, and the lip is nearly flat and is spotted on a pale centre.

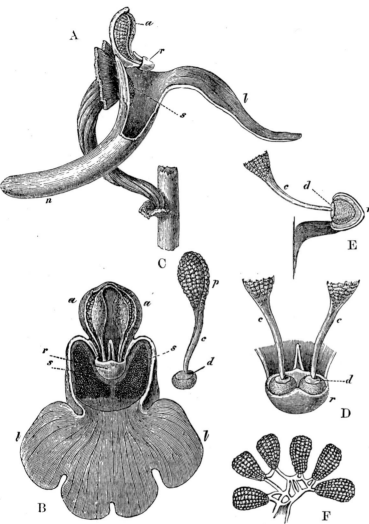

The mechanisms of *Orchis mascula*

a. anther, consisting of two cells; *r*. rostellum; *s*. stigma; *l*. labellum; *n*. nectary; *p*. pollen mass; *c*. caudicle of pollinium; *d*. viscid disc of pollinium. *A*. Side view of flower, with all the petals and sepals cut off except the labellum. *B*. Front view of flower. *C*. One pollinium, showing the packets of pollen-grains, the caudicle and viscid disc. *D*. Front view of the caudicles of both pollinia with the discs lying within the rostellum, its lip being depressed. *E*. Section through one side of the rostellum, with the included disc and caudicle of one pollinium, lip not depressed. *F*. Packets of pollen-grains, tied together with elastic threads

Let us see, as Darwin said, how it works.

The pollinating insect is a bumble-bee, though he is not aware of his mission—he comes only for the nectar. When he alights on the lip (which forms a good landing-place) he pushes his head into the flower, in order to reach the end of the nectary with his proboscis, or tongue. You can simulate the bee by very gently using a sharp-pointed pencil. Across the gangway of the nectary is the pouch-shaped rostellum, and it is almost impossible for the bee to avoid it. At a touch the rostellum splits across, revealing the two pollinia with their discs nestling in the pollen-cells. Out of each disc grows a stem or caudicle supporting a packet of pollen, for unlike other flowers the pollen is not a fine powder but grains stuck together in masses. So sticky are these discs that they adhere to anything that touches them, and within a few seconds their glue sets hard and dry like cement. So when the bee, having feasted, withdraws his head (or you withdraw your pencil) one of the pollinia or usually both will be firmly attached to it, and standing upright like a pair of horns.

We will follow the flight of the bee. He visits another *Orchis mascula* and pushes his head into the flower. Now, but for what Darwin called 'a beautiful contrivance' the pollinium would strike against the anther-cell. The flower, so that it can be pollinated, requires that it be pushed into the stigma, which is lower down, and this happens because the apparently insignificant disc is endowed with a remarkable power of contraction which forces the pollinium to sweep through a downward right-angle so that it is in the position for ramming the stigma. It takes about thirty seconds for the pollinium to make this 45° angle, and Darwin found that on average the time it took for a bee to leave one flower and fly to another—was thirty seconds.

Now another adaptation comes into play. The stigma is sticky, but not sticky enough to pull the whole pollinium off the bee's head, yet sufficiently so to break the elastic threads tying the packets of pollen-grains together. So only some are left on the stigma. Hence a pollinium attached to a bee can be applied to many stigmas and will pollinate them all.

Having said that the glue coating the disc sets hard when exposed to the air, you may wonder how those remaining in the open pollen-cells do not also harden. The answer is that in the pouch which encloses them there is a fluid which keeps them damp, sealing them from the air. The flower has another safeguard: should the bee withdraw before the discs glue themselves to it, the lip of the pouch will spring back and enclose them again, ready for the next visitor.

One of the orchids which most delighted Darwin was *Orchis pyramidalis*. 'In no other plant,' he wrote, 'or indeed in hardly any animal, can adaptations of one part to another, and of the whole to other organisms widely remote in the scale of nature, be named more perfect than those presented by this Orchis.' Its flowers are visited both by butterflies and moths, and Darwin thought it not fanciful to believe that its bright-purple blossoms attracted the day-fliers, and its strong foxy odour the night-flyers.

The Pyramidal Orchid, he found, has its parts arranged very differently from *Orchis mascula*. First, there are two quite distinct rounded stigmatic surfaces. These are placed on each side of the pouched rostellum which, instead of standing above the nectary, overhangs and partially blocks the entrance to it. The flowers of this orchid have only one disc, and this, shaped like a saddle, carries on its seat the two pollinia. When the moth withdraws its tongue the saddle with its riders on top becomes attached to it, and as soon as the saddle is exposed to the air a rapid movement takes place—the two flaps curl inwards and embrace the moth's tongue. Darwin timed this movement and found that the flaps curled inwards so as to touch each other in nine seconds, and in nine more seconds the saddle was converted into a solid ball, by the flaps curling still more inwards.

As with the Early Purple Orchid, the pollinia of the Pyramidal Orchid ·go through the same 45° angle to ram the stigma of the next flower. 'I have shown this little experiment to several persons,' Darwin wrote, 'and all have expressed the liveliest admiration at the perfection of the contrivance by which this Orchid is fertilised.'

Perfection indeed. When the moth alights it finds two prominent ridges on the lip, sloping down and narrowing. These not only guide it to the nectary but prevent its tongue from being inserted obliquely. For, if it were, the saddle would become attached obliquely and the pollinia would not strike exactly the two stigmatic surfaces of the next flower.

It was fortunate that Darwin had so many friends among naturalists and horticulturists. When an orchid was rare in Britain he was able to get it elsewhere, as with *Neotina intacta* which John Traherne Moggridge sent him from the north of Italy, informing him that it was remarkable for producing seeds without the help of insects. To prove this Darwin netted the pinkish-white flowers, finding that the pollen was not sticky and dropped easily on to the stigma. Sure enough, they set seed. He tried the same experiment with the Green-winged Orchid (*Orchis morio*) and *Orchis mascula*, covering some plants with a bell glass and leaving others uncovered. None of the covered ones set seed nor, when he took the bell glasses off, were the pollinia carried away by insects. From this he came to the conclusion that there was a proper season for each kind of orchid, and that insects cease their visits when the proper season has passed.

Similarly a wet and cold summer, like that of 1860, produced very few seed-capsules, for fewer insects were flying. A normal summer showed how well moths and butterflies performed their office of marriage-priest, as he called it. He examined six spikes of the Pyramidal Orchid and every single expanded flower had its pollinia removed.

When Robert Brown had long ago recommended Darwin to read Sprengel's *Secret of Nature Displayed*, that young man, we remember, found in the book 'some little nonsense'. Part of the nonsense was Sprengel's attitude to insects: he credited them with having little intelligence. He had studied the flowers of the Marsh Orchid and the Green-winged Orchid and declared that he could never find a drop of nectar, despite the well-developed spur in both species. *Scheinsaftblumen* — sham-nectar-producers, Sprengel called them, believing that the plants must exist by some organised system of deception. But was it credible that the flowers could play such a trick on the insects they needed to pollinate them?

Darwin did not think so.

He decided to put the Green-winged Orchid to the test. As soon as many flowers were open he examined them for twenty-three consecutive days: after hot sunshine, after rain, and at all different hours including midnight and early morning. He kept the spikes in water, irritated the nectaries with a bristle, and exposed them to irritating fumes. In some flowers the pollinia were intact. In others they had been removed, proving that insects had visited them. Yet in none of their nectaries could he detect the minutest drop of honey, even under the microscope. It was the same with the nectaries of *Orchis maculata*, the Spotted Orchid. Yet repeatedly he saw Empis flies keeping their tongues inserted into them. He could only conclude that the nectaries of these orchids do not ever contain nectar.

Then how did they attract insects? He did another experiment, this time dissecting the nectaries of *morio* and *maculata*. He was astonished to find first, a considerable cavity between the inner and outer membranes; second, that the inner membrane could be penetrated very easily; and finally that a quantity of fluid was contained between them. So copious was it that he could squeeze out large drops.

He now examined the nectaries of *Gymnadenia conopsea*, the Sweet-scented Orchid, and *Platanthera chlorantha* (*Habenaria chlorantha*), the Butterfly Orchid, and, as he suspected, there was no fluid between the membranes, which in fact were closely united. He concluded that in orchids having no nectar in the nectary, insects pierced the membrane to find the fluid. This was a bold hypothesis, for at that time no case was known of insects piercing even the most delicate membrane.

The helleborine *Cephalanthera grandiflora* was a remarkable orchid, for it attracted insects by offering them solid food. The end part of the lip was frosted with tiny orange-coloured globules, and inside the cup of the flower were wrinkled ridges of a darker orange. These ridges, Darwin found, were often gnawed, as if by some animal, and sometimes bitten-off fragments were lying at the bottom of the cup. He discovered that the movements made by the insects in gnawing helped the whole operation of removing the pollen and dropping it on the stigma below.

There seemed no end to the beautiful contrivances. In the two Australasian orchids *Pterostylis trullifolia* and *P. longifolia* two of the petals and one of the sepals formed a hood enclosing the column (which in all common orchids is composed of a single well-developed stamen with the pistils). When an insect landed on the lip it sprang up like a drawbridge and shut the insect inside, staying shut for up to an hour and a half, when the drawbridge was lowered. The prisoner could now escape but only by crawling through a narrow passage formed by two projecting shields, in so doing removing the pollinium.

The pretty Lady's Tresses Orchid (*Spiranthes spiralis*) had a disc shaped like a boat, which was held in place by a fork at the top of the rostellum. A membrane covered the boat like a deck, and safely stowed in the bow was the precious cargo of pollen-masses whose elastic threads were grouped together on each side like a pair of oars. Darwin noticed a slight furrow running down the middle of the boat, which a visiting bee would inevitably touch. He simulated its proboscis by laying a bristle along it, and instantly a chain-reaction was set in motion. The rostellum split along its whole length and freed the boat, a milky fluid began oozing out, and when Darwin withdrew the bristle the boat was stuck fast to it. Thus a visiting bee would lift the little boat from its stocks and bear it away with its cargo of pollen.

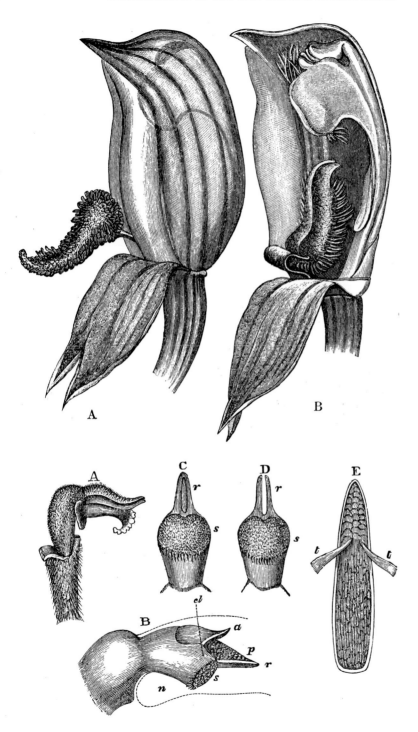

The mechanisms of the orchid *Pterostylis longifolia*. *A*. Flower in its natural state: the outline of the column is dimly seen within. *B*. Flower with the near lateral petal removed, showing the column with its two shields, and the labellum in the position it occupies after having been touched

The mechanisms of the orchid *Spiranthes spiralis (autumnalis)*, Lady's Tresses *a*. anther; *p*. pollen-masses; *t*. threads of the pollen-masses; *cl*. margin of clinandrum; *r*. rostellum; *s*. stigma; *n*. nectar receptacle. *A*. Side view of flower in its natural position, with the two lower sepals removed. The labellum can be recognised by its fringed and reflexed lip. *B*. Side view enlarged of a mature flower, with all the sepals and petals removed. The positions of the labellum and the upper sepal are shown by the dotted lines. *C*. Front view of the stigma, and of the rostellum with its embedded, central, boat-formed disc. *D*. Front view of the stigma and rostellum after the disc has been removed. *E*. Disc, removed from the rostellum, greatly enlarged, viewed from the back, with the attached elastic threads of the pollen-masses. The pollen-grains have been removed from the threads

Listera ovata (the Twayblade) was one of the most remarkable of all orchids. We remember that Joseph Hooker described it but did not mention the part insects play in its fertilisation. The lip of this flower exudes nectar on which small insects enjoy feasting. They lick their way slowly up the narrowing surface until their heads are directly below the over-arching crest of the rostellum. When they raise their heads they touch the crest. This instantly explodes and the pollinia are shot out and firmly cemented to the insect's head. There was a bed of Twayblades some miles away from Down, where Darwin's son George watched it all happen. He was struck by the number of spider-webs spread over the plants, as if the spiders were aware of how attractive the *Listera* was to the little flies.

After observing the behaviour of fifteen genera of British orchids, Darwin turned to the great exotic tribes that ornament tropical forests, and of these he examined fifty genera. Some of these orchids were rare. They came to him from Kew, from James Veitch junior, from John Lindley and others. Lady Dorothy Nevill sent him 'a lot of treasures' from her magnificent collection. Sir Robert Schomburgk, discoverer of the huge saucer-leaved Amazon Lily, was another contributor. These orchids were housed in the small glass lean-to behind the kitchen.

When he came to examine the first one he found a most extraordinary flower. The plant's modern name is *Cryptophoranthus*, meaning 'hidden blossom'. The species was *atropurpureus*. In Darwin's day the orchid's name was *Masdevallia*, named for José Masdevall, a Spanish botanist and physician, and its species name *fenestrata*, which exactly described the two minute oval windows on either side of the flower. These little windows were the only way into it, and so far as Darwin could see the insect would have to be very small, for it seemed improbable that a larger insect would poke its proboscis through one of the windows. Dissecting the flower he found a cluster of eggs that some insect had laid near the base. This was a surprise, but Darwin had to admit that he never discovered how this orchid was fertilised. To this day the answer is the same: the pollinator is not known for any *Cryptophoranthus*.

The *Angraecum sesquipedale* orchid of large six-rayed flowers like stars formed of snow-white bracts, had excited the admiration of travellers in Madagascar. A green whip-like nectary of astonishing length hung down beneath the lip. In several flowers sent him by James Bateman, who wrote on Mexican orchids, Darwin found them to be eleven and a half inches long but with only the lower inch and a half filled with nectar. He thought that the only insect capable of pollinating it must be some kind of Sphinx or Hawk moth (which of all insects has the longest tongue) with a proboscis of that length. This belief was ridiculed by the entomologists, but Darwin wrote off to his friend Fritz Müller in Brazil who informed him that there was one with a proboscis of nearly sufficient length—between ten and eleven inches. Its Madagascar counterpart (with a proboscis of exactly the right length) is *Xanthopan morgani praedicta*.

It seemed that each orchid was more remarkable than the last. This was true of the *Catasetum*. To start with it had a weird-looking flower. Darwin thus described it: 'The dull coppery and orange-spotted tints,—the yawning cavity in the great fringed labellum—the one antenna projecting with the other hanging down—give to these flowers a strange, lurid, and almost reptilian appearance.'

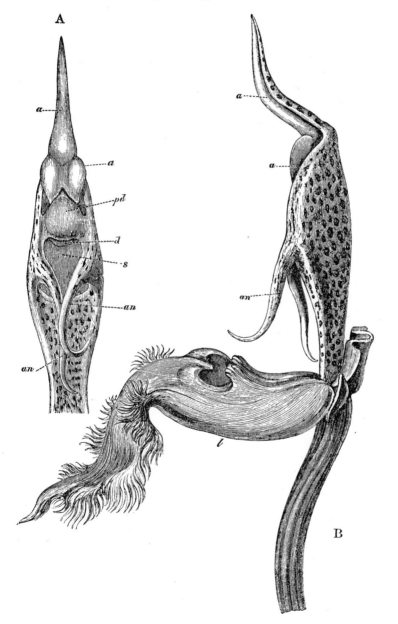

The mechanisms of the orchid *Catasetum saccatum*
A. Front view of column. *a.* anther; *an.* antennae of the rostellum; *d.* disc of pollinium; *l.* labellum; *pd.* pedicel of pollinium; *s.* stigmatic chamber. *B*. Side view of flower with all the sepals and petals removed, except the labellum

EXOTIC ORCHIDS examined by Darwin:

(*left*) *Cycnoches ventricosum*, the Swan Orchid

(*above*) *Mormodes histrio*, the Goblin Orchid of curious twisted flowers

(*below left*) *Angraecum sesquipedale* whose 11½ in.-long nectaries require a pollinator with a proboscis of the same length

(*below right*) *Odontoglossum grande*, with flowers up to 7 in. across

The antennae he mentions are the most curious organ of the flower, and occur in no other genus. They form rigid curved horns tapering to a point, with a slit like an adder's tongue. When the left-hand one is touched it flirts the heavy disc out of the stigmatic chamber with such force that the whole pollinium is ejected, bringing away with it two balls of pollen and tearing the loosely attached anther from the top of the column. They were the snipers of the orchid world: friends who had them in their hot-houses were liable to be shot in the face.

It was such a fascinating subject for experiment that Darwin had to beg more plants. He wrote to Hooker: 'If you really can spare another Catasetum, when nearly ready, I shall be most grateful. Had I not better send for it? . . . A cursed insect or something let my last flower off last night.'

But there was something even more queer about the *Catasetum*. 'It is exclusively a male form,' Darwin pointed out, 'so that the pollen-masses must be transported to the female plant, in order that seed should be produced.' This sounds a contradictory statement. But Darwin had a floral joke up his sleeve. Sir Robert Schomburgk announced that he had found three forms of orchid which he believed to be of three different genera growing on the one plant. They were *Catasetum tridentatum, Monacanthus viridis,* and *Myanthus barbatus.* John Lindley's horrified reaction to this was that 'such cases shake to the foundation all our ideas of the stability of genera and species.' Sir Robert told Darwin that he had seen hundreds of catasetums in Essequibo without ever finding one specimen with seeds, whereas he had found enormous seed-vessels on the *Monachanthus.* Dr Hermann Crüger, director of the Trinidad botanic garden, said much the same thing.

It was left to Darwin to solve the riddle. With his usual care he examined the flowers of all three. He dissected them. He kept them in spirits and examined their cells. He compared their parts. Finally he was able to announce that the *Catasetum* was a male, that *Monacanthus* was the lady, and that *Myanthus* was a hermaphrodite. Since Darwin's discovery *Catasetum tridentatum* (whose species name is now *macrocarpum*) has reigned supreme, and *Monacanthus* and *Myanthus* have been banished from taxonomy.

Another puzzle was *Mormodes ignea,* one of the Vanda orchids and a native of Central America, gorgeous with its trumpet streaked with scarlet, and bright orange lip. A plant with two spikes of flowers was lent to Darwin by Sigismund Rucker, an orchid fancier of West Hill, Wandsworth, to whom Darwin recorded his thanks for being allowed to keep the plant for a considerable time. He had need of time, for if the flower was extraordinary to look at its mechanism was even more curious. Every part of the flower was twisted, so that all the organs—anther, rostellum, and upper part of the stigma—on the left side of the spike faced the left, and those on the right faced the right. The lip was twisted and arched over the twisted column. This positioning was important, for had the column been straight the pollinia would have struck the lip and rebounded. And had the organs not been twisted so as always to face the outside, there would not be a clear space for the pollinia to eject and strike visiting insects.

But how were the pollinia ejected? Where was the trigger that set off the mechanism? He pressed the rostellum with variously shaped objects, but the disc never once stuck properly to it. Using a broad spatula, the pollinium struggled against it and sometimes coiled itself up spirally, and the disc did not adhere at all. After twelve tries he was in despair. And then he remembered a tiny hinge he had seen, no bigger than a pin's head, articulating the top of the anther-case to the column below its bent summit. Suppose, he thought, an insect were to alight on the crest (and there was no other convenient landing place), and suppose it leaned over the front of the column to suck the sweet sticky petals? Gently Darwin touched the minute hinge with a very fine needle, and instantly the pollinium was shot upwards and fell on the crest of the lip where the insect would be.

There were other Vandas he experimented with to find the different triggers. *Cycnoches ventricosum* with its winged petals was, as its Greek name tells us, like a beautiful flying swan. James Harry Veitch twice sent him flowers and buds to study and Darwin found that its trigger was a filament between little leaf-like appendages. In *Mormodes luxata* it was not the anther-hinge but the apex of the column.

The study of Vandas, Darwin wrote, and their wonderful and often beautiful flowers, with all their many adaptations, with parts capable of movement, and other parts endowed with something so like, though no doubt different from, sensibility, had been to him most interesting.

In his examination of orchids hardly any fact had struck him so forcibly as these 'endless diversities of structure,—the prodigality of resources,—for gaining the very same end, namely, the fertilisation of one flower by pollen from another plant.' And he added: 'This fact is to a large extent intelligible on the principle of natural selection.' Orchids had provided him with one of the finest of all test-cases for the theory of adaptive evolution.

The fundamental framework of the flower consisted of the remnants of the fifteen primary organs—remnants but still visible: the stigma that had become the rostellum, the column that was a compacted stamen and pistils. 'So,' said Darwin, 'almost everyone who believes in the gradual evolution of species will admit that their presence is due to inheritance from a remote parent-form.'

12

Hops, Hooks and Tendrils

'Twiners entwining twiners . . .'

Plants require light in order to obtain chlorophyll (literally green leaf) necessary for forming starch and energy. Every leaf turns itself to the light. Trees obtain it by soaring above other plants. Climbers such as the clematis, ivy, bramble, rose and White Bryony use leaf-petioles, tendrils and hooked thorns as ropes and grappling irons to hoist themselves upward.

THE ORCHID BOOK was published in May 1862, the same year that a *Memoir of Henslow* appeared under the pen of his old friend and brother-in-law Leonard Jenyns.

Henslow had died in May the year before, after a lingering illness, and as one 'who could so thoroughly appreciate the excellence of his disposition' Darwin was asked to contribute. 'I took pleasure in writing my impression of his admirable character,' he told Joseph Hooker to whom the loss of Henslow was 'a blank in my existence never to be replaced'. On his marriage to Frances Henslow, Joseph and his father-in-law had become close friends. He thought of Henslow as 'one of those friends formed *late in life*—to be a lamp unto our path who we never go ahead of as we do the instructors of our youth.'

Darwin had gone ahead of him, but he was deeply conscious of the debt he owed. His contribution to the *Memoir* recaptured the days at Cambridge, the gatherings at Henslow's house presided over by that rare spirit of charm and astonishing depth of knowledge. He recalled with nostalgia the excursions to the fens. More than ten years later, in his *Autobiography*, he again wrote of his good friend, speaking of him as the man who had 'influenced my whole career more than any other.'

It was a pity that Henslow did not live to see the publication of the *Fertilisation of Orchids*, for this was the first book Darwin had written purely on plants. He would have been proud of the reviews.

'The Botanists praise my Orchid-book to the skies,' its author wrote to John Murray. 'Some one sent me (perhaps you) the "Parthenon", with a good review. The *Athenaeum* treats me with very kind pity and contempt; but the reviewer knew nothing of his subject.'

Five days later, after more notices, he wrote: 'There is a superb, but I fear exaggerated, review in the "London Review". But I have not been a fool, as I thought I was, to publish; for Asa Gray, about the most competent judge in the world, thinks almost as highly of the book as does the "London Review". The *Athenaeum* will hinder the sale greatly.' It did not. The book ran happily into two editions and seven impressions.

Joseph Hooker 'thought it very well done indeed', and that he, George Bentham, Miles Joseph Berkeley, Kew's Daniel Oliver, and Asa Gray should approve of his work heartened Darwin who had begun to think he was making a complete fool of himself by publishing in a semi-popular style. Gray's review in *Silliman's Journal* made a strong point of this when he wrote enthusiastically of the 'fascination it must have for even slightly instructed readers.'

Darwin told him: 'Now I shall confidently defy the world.'

George Nicholson, a Kew man whose great work was *The Dictionary of Gardening* (1885–1901), referred his readers to the authority on 'Orchid Fertilisation' in these words:

The great source of information on this, as on so many other important and interesting questions in Natural Science, is to be found in the writings of Charles Darwin who has written, upon this subject, the well-known work 'On the Various Contrivances by which British and Foreign Orchids are Fertilised by Insects'. This book must always be referred to by those who wish to understand the very curious structures that adapt many Orchids, in a very peculiar degree, to benefit by the visits of insects, while a smaller number are adapted for self-fertilisation alone.'

It was Hooker who paid him the greatest tribute. In December 1862 he wrote to his friend Brian Hodgson of Darjeeling for whom he named *Rhododendron hodgsonii*:

Darwin still works away at his experiments and his theory, and startles us by the surprising discoveries he now makes in Botany; his work on the fertilisation of orchids is quite unique—there is nothing in the whole range of Botanical Literature to compare with it.

This book with his other works raised him in Hooker's estimation to the position of the 'first Naturalist in Europe', and he added:

Indeed I question if he will not be regarded as great as any that ever lived; his powers of observation, memory and judgement seem prodigious, his industry indefatigable and his sagacity in planning experiments, fertility of resources and care in conducting them are unrivalled, and all this with health so detestable that his life is a curse to him and more than half his days and weeks are spent in inaction—in forced idleness of mind and body.

He reviewed the book for the *Gardeners' Chronicle*. Acknowledging the notice, Darwin thought that 'Perhaps I am a conceited dog; but if so, you have much to answer for; I never received so much praise, and coming from you I value it much more than from any other.'

Variation was crawling along, always with some interruption. The 7th of October saw him dealing with 'Facts of Varieties of Plants', as he recorded in his diary, the 21st of December with bud-variation. Three days later he was writing to Hooker: 'And now I am going to tell you a *most* important piece of news!! I have almost resolved to build a small hot-house, my neighbour's really first-rate gardener has suggested it, and offered to make me plans, and see that it is well done.' Darwin thought it would be 'a grand amusement for me to experiment with plants.' Sir John Lubbock, his neighbour, came to regard Darwin as his 'father in science' and did some of his earliest work on Darwin's collections. His gardener was Horwood.

The new hot-house, which was built against the high wall of the kitchen garden, was ready by the middle of February, 'and I long to stock it, just like a schoolboy,' he wrote to Hooker. 'Could you tell me pretty soon what plants you can give me; and then I shall know what to order?' And a week later: 'You cannot imagine what pleasure your plants give

The greenhouse at
Down House, as it was
in Darwin's time

me.' In March: 'A few words about the Stove-plants; they do so amuse me. I have crawled to see them two or three times.'

Illness was dogging him. It took him six and a half weeks to write a chapter on Inheritance, eight weeks to write one on Crossing and Sterility, a month to do Selection. In February he managed to spend ten days in London with Erasmus and read a paper on *Linum* to the Linnean Society.

Then 'about April 20th' he 'Began to count seeds of Lythrum.' He counted 20,000 of them under the microscope and polished off a Linnean Society paper on *Lythrum salicaria* which with others was to develop into *The Different Forms of Flowers*, a book he dedicated to Asa Gray and published in 1877.

Asa Gray was much in Darwin's mind at this time, for a new interest had seized him, and he had been 'led to take up the subject', as told in his *Autobiography*, 'by reading a short paper by Asa Gray published in 1858.'

The subject was climbing plants.

The paper was on 'The coiling of Tendrils', in which Asa Gray quoted Hugo von Mohl as saying that this movement resulted from 'an irritability excited by contact'. Gray thought that this might fit into Darwin's *Variation* studies. He sent seeds of *Echinocystis lobata* as a good example of a tendril-bearer. It was a rapid grower. In May Darwin wrote to Hooker.

> I have been observing pretty carefully a little fact which surprised me; and I want to know from you and Oliver whether it seems new or odd to you, so just tell me whenever you write; it is a very trifling fact, so do not answer on purpose.
>
> I have got a plant of *Echinocystis lobata* to observe the irritability of the tendrils described by Asa Gray, and which of course, is plain enough. Having the plant in my study, I have been surprised to find that the uppermost part of each branch (i.e. the stem between the two uppermost leaves excluding the growing tip) is *constantly* and slowly twisting round making a circle in from one and a half to two hours; it will sometimes go round two or three times, and then at the same rate untwists and twists in opposite directions. It generally rests for half an hour before it retrogrades. The stem does not become permanently twisted. The stem beneath the twisting portion does not move in the least, though not tied. The movement goes on all day and all early night. It has no relation to light, for the plant stands in my window and twists from the light as quickly as towards it. This may be a common phenomenon for what I know, but it confounded me quite, when I began to observe the irritability of the tendrils. I do not say it is the final cause, but the result is pretty, for the plant every one and a half or two hours sweeps a circle (according to the length of the bending shoot and the length of the tendril) of from one foot to twenty inches in diameter, and immediately that the tendril touches any object its sensitiveness causes it immediately to seize it; a clever gardener, my neighbour, who saw the plant on my table last night, said: 'I believe, Sir, the tendrils can see, for wherever I put a plant it finds out any stick near enough.' I believe the above is the explanation, viz. that it sweeps slowly round and round. The tendrils have some sense, for they do not grasp each other when young.

The *Echinocystis* delighted him. But now he wanted more plants to observe. He wrote to Hooker in July that he was 'getting very much amused by my tendrils, it is just the sort of niggling work which suits me, and takes up no time and rather rests me while writing. So will you just think whether you know any plant, which you could give or lend me, or I could buy, with tendrils, remarkable in any way for development, for odd or peculiar structure, or even for an odd place in natural arrangement.'

He had made a discovery, but a letter to Asa Gray brought an unexpected reply. Darwin wrote back on the 4th of August: 'My present hobby-horse I owe to you, viz. the tendrils: their irritability is beautiful, as beautiful in all its modifications as anything in Orchids.' ('All my geese are swans,' he once cheerfully admitted.) 'About the *spontaneous* movement (independent of touch) of the tendrils and upper internodes, I am rather taken aback by your saying, "is it not well known?" I can find nothing in any book which I have. . . .'

Asa Gray told him that two books had been written on the subject, both in 1827, *Uber das Winden der Pflanzen* by Ludwig H. Palm, and *Uber den Bau und das Winden der Ranken und Schlingpflanzen* by Hugo von Mohl. There was also von Mohl's treatise on 'The Vegetable Cell' which he had mentioned in his paper.

Darwin read them—and continued happily with his observations and with the writing of a paper he had begun in May. Modestly he told Hooker that he had 'a good deal of new matter.' In fact, as he added: 'It is strange, but I really think no one has explained simple twining plants.'

He went on to examine a hundred climbers of all different kinds, for he was determined now to investigate the subject completely. 'I was all the more attracted to it,' he says in his *Autobiography*, 'from not being at all satisfied with the explanation which Henslow gave us in his Lectures, about Twining plants, namely, that they had a natural tendency to grow up in a spire. This explanation proved quite erroneous.'

So this was the next interruption to *Variation*, and again the paper became a book, *The Movements and Habits of Climbing Plants*, published by the Linnean Society in 1865 and ten years later by Murray. It became a standard work and remains so today.

Like the Orchid book it was another proving-ground for his theory. In it he was able to demonstrate two of its principles—the struggle for existence, and modification, and added this evidence to the fourth edition of the *Origin* which was published in 1866. In the wild, climbing plants are always to be found among the dense vegetation of hedges, thickets, and forests, where competition for light and air is keenest. To aid their ascent they use hooks and tendrils, which are modifications of other organs. A tendril was once a leaf.

Down was in Kent, and Kent is the Garden of England—the Hop Garden, it could be called. All over the county are to be seen the queer lop-sided towers of the kilns where the hops are dried. In spring the men on stilts stringing the hop poles are long-legged giants stalking from row to row. In summer every inch of string and pole is covered with sheets of green which spread over square miles of countryside.

How plants climb:

The petiole of a clematis leaf acts as a hook

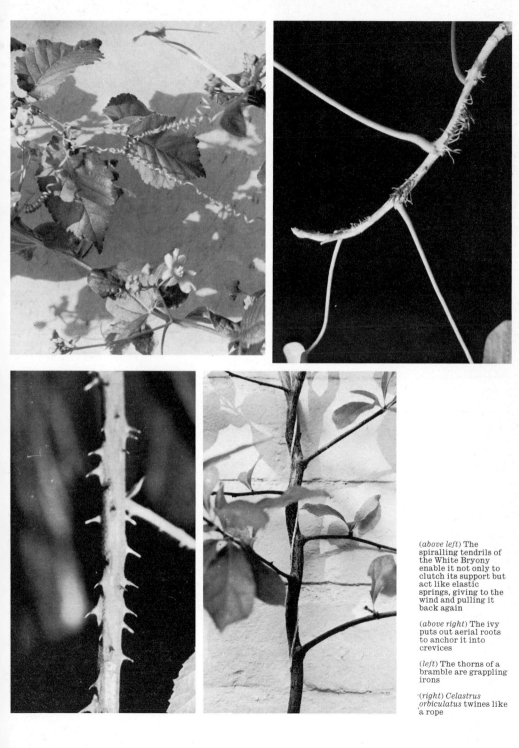

(*above left*) The spiralling tendrils of the White Bryony enable it not only to clutch its support but act like elastic springs, giving to the wind and pulling it back again

(*above right*) The ivy puts out aerial roots to anchor it into crevices

(*left*) The thorns of a bramble are grappling irons

(*right*) *Celastrus orbiculatus* twines like a rope

The Hop (*Humulus lupulus*) was the first climbing plant Darwin described.

They began growing in April. When the first shoot rose from the ground, the two or three first-formed internodes (the part of the stem between two nodes or leaf-axils) were straight and remained stationary. The next-formed was seen to bend to one side and to travel slowly round towards all points of the compass—moving, like the hands of a watch, with the sun. The movement very soon acquired its full speed. From seven observations made during April, and during August on shoots coming from a Hop that had been cut down, the average rate during hot weather and during the day was 2 hours 8 minutes for each revolution. The revolving movement continued as long as the plant went on growing; but each separate internode as it became old ceased to move.

To ascertain more precisely the amount of movement each internode made, 'I kept a potted plant, during the night and day, in a well-warmed room to which I was confined by illness,' Darwin wrote. A long shoot projected beyond the end of the supporting stick, so that only a very young internode 1¼ of an inch long was left free. He assumed that it made at least one revolution during the first twenty-four hours. Early the next morning its position was marked and it made a second revolution in nine hours, this time moving much faster. The third circle was performed in the evening in a little over three hours. After thirty-seven revolutions the internode became upright and after moving to the centre became motionless. He tied a weight to its tip, so as to bow it slightly and thus detect any movement. There was none. Some time before the last revolution was half performed, the lower part of the internode ceased to move.

He had been measuring its growth. After the seventeenth revolution the internode had grown from 1¾ to 6 inches in length and carried an internode 1⅞ inches long which was just perceptibly moving, and this carried a very minute internode. At the twenty-seventh revolution the lower and still moving internode was 8⅜, the middle one 3½, and the last one 2½ inches in length. He measured, too, the inclination of a whole shoot: it had swept a circle of 19 inches in diameter. When the movement ceased, the lower internode was 9 inches long and the middle one 6 inches; so that, from the twenty-seventh to the thirty-seventh revolution, three internodes were revolving at the same time. By now the whole shoot had become twisted three times round its own axis, in the line of the sun's course.

The lower internode, when it ceased revolving, became upright and rigid. The main shoot, now grown beyond the stick, became bent and nearly horizontal, but it continued to revolve in a slight and slow swaying movement. As it grew it hung down more and more, its growing tip revolving and turning itself up more and more. There was an interesting point about the surface of the internode: one side of it from being convex became concave. Darwin was able to show that this was a sure sign of the revolving movement.

Hoya carnosa was his next subject, the beautiful Wax Flower from

Queensland. A hanging shoot 32 inches long swayed from side to side in semicircles while its extreme internodes made complete revolutions. This swaying was due to the movement of the lower internodes which had not enough force to swing the whole shoot round the central supporting stick.

Another waxy-flowered greenhouse climber was *Ceropegia gardnerii* which had been introduced from Ceylon only recently, in 1860. It had curious flowers: the creamy-white petals blotched with purple curled round to make cups. Darwin had one on his study table. He allowed the top to grow out almost horizontally to a length of 31 inches, finding that it revolved in an anticlockwise direction, the opposite to the Hop, at rates between 5 hours 15 minutes and 6 hours 45 minutes for each revolution. The extreme tip thus made a circle of more than 5 feet in diameter and 16 feet in circumference, travelling at the rate of 32 or 33 inches per hour. 'It was an interesting spectacle,' he wrote, 'to watch the long shoot sweeping this grand circle night and day, in search of some object round which to twine.'

The *Ceropegia* was his most willing subject. When he placed a tall stick so as to arrest the lower and rigid internodes, the straight shoot slowly and gradually slid up the stick but did not pass over the top. Then, after an interval sufficient to have allowed half a revolution, the shoot suddenly bounded from the stick, fell over and then resumed its climb. It now began revolving again, after half a turn coming into contact again with the stick, again slid up it and again bounded from it and fell over. 'This movement of the shoot had a very odd appearance,' Darwin wrote, 'as if it were disgusted with its failure but were resolved to try again.'

Many odd facts came out of these experiments. He found that a greater number of twiners revolve in an anticlockwise direction, and that the woody shoots of the wistaria move faster than those of the flexible Morning Glory and thunbergia. He raised seventeen plants of *Loasa eurantiaca*, and of these eight revolved left to right, five right to left, while four revolved and twined first in one direction and then reversed their course. It was interesting that of our English twiners, only the honeysuckle twined round trees, while tropical twiners could easily ascend them, which they would have to do in order to reach the light.

It was when he was half-way through the observations for his book that he learnt that the surprising phenomenon of these spontaneous revolving movements had been observed by the two German botanists. But von Mohl had thought that it was the twisting of the axis that caused the revolving movement. Darwin proved that this was not so. 'If we take hold of a growing sapling, we can of course bend it to all sides in succession, so as to make the tip describe a circle, like that performed by the summit of a spontaneously revolving plant. By this movement the sapling is not in the least twisted round its own axis. I mention this because if a black point be painted on the bark, on the side which is uppermost when the sapling is bent towards the holder's body, as the circle is described, the black point gradually turns round and sinks to the lower side, and comes up again when the circle is completed; and this

gives the false appearance of twisting. Which, in the case of spontaneously revolving plants, deceived me for a time.' He did many
experiments to prove these spontaneous revolving movements, by
painting black lines and coloured streaks along the surfaces of stems.

Having investigated twining plants he turned to the leaf-climbers
and tendril-bearers. Few people, perhaps, who grow clematis in their
gardens realise that these plants climb by their leaves, using the leaf-
stem or petiole as a hook to latch round the nearest support. Again it is
the searching revolving movement which helps it to find a support.
Darwin experimented with eight species of clematis and with seven
species of *Tropaeolum*, first with the *grandiflorum* variety of *T. tricolorum*,
finding that this plant had a belt-and-braces technique of climbing not
only by hooked petioles but by the aid of tendril-like clasping filaments,
as well as twining its stem spirally.

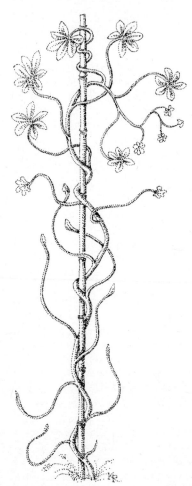

Tropaeolum tricolorum,
the Three-coloured
Indian Cress, has three
mechanisms for
climbing. It not only
twines its stem but
makes hooks of its
leaf-stalks and clasps
of its tendril-like
filaments. Leaf-
climbers, Darwin
found, were
intermediate between
twiners and tendril-
bearers. This
Tropaeolum showed
him every step of the
gradation. The plant
first puts out tendril-
like leaves (shown at
the bottom of the
drawing). As the plant
grows taller these
have flattened ends.
Finally the plant
grows proper leaves
with fully climbing
petioles, and the
tendril-like leaves
then drop off

Leaf-climbers, Darwin pointed out, are intermediate between twiners and tendril-bearers, and *Tropaeolum* showed every step of the gradation. Its young stem twined regularly round a thin vertical stick: on one plant Darwin counted eight spiral turns in the same direction. Grown older, the stem often ran straight until it was arrested by the clasping petioles, when it made one or two turns in the reverse direction. Until, however, the plant grew to a height of two or three feet no true leaves were produced but, first of all, filaments with pointed ends, a little flattened, and furrowed on the upper surface. As the plant grew taller new filaments were produced with slightly enlarged tips, then others with a rudimentary segment of a leaf. Soon other segments appeared, and at last a perfect leaf was formed. 'So that on the same plant you may see every step,' Darwin said, 'from the tendril-like clasping filaments to perfect leaves with clasping petioles.' After the plant had grown to a considerable height and was securely attached to its support by the petioles of the true leaves, the filaments withered and dropped off.

Other tropaeolums behaved differently. The species *azureum* had no filaments or rudimentary leaves. *T. pentaphyllum* did not possess the power of spirally twining—due, Darwin thought, to continual interference from the clasping petioles. The internodes of a nine-inch plant of *tuberosum* did not move at all. On an older plant they sometimes stood still for hours, during some days moving only in a crooked line, on other days making small irregular spires or circles. The internodes of a variety named 'Dwarf Crimson Nasturtium' moved irregularly during the day towards the light, and from the light at night.

A purple variety of the Mexican *Maurandia lophospermum* presented Darwin with an interesting case. Its young internodes were sensitive to the touch. When a petiole of this species clasped a stick, it drew the base of the internode against it; and then the internode itself bent towards the stick, which was caught between the stem and the petiole as if by a pair of pincers. The internode afterwards straightened itself, except for the part in actual contact with the stick. He did fifteen experiments, lightly rubbing several of them with a thin twig, and after about two hours they were all bent, straightening again four hours later. He rubbed some internodes one day on one side and the next day on the opposite side or at right angles, and the curvature was always towards the rubbed side.

There was a small but interesting group of plants which climbed by means of their leaf-tips or extended midribs of their leaves. One of these was the liliaceous *Gloriosa plantii*, a native of Mozambique with reddish-yellow flowers. The young leaves when first developed stood up nearly vertically, but soon bent over until they were horizontal. The end of the leaf formed a narrow ribbon-like thickened projection which at first was nearly straight. By the time the leaf was horizontal it was a well-formed hook strong enough and rigid enough to catch any object, anchor the plant and stop the revolving movement. At its early stage of growth the plant could support itself, and no hooks were formed on the top leaves of a fully grown plant in flower, which did not need to climb any higher. 'We thus see,' wrote Darwin, 'how perfect is the economy of nature.'

He visited Veitch's Royal Exotic Nursery in the King's Road, Chelsea, to see how Pitcher Plants climbed. They did so by the midrib—which extended from the end of the leaf to the pitcher, coiling round any support. The twisted part became thicker and the stalk often made a turn when not in contact, this twisted part likewise becoming thickened. Darwin thought that the main purpose of the coiling was to support the pitcher with its load of honey.

He now came to his 'hobby-horse', the tendrils. These were modified leaves with their petioles or stems, modified flower-stalks, and even modified branches. In *Bignonia venusta* they were modified leaves. Each tendril had three fingers, and they lay in different planes. Their tips were hooked, so that the whole tendril made an excellent grapnel.

Bignonia capreolata (now *Doxantha capreolata*) had tendrils which always groped towards the dark, no matter how many tricks Darwin played on it by turning the pot to the light or placing it in a box open only on one side and obliquely facing the light. 'Six wind-vanes could not have more truly shown the direction of the wind,' he wrote, 'than did these branched tendrils the course of the stream of light which entered the box.' He gave them a glass tube blackened inside, and a well-blackened zinc plate. They bent themselves round the edges of the plate and curled round the tube. 'But they soon recoiled from these objects with what I can only call disgust,' Darwin related, 'and straightened themselves.'

He then placed a post with a very rugged bark close to a pair of tendrils. Twice they touched it for an hour or two and twice they withdrew. At last one of the hooks curled round and firmly seized a minute point of bark, and immediately the other branches spread themselves out, following every minute inequality of the surface. Darwin now placed near the plant a post without bark but much fissured, and the points of the tendrils crawled into all the crevices in a beautiful manner.

The final process was even more surprising. He discovered it by accidentally having left a piece of wool near a tendril. And when he saw what was happening he bound a quantity of flax, moss, and wool loosely round sticks and offered them. Immediately there was great excitement. Hooked points caught hold of even loosely floating fibres. The tips of the hooks began to swell and after a few days were changed into whitish irregular balls about one-twentieth of an inch in diameter which began to secrete a sticky resin, catching at the fibres around and enveloping them. He saw one little ball with between fifty and sixty fibres of flax crossing it at various angles and all embedded in it. On one tendril eight of these balls formed, and one supported a weight of nearly seven ounces. This bignonia in its native southern United States home was accustomed to climbing trees clothed with lichens and mosses. No wonder it was happy with Darwin's flax. The turning to the dark told of the ancestry of the tendrils, which were metamorphosed leaves. Forest climbers from the moment of emerging from the seed will always reach towards the dark in their effort to find a support: that is, they are reaching towards the shadow that must denote a tree.

Cobaea scandens was in Darwin's estimation 'an excellently constructed climber'. Each of its tendrils had a network of branches, the finer ones so thin and flexible that they were blown about by a breath of air. Yet they were strong and elastic. At the extremity of each branch was a tiny double hook. On one tendril eleven inches long Darwin counted ninety-four of these beautifully constructed little hooks which, he found, 'readily catch soft wood, or gloves, or the skin of the naked hand.' So if, by revolving, the tendril failed to find a twig to cling to, the wind would ensure that its filmy branchlets were wafted to a support.

Echinocystis lobata, the Wild Cucumber, was the plant he had grown from seeds sent him by Asa Gray. Its tendrils, always forming an acute angle with the stem, were endowed fantastically with the power of stiffening when they had to pass over the shoot in its circular movement. If the tendril had not possessed this power it would have struck against the shoot and been stopped.

Darwin found the tendrils so sensitive that they hardly ever failed to seize a round stick placed in their path. He put a pencil so far from one tendril that it could curl only half-way, and was amazed when after a few hours it was curled twice or even three times round it. He thought first of all that it had grown rapidly, but coloured dots and measurements proved that it had hardly moved at all. Watching it through a lens he discovered that it dragged itself onwards by a slow alternate movement like 'a strong man suspended by the ends of his fingers to a horizontal pole, who works his fingers onwards until he can grasp the pole with the palms of his hands.'

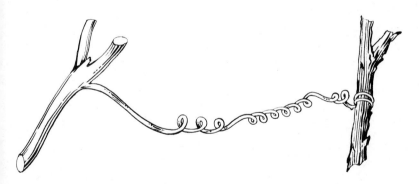

A caught tendril of *Bryonia dioica*. Note the straightened portion where the spiralling direction is reversed, a device, Darwin explained, by which self-twisting is avoided

Tendrils on such climbers as the Passion Flower and the White Bryony grew spiralled and then straight, then spiralled again in the reverse direction. This enabled the tendril to drag up the shoot and, when the shoot had grown, contract. It was this elasticity, too, that prevented the plant from being loosed from its support: it was able to give to the wind and spring back. Darwin must have been remembering the days of the *Beagle* when he wrote the following:

> I have more than once gone on purpose during a gale to watch a Bryony growing in an exposed hedge, with its tendrils attached to the surrounding bushes; and as the thick and thin branches were tossed to

and fro by the wind, the tendrils, had they not been excessively elastic, would instantly have been torn off and the plant thrown prostrate. But as it was, the Bryony safely rode out the gale, like a ship with two anchors down, and with a long range of cable ahead to serve as a spring as she surges to the storm.

There were the plants that climbed by the grappling irons of hooks—as do roses and brambles; by rootlets, like ivy. As an oddity there was the epiphytal philodendron of Southern Brazil with its aerial roots that wound spirally downwards round its tree, instead of upwards. The hook-climbers were the least efficient. As Darwin pointed out, they can climb only by clambering over a tangled mass of vegetation, while root-climbers can ascend even naked faces of rock.

The leaf-climbers and tendril-bearers far exceeded them in number and in the perfection of their mechanisms. They were related to each other and to twiners, for some of the leaf-climbers had twining relatives within the same family and even in the same genus. In Mikania there were leaf-climbers and twining species. The leaf-climbing species of clematis were very closely allied to the tendril-bearing Naravelia. The Fumariaceae included closely allied genera which were leaf-climbers and tendril-bearers. A species of bignonia was both a leaf-climber and a tendril-bearer, and closely allied species were twiners. And nearly all of them had the same remarkable power of spontaneously revolving.

'If we can consider leaf-climbers alone,' said Darwin, 'the idea that they were primordially twiners is forcibly suggested.' And it was probable, he added, that tendril-bearers were primordially twiners. Why then had they not remained twiners? The answer was simple. Without its tendrils could the bryony have survived the gale?

Darwin wrote in summing up his work:

It has often been vaguely asserted that plants are distinguished from animals by not having the power of movement. It should rather be said that plants acquire and display this power only when it is of some advantage to them; this being of comparatively rare occurrence, as they are fixed to the ground, and food is brought to them by the air and rain. We see how high in the scale of organisation a plant may rise, when we look at one of the more perfect tendril-bearers. It first places its tendrils ready for action, as a polypus places its tentacles. If the tendril be displaced, it is acted on by the force of gravity and rights itself. It is acted on by the light and bends towards or from it, or disregards it, whichever may be most advantageous. During several days the tendrils or internodes, or both, spontaneously revolve with a steady motion. The tendril strikes some object, and quickly curls round and firmly grasps it. In the course of some hours it contracts into a spire, dragging up the stem, and forming an excellent spring. All movements now cease. By growth the tissues soon become wonderfully strong and durable. The tendril has done its work, and has done it in an admirable manner.

13

Man's Gigantic Experiment

From wild flowers, sometimes insignificant, the florist developed our modern border blooms, by selecting the biggest and strongest seedlings and breeding from them, next year repeating the process, and next year again.

This, however, was a wasteful hit-or-miss method. It was not until Darwin gave the horticulturist laws by which to work that hybridising was put on a scientific basis.

I T WAS *VARIATION* at last, a book that emerged in 1868 as two thick volumes of 900 pages. The task of writing it took Darwin until the beginning of 1867, for during much of the time he was battling with illness, from the 22nd of April 1865 'unable to do anything (except the origin for 2nd French Edit.) till early in December when I began correcting Homomorphic seeds.'

The object of *Variation* was to show how animals and plants had changed under domestication and cultivation, and what bearing these changes had on the general principles of variation.

Variation was a corner-stone of Darwin's evolutionary theory, and in his Introduction he made it clear that man could not tamper with nature and so cause variability: he could only allow elective affinities to come into play. If organic beings did not possess an inherent tendency to vary, man could do nothing. He unintentionally exposed his animals and plants to various conditions of life and variability supervened, which he could not prevent or check.

Darwin considered the simple case of a plant which has been cultivated for a long time in its native country, and which consequently has not been subjected to any change of climate.

It has been protected to a certain extent from the competing roots of plants of other kinds; it has generally been grown in manured soil; but probably not richer than that of many an alluvial flat; and lastly, it has been exposed to changes in its conditions, being grown sometimes in one district and sometimes in another, in different soils. Under such circumstances, scarcely a plant can be named, though cultivated in the rudest manner, which has not given birth to several varieties. It can hardly be maintained that during the many changes which this earth has undergone, and during the natural migrations of plants from one land or island to another, tenanted by different species, that such plants will not often have been subjected to changes in their conditions analogous to those which almost inevitably cause plants to vary. No doubt man selects varying individuals, sows their seeds, and again selects their varying offspring. But the initial variation on which man works, and without which he can do nothing, is caused by slight changes in the conditions of life, which must often have occurred under nature. Man, therefore, may be said to have been trying an experiment on a gigantic scale; and it is an experiment which nature during the long lapse of time has incessantly tried. Hence it follows that the principles of domestication are important for us. The main result is that organic beings thus treated have varied largely, and the variations have been inherited. This has apparently been one chief cause of the belief long held by some few naturalists that species in a state of nature undergo change.

He hoped that his readers would obtain some light on the causes of variability—on the laws which govern it, such as the direct action of climate and food, the effects of use and disuse, and of correlation of growth—and of the amount of change to which animals and plants are liable under domestication.

In relating these topics to Selection, Darwin had to demonstrate that his various theories could provide a consistent explanation for the facts of variation, and that these facts could not otherwise be explained.

Cultivated plants involved Darwin in some initial difficulties. In some cases the wild prototype was unknown or doubtfully known. In other cases it was hardly possible to distinguish between escaped seedlings and truly wild plants, so that there was no safe standard of comparison by which to judge any supposed amount of change. Many botanists believed that some of our anciently cultivated plants had become so profoundly modified that it was not possible to recognise their aboriginal parent-forms; and in any case, said Darwin, botanists had generally neglected cultivated varieties as beneath their notice.

Equally perplexing was the question of whether some of them were descended from one species or from several inextricably commingled by crossing and variation. Variations often passed into, and could not be distinguished from, monstrosities, or abnormal flowers. Many varieties were propagated by grafts, buds, layers, and bulbs, and often it was not known how far their peculiarities could be transmitted by seed.

On the other hand, there was the evidence supplied by Alphonse de Candolle who in 1855 in his *Géographie botanique raisonnée* had listed 157 of the most useful cultivated plants, of which, he believed, 32 were unknown in their wild state. But, Darwin argued, de Candolle did not include several plants with loosely defined characters, namely the various forms of pumpkins, millet, sorghum, kidney-bean, dolichos, capsicum and indigo. Nor did he include ornamental flowers. Anciently cultivated flowers such as certain roses, the common Imperial Lily, the Tuberose, and even the Lilac, were said not to be known in the wild.

De Candolle had concluded that plants have rarely been so much modified by culture that they could not be identified with their wild prototypes. 'But on this view,' said Darwin, 'considering that savages probably would not have chosen rare plants for cultivation, that useful plants are generally conspicuous, and that they could not have been inhabitants of desert or of remote and recently discovered islands, it appears strange to me that so many of our cultivated plants should still be unknown or doubtfully known in the wild state.' This, however, would be understandable if the plants had been greatly modified under cultivation or if they had become extinct. Darwin in his travels had seen the sort of food savages ate. There was no reason to suppose that our cereal plants originally existed in their present state so valuable to man.

Comparing the fine vegetables and luscious fruits in our kitchen gardens and orchards, it was difficult to believe that the stringy roots of a wild carrot and parsnip, or the wild crabs and sloes, should ever have been valued. Yet, from what he knew of Australian and South African savages, he could not doubt it. Savage man after many and hard trials had found what plants were harmless and good to eat, and taken the first step in cultivation by sowing the seeds near his dwelling. Darwin reminded his readers that the selection of seed-corn was strongly recommended in ancient times by Columella and Celsus. He quoted Virgil as saying

> I've seen the largest seeds, tho' view'd with care,
> Degenerate, unless th' industrious hand
> Did yearly cull the largest.

Since the days of the ancient Greeks countless varieties of cabbage had been formed. Theophrastus distinguished only three, de Candolle more than thirty. It is a very variable plant, so great care must be taken to prevent crossing. To prove this, Darwin raised 233 seedlings from cabbages of different kinds which he purposely planted near each other, and of these not less than 155 were deteriorated and mongrelised. None of the remaining 78 was perfectly true. Culture and climate greatly affected the growth of cabbages. In Jersey they often grew to a height of 10 or 12 feet, and their woody stems were used as rafters and as walking sticks—a specimen was in the Museum at Kew. One on the island, which attained 16 feet, had its spring shoots at the top occupied by a magpie's nest.

But that was going a bit too far. Nor was it what we looked for in our kitchen gardens.

There were numerous varieties of the common garden pea. Darwin planted 41 English and French varieties. They differed in height from between 6 and 12 inches to 8 feet. It is interesting, incidentally, that one of them was the Sugar Pea or *Pois sans parchemin*, now again in fashion. The peas, Darwin found, kept true, because they were self-fertilising.

There was little doubt about the parentage of the potato. In the Chonos Archipelago and in the Andes he had seen them growing wild and had recognised them instantly. This species was *Solanum maglia* 'or the Darwin Potato, as we might suitably call it in English,' wrote George Nicholson in his *Dictionary*, recommending it as better fitted for growing in England and Ireland than *Solanum tuberosum* which was a plant of a comparatively dry climate. There was indisputable evidence that it yielded an abundant supply of eatable potatoes.

The great nurseryman of the day specialising in hybridising fruit trees was Thomas Rivers of Sawbridgeworth. In December 1862 Darwin was corresponding with him about the origin of the peach and nectarine. 'I am collecting all accounts of what some call "Sports", that is, of what I shall call "bud-variations", ie a moss-rose suddenly appearing on a provence rose—a nectarine on a peach &c &c.—Now what I want to know, & what is not likely to be recorded in print, is whether very slight differences, too slight to be worth propagating, thus appear suddenly by *buds*.' He learned that Rivers had raised three varieties of nectarine from

I.   II.   III.   IV.

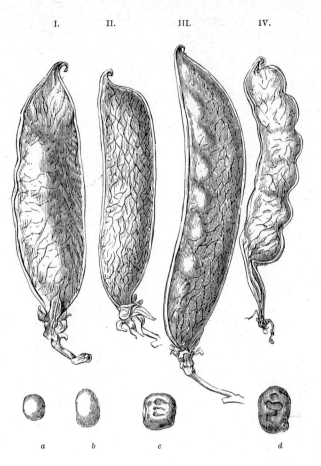

a   b   c   d

The pea is perfectly
fertile without the aid
of insects, and if
several varieties are
grown side by side
each breeds true. Man,
however, by cross-
pollinating can obtain
shorter or taller peas,
wrinkled or smooth.
Darwin grew forty-one
different kinds of peas
in his garden!

stones of three distinct varieties of the peach, and conversely peaches from nectarine stones, with some that were half-and-half.

In 1838 J. C. Loudon in his *Arboretum et Fruticetum Britannicum* had classed (under the family Amygdaleae) the peach as *Persica vulgaris* and the 'Smooth-skinned Peach or Nectarine Tree' as *Persica laevis*. By Darwin's day they had become *Amygdalis persica*, but Loudon thought they might only be varieties and even spoke of 'the different modific-ations which the tree undergoes'. Today we know that Darwin was right, for the nectarine is a smooth-skinned sport of the peach.

Thomas Rivers called his attention to several varieties connecting the almond and the peach, such as the Peach-Almond. From this gradation and from the fact that the peach has never been found growing wild, it seemed to Darwin that the peach was the descendant of the almond, improved and modified in a marvellous manner. Since his day, because of their special similarities, the peach, nectarine and almond have been classed together under the Amygdalus group of the genus *Prunus*.

In the *Gooseberry Grower's Register* for 1862 there were 243 distinct varieties listed as winning prizes at various times, so that a vast number must have been exhibited. It was a good thing that the Down House garden was fairly large, for Darwin tells us that he cultivated fifty-four varieties. He was struck by the fact that although the fruit differed greatly the flowers in all the kinds were very much the same. But what was really astonishing was the increase in the weight of the berry. According to Downing, the American fruit-grower, the wild gooseberry weighed only about a quarter of an ounce or 5 dwts. By 1786 the weight had doubled, as the Gooseberry Register recorded. In 1852 a Staffordshire gooseberry attained 37 dwts 7 grains. The increase was partly due to the careful training of branches and roots, and to composting and mulching, but mainly to the continued selection of the best seedlings.

Useful and ornamental trees came into Darwin's discussions. Lawson's of Edinburgh listed twenty-one varieties of the common ash in their catalogue. Paul's nursery at Waltham Cross boasted eighty-four varieties of holly. 'In the case of trees,' Darwin wrote, 'all the recorded varieties, as far as I can find out, have been suddenly produced by a single act of variation. The length of time required to raise many generations, and the little value set on the fanciful varieties, explains how it is that successive modifications have not been accumulated by selection.'

Our useful trees, Darwin pointed out, 'have seldom been exposed to any great change of conditions. Yet in examining extensive beds of seedlings in nursery-gardens considerable differences may be generally observed.' While touring in England he had always been surprised at the difference between the same kind of tree, according to whether it grew in a hedgerow or woodland, and he believed that hedgerow trees varied more than those growing in a primeval forest. Exposed to conditions not strictly natural, they had yielded a greater number of strongly marked and curious variations of structure.

The hawthorn had varied much in the form of its leaves, and in the size, colour and shape of its berries, some having golden-yellow, some black or whitish, others woolly berries. Loudon, who described twenty-nine well-marked varieties, attributed the reason why the hawthorn had yielded more varieties than most other trees to the fact that nurserymen selected any remarkable variety from the immense beds of seedlings annually raised for making hedges.

The rapidity with which new garden flowers could be bred was instanced by Darwin when he wrote of the pansy and the little white rose of Scotland (*Rosa spinosissima*), the Burnet Rose. In 1793 Robert Brown, a partner in the firm of Dickson & Brown of Perth, dug up some of these wild roses from the Hill of Kinnoull and transplanted them into his nurseries. One of these bore flowers slightly tinged with red, from which he raised a plant with semi-monstrous flowers also tinged with red. Seedlings from this flower were semi-double, and by continuous selection eight sub-varieties were raised in about nine or ten years. In less than twenty years these double roses had so much increased in number and kind that twenty-six distinct varieties, classed in eight sections, were described by

Joseph Sabine, secretary of the Horticultural Society. In 1841 three hundred varieties could be bought in Scottish nurseries. These were blush, crimson, purple, red, marbled, bicoloured, white, and yellow, and differed as much in the size and shape of the flower as in colour.

In 1687 John Evelyn the diarist, who was a great horticulturist, was growing the Heartsease in his garden at Sayes Court near Deptford. But it was not until about 1810 that the flower was looked at as more than a pretty little tricoloured wildling. Fortunately for the future of the beautiful velvety pansy, Lady Monke, who was then Lady Mary Bennet, began collecting them from various parts of her father's estate at Walton-on-Thames. She made a small garden in the shape of a heart, into which she transplanted all the different kinds of Heartsease she could find. William Richardson, her father's gardener, helped her and by crossing and cultivating them obtained some improved varieties. These came to the notice of James Lee, the famous Hammersmith nurseryman, who imported a large blue variety from Holland and began hybridising them with the Walton Heartsease. In a few years twenty varieties were in commerce. Others entered the field. The result was the modern pansy. Darwin tells us that the first great change was the conversion of the dark lines in the centre of the flower into the dark eye which is now considered one of the chief points of a first-rate flower. By 1835, four hundred named varieties were on sale.

Those who pronounce the dahlia as dahl-ia (and not day-lia) are historically correct, for the plant was named in honour of the Swedish botanist and pupil of Linnaeus, Andreas Gustav Dahl. It was discovered by Nicholas de Menonville in Mexico where it was called the Acoctli or Cocoxochitl, meaning water tubes, the reference being to the plant's hollow stems. Darwin was correct in believing that all the varieties of his day were descended from a single species. The flower was introduced into France in 1802, and to John Fraser of Sloane Square we owe its successful introduction from there into England in the same year. Darwin quoted Joseph Sabine as remarking that: 'It seems as if some period of cultivation had been required before the fixed qualities of the native plant gave way and began to sport into those changes which now so delight us.'

Sabine was writing in 1820 and Darwin more than forty years later. In the interval, as Darwin tells us, the flowers were greatly modified from a flat to a globular form. Anemone- and ranunculus-like races differing in the form and arrangement of the florets had arisen, also dwarf races, one of which was only eighteen inches tall. The petals, uniformly coloured or tipped or striped, were of an almost infinite diversity of tints. Seedlings of fourteen different colours had been raised from the same plant, though many of the seedlings were the same colour as their parents. The period of flowering had been considerably hastened, and this, as Darwin wrote, had probably been brought about by continual selection. In 1808 the flowering period of dahlias was brief, September to November. By 1828 new dwarf varieties began flowering in June, the dwarf purple 'Zelinda' sometimes blooming earlier.

The hyacinth was introduced into England sometime before 1596 (John Gerard was growing it in his London garden at that date) from the Levant, the petals of the original flower being narrow, wrinkled and pointed, and of a flimsy nature, where now they are broad, smooth, solid and rounded. The nurseryman William Paul was an authority on their history, and he must have read the *Origin of Species* thoroughly. Darwin quoted him as writing the following in 1864: 'This single flower serves well to illustrate the great fact that the original forms of nature do not remain fixed and stationary, at least when brought under cultivation. While looking at the extremes, we must not, however, forget that there are intermediate stages which are for the most part lost to us. Nature will sometimes indulge herself with a leap, but as a rule her march is slow and gradual.'

This was Darwin's theory in a nutshell. He had written: 'As natural selection acts solely by accumulating slight, successive, favourable variations, it can produce no great or sudden modification; it can act only by short and slow steps. Hence, the canon of "Natura non facit saltum", which every fresh addition to our knowledge tends to confirm, is on this theory intelligible.'

We remember that in his letter to Thomas Rivers, Darwin allowed ' "Sports", that is . . . what I shall call "bud-variations" '. Why then did he not allow that nature did sometimes make a *saltus* or sudden leap, what today we would call a mutation? By now he was a thorough-going botanist, at least in the realm of plant morphology, and already he was aware that a simple and common metamorphosis such as that of stamens into carpels, leaves into tendrils, stigma into rostellum, can deceive even a trained botanist. We recall his classic discovery that the *Catasetum* was three different forms and not, as the botanists had thought, three different kinds of plants. Were Darwin living today he would know that even a single gene can alter a plant's structure or appearance, and that several genes can alter it greatly. He would have taken this in his stride. To him the concert of genes would have been the accumulation of slight variables, again acting under the laws of Natural Selection.

In the last chapter of Volume I of *Variation* he dealt with bud-variation in detail, instancing trees, shrubs, and even herbaceous plants such as the phlox, in which this commonly occurred. Of the barberry (berberis) he said: 'There is a well-known variety with seedless fruit, which can be propagated by seedlings or layers; but suckers always revert to the common form, which produces fruit containing seed.' He added: 'My father repeatedly tried this experiment, and always with the same result.'

So horticultural experiments were nothing new in the Darwin family. Yet it has been wondered whence Charles Darwin got his interest in plants.

He now asked himself the reasons for plant-variation. He had to look for the factors of inheritance. 'The great principle of inheritance,' he began, 'has been recognised by agriculturists and authors of various nations, as shown by the scientific term *Atavism*, derived from atavus, an ancestor; by the English terms of *Reversion*, or *Throwing-back*; by the French *Pas-en-arrière*; and by the German *Rückschlag*, or *Rückschritt*.' An example of reversion was sometimes found in the Heartsease: plants perfectly wild in their foliage and flower frequently grew from seeds gathered from the finest cultivated varieties, though in their case reversion was not to a very ancient period, because the development of this flower was of modern origin. With most of our cultivated vegetables there was some tendency to reversion and this would be more evident if gardeners in looking over their beds of seedlings did not pull up the 'rogues'. In turnip and carrot beds a few plants would often 'break'—that is, flower too soon, and their roots were generally hard and stringy, like those of their ancestors. James Buckman, the expert on agricultural plants, had proved how easy it was to bring back cultivated plants to their wild or nearly wild condition: he had done this with the parsnip, selecting back for only a few generations. Hewett C. Watson had done the same thing with Scottish kale: in the third generation some of his plants closely resembled the indigenous kale found growing about old castle walls.

These were cases of cultivated plants reverting completely to their wild forms.

There was another type of reversion, where hybrids reverted to both or to one of their parent-forms after several or even many generations. Darwin had often seen this with plants and he believed that characters 'exist in a latent state, ready to be evolved under certain conditions.' These might be a different environment, climate or soil.

Abnormal flowers (the botanical 'monsters') often showed reversion clearly. The corydalis, linaria, and antirrhinum all have irregular flowers—that is, some petals form a hood, others a lip, unlike a daisy which is symmetrical. Darwin did many experiments with these flowers. Sown from seed, a regular or peloric flower sometimes appeared, having a symmetrical cup of petals on top of the calyx instead of the hood and lip. Examining them he discovered a fifth stamen. The normal flower has only four but it does have a minute rudiment of a fifth. This led him (correctly as we now know) to believe that at some ancient epoch the flower of the antirrhinum resembled the flower of the peloric form he had grown in his garden.

'We must believe,' he wrote, 'that a vast number of characters, capable of evolution, lie hidden in every organic being.' He added: 'On this view of the nature of peloric flowers, and bearing in mind certain monstrosities in the animal kingdom, we must conclude that the progenitors of most plants and animals have left an impression, capable of redevelopment, on the germs of their descendants, although these have since been profoundly modified.' This germ was perhaps the most wonderful object in nature. But on the doctrine of reversion it became a far more marvellous object, for we must believe, said Darwin, that it is crowded with invisible characters related to a long line of ancestors separated by hundreds or even thousands of generations from the present time, 'and these characters, like those written on paper with invisible ink, lie ready to be evolved whenever the organisation is disturbed by certain known or unknown conditions.'

Characters of all kinds, then, tended to be inherited, and those which have withstood all counteracting influences and been truly transmitted would generally continue to withstand them, and consequently be faithfully inherited. But sometimes there was a difference in the strength or 'prepotency' of transmission in the two parent-forms. Or the two parents might be equally prepotent. He raised a large bed of plants from the peloric Snapdragon (*Antirrhinum majus*) artificially fertilised by its own pollen. Sixteen plants survived the winter and they were all as perfectly peloric as their parent. He had also crossed the peloric Snapdragon with pollen of the common form, and the latter reciprocally with peloric pollen. Of these two large beds of seedlings not one was peloric. He now allowed these crossed plants to sow themselves, and out of 127 seedlings 88 proved to be common Snapdragons, 2 were in an intermediate condition, and 37 were perfectly peloric.

Finally he considered one of the most important facts in biology: the difference between the embryo and the adult. The explanation was that although variations are inherited these do not necessarily or generally appear at a very early period of growth. Because of this the embryo, even after the parent-form has undergone great modification, is left only slightly modified. Many seedlings of widely different plants descended from a common progenitor resembled each other and probably resembled their common ancestor. 'We can thus understand,' said Darwin, 'why embryology throws a flood of light on the natural system of classification, as this ought to be as far as possible genealogical.'

At the end of his book Darwin put forward a provisional theory to account for the factors of inheritance. He called these factors 'gemmules' and supposed that they were 'thrown off' from each different unit throughout the body, finally to be 'contained within each bud, ovule, spermatozoon, and pollen-grain'. 'My well-abused theory of Pangenesis,' he later called it. Professor John Heslop-Harrison has a kinder comment. 'As we see it today,' he wrote in *A Century of Darwin*, 'the major error of pangenesis lay in giving weight to a mechanism which would allow direct registration of somatic effects on the germ-plasm: but it was only "some such" theory which Darwin hoped to see established, and in the conception of "gemmules" as the carriers of inheritance he was not so distant from the twentieth-century conception of the gene.'

At the same time as Darwin was drawing up his conclusions on inheritance, Gregor Mendel, an Augustinian monk who taught science at the monastery at Brünn, was hybridising peas and keeping exact records of the various features—dwarfness and tallness, yellow seeds and green seeds, round peas and wrinkled peas—as they appeared in succeeding generations. The constant numerical ratios he obtained from some 10,000 plants told him that there must be some natural law of inheritance.

Darwin also was methodical. He, too, tabulated his results, and it is an astonishing fact that with his cabbages and with his antirrhinums, as with other plants, he obtained Mendelian 3:1 ratios.

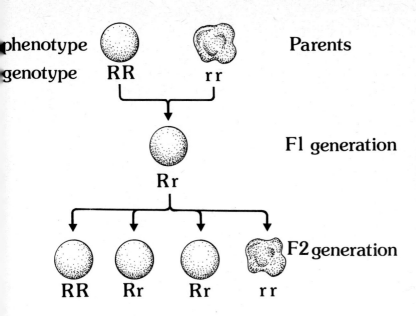

phenotype / genotype — Parents — RR — rr — F1 generation — Rr — F2 generation — RR — Rr — Rr — rr

The 3:1 ratio which Gregor Mendel worked out in his studies of the behaviour of hereditary factors, and which Darwin also obtained. These and other ratios gave birth to the science of genetics. Here a pure-bred pea of the round phenotype (genotype RR) is crossed with a pure-bred wrinkled phenotype (rr). In the first filial generation (FL) all the progeny are round phenotypically (Rr genotypically), as R is dominant over r. These FL generation offspring are then selfed and the second filial generation (F2) will be in the ratio of 3 round to 1 wrinkled

Mendel found that where the parents showed a marked difference in special characters, such as tallness and dwarfness, the hybrid offspring in the first generation was always tall. This was what Darwin called 'prepotency'. ('In some cases prepotency apparently depends on the same character being present and visible in one of the two breeds which are crossed, and latent or invisible in the other breed.') Mendel called the first a dominant characteristic and the other recessive.

There were other laws of inheritance which Mendel found, which Darwin also discovered, and it has always been considered a tragedy that Darwin did not draw the inference from his results which would have enabled him to come to the same conclusions. But he was no mathematician. Otherwise, such repetitive results as 3:1, or 9:3:3:1, or 1:2:1 would have awakened his curiosity—and he could have been Mendel. But Mendel could never have been Darwin.

As it was, Mendel read the *Origin of Species* and took careful note of all Darwin had to say on hybridising, and the paper he read on hybridism to the Brünn natural history society in 1865 lay forgotten until 1900 when it was independently and simultaneously unearthed by three botanists —Hugo de Vries, working in Holland, and Carl Erich Correns and Erich von Tschermak-Seysenegg in Germany—and presented to the world as the science of genetics. Long before then Darwin, the founder of it all, had made other discoveries.

He was not even finished with *Variation*. Crossing was a vast field of inquiry he could barely touch on, even in three chapters closely reasoned and packed with information.

It was to be the subject of one of his greatest books—*The Effects of Cross- and Self-Fertilisation.*

14

Mayhem and Murder in the Plant World

It was Darwin's discovery that some plants, like the sundew, not only capture insects in order to feed upon them but possess 'nervous matter' analogous to the nerves of animals.

Grandfather Erasmus Darwin supposed that the Venus's Fly-trap surrounded itself with snares to prevent insects making depredations upon its flowers. Other naturalists had found out curious facts about other insectivorous plants.

It was left to Charles Darwin to investigate the whole subject and, with the authority of thousands of complex experiments, tell the complete story.

IN 1864 DARWIN WAS given the greatest honour British science can confer, the Copley Medal of the Royal Society. It was awarded 'for his important researches in Geology, Zoology and Botanical Philosophy'.

He had a busy time replying to letters of congratulation. To William Darwin Fox: 'I was glad to see your handwriting. The Copley being open to all sciences and all the world, is reckoned a great honour; but excepting from several kind letters, such things make little difference to me. It shows, however, that Natural Selection is making some progress in this country, and that pleases me. The subject, however, is safe, in foreign lands.' In this year the *Origin of Species* was translated into Dutch and Russian. The first foreign editions were the American and German in 1860, followed by the French in 1862.

To Joseph Hooker he wrote: 'How kind you have been about this medal; indeed, I am blessed with many friends, and I have received four or five notes which have warmed my heart. I often wonder that so old a worn-out dog as I am is not quite forgotten.' He was fifty-five.

And to Huxley: 'I must and will answer you, for it is a real pleasure for me to thank you cordially for your note. Such notes as this of yours, and a few others, are the real medal to me, and not the round bit of gold.'

He thought Hugh Falconer should have the Medal, as ought John Lubbock who 'tells me that some old members of the Royal are quite shocked at my having the Copley.' In fact the malcontents had prevented the award being made in the previous year—to Lyell's disgust. Darwin was not present at the ceremony and so did not hear Lyell's after-dinner speech when he made 'a confession of faith as to the "Origin".' His Scottish reserve put it in these words: 'I said I had been forced to give up my old faith without thoroughly seeing my way to a new one. But I think you would have been satisfied with the length I went.'

The times were moving with the pace Darwin had set, though Lyell's book *The Antiquity of Man*, published the year before, had been a great disappointment to him, for 'he has not spoken out on species, still less on man.' Huxley in the same year had published *Man's Place in Nature*, a work important not only for the evolutionary links between *homo sapiens* and his ancestors but for the problems it posed of man's place in the world of tomorrow.

Huxley was by now a frequent visitor at Down. A family friendship had sprung up after the death from scarlet fever of his four-year-old son Noel in 1860. Two other children in danger and his wife six months pregnant were poignant reminders to the Darwins of the nightmare of 1851. They begged Henrietta to come to Down and bring the children. The visit forged bonds of lasting affection.

Lyell's and Huxley's books were followed by a paper Wallace published in 1864 in the *Anthropological Review*, then by a massive tome on the subject by Ernst Häckel, and finally by Darwin's own *Descent of Man* which occupied him intermittently from 1867. Published four years later it ran into its fourth edition in the December. Immediately it was off his hands he began another anthropological work, *The Expression of the Emotions in Man and Animals*, a book that had really started with the birth of his first son, William, in 1839, when he had 'at once commenced to make notes on the first dawn of the various expressions which he exhibited'. Day by day he had posted himself by his baby's cot. It was no mere fond father observing and recording, for the interpretation of his expressions led directly to his interpretation of the origin of human nature. The book was published in 1872.

William was now thirty-two and a partner in the Southampton and Hampshire Bank, later incorporated in Lloyd's. All the Darwin boys had left the family nest and were doing well. From Cambridge in January 1868 had come the news that George was Second Wrangler and a Fellow of Trinity, in February consolidating his position among the mathematical dons by winning the second Smith's Prize. Leonard was at Woolwich, having come second in the entrance examination for officers when he was eighteen. Francis who had taken his B.A. at Cambridge was studying medicine at St George's Hospital, but was soon to become assistant to his father. Horace, the youngest, was still at Trinity. In the late spring of 1871 Henrietta was in London taking courses in geometry and other subjects. In August she married Richard Buckley Litchfield, scholar, philosopher and philanthropist, who was a founder of the Working Men's College. Bessy remained unmarried.

So life at Down House was quiet, but never dull. Emma took to gardening, a true Wedgwood pursuit, for after all, her uncle John Wedgwood was the first founder of the Horticultural Society. And she much enjoyed evening walks with her husband. She was in her element when friends came to stay, and sometimes she succeeded in stealing Charles away for a holiday.

In 1868 they were at Freshwater Bay in the Isle of Wight, staying in Dumbola Lodge, owned by Mrs Julia Margaret Cameron who was already famous for photographs of the great. The well-known photograph she took of Darwin on this holiday shows him with the beard he had grown two years before. There were visits to Cambridge to see the boys, now and again a week in London with Erasmus, and a stay with William at Basset, Southampton, where Darwin was always happy. On the summer holiday of 1869 he made a sentimental visit to Shrewsbury. They called at The Mount and the new owner showed them over the house. This was meant kindly but Darwin was deeply disappointed. 'If I could have been left alone in that greenhouse for five minutes,' he said afterwards, 'I know I should have been able to see my father in his wheel-chair as vividly as if he had been there before me.'

Family ties were loosening. Three years before, his two sisters Catherine and Susan had died. But the immediate ties between Darwin and his children were, if possible, stronger. He became 'F' to them, no longer 'Your Father' or 'Papa'. When the boys were at home on holiday they helped with whatever experiments were going on. William, Frank and George did illustrations for *Climbing Plants*, *Insectivorous Plants* and others of his books—George by far the most.

Since 1860 a pile of notes had been awaiting his attention. They were to do with 'My beloved Drosera'. He finished the last proofs of *Expression* on the 22nd of August 1872 and next day 'Began working at Drosera', as his diary records.

What emerged in June 1875 under the usual Murray imprint was *Insectivorous Plants*, a drama of plant life. For, as the author wrote to Asa Gray of his first subject in the book, *Drosera rotundifolia*, the Round-leaved Sundew, 'It is a wonderful plant, or rather a most sagacious animal.' That a plant could indeed behave like an animal was an astonishing fact. That a plant had what he could only call 'nervous matter' analogous to the nerves of animals was even more astonishing, and this was Darwin's unique discovery.

Linnaeus had known that the *Dionaea*, called Venus's Fly-trap, caught insects. His correspondent John Ellis had sent him a description of it and a drawing. Grandfather Erasmus Darwin supposed that the *Dionaea* surrounded itself with insect traps to prevent depredations upon its flowers. The Rev. Dr Curtis of North Carolina discovered the plant's digestive fluid. Other naturalists had found out curious facts about other insectivorous plants. But the facts were scattered and disconnected. It was left to Darwin to investigate the whole subject and, with the authority of thousands of complex experiments, tell the complete story.

It was a sensational story of macabre happenings, of mayhem and murder. Who could believe that the pretty little Common Butterwort with its golden star of leaves and springing violet flower had such a sinister nature? Or that under the quiet waters of a pond there were plants trawling nets to catch fish? And, worse, that there existed a perfect chamber of horrors involving lures, anaesthesia and drownings?

Darwin told the story superbly, missing not a detail.

The sundews he first examined were growing, we remember, on heathland in Sussex. He gathered a dozen plants. On thirty-one of their fifty-six fully expanded leaves were dead insects or remnants of them. On one plant all six leaves had caught their prey, and on several plants many of the leaves had caught more than a single insect. One large leaf bore the remains of thirteen. The largest insect Darwin ever saw caught in this way was a Small Heath butterfly.

He was struck by the stickiness of the leaves. Other plants had this feature: the buds of the Horse Chestnut, for instance, and these too caught flies, but without any advantage to the tree. The difference was that *Drosera* was especially adapted for the purpose of catching insects.

Drosera rotundifolia was a ground-hugging plant whose half-dozen rounded leaves made a circle. The whole of their upper surface was covered with filaments—'or tentacles, as I shall call them,' wrote Darwin, 'from their manner of acting.' Each tentacle bore a gland on its summit, surrounded by a large drop of a very sticky secretion. They glittered in the sun, giving rise to the plant's poetical name of the sun-dew.

The tentacles on the centre part of the leaf were short and stood upright, the pedicels attaching them to the leaf being green. Towards the edge of the leaf they became longer and more inclined outwards, their pedicels being purple.

The structure of the glands was remarkable and their functions complex. Darwin found that they secreted, absorbed, and were acted on by various stimulants. They consisted of an outer layer of small many-sided cells containing purple granular matter or fluid. Within this layer was an inner one of differently shaped cells, also filled with purple fluid but of a slightly different tint. In the centre was a group of elongated cylindrical cells of unequal length, closely pressed together and surrounded by a spiral line which he was able to separate as a fibre. These cells were filled with a clear fluid and were connected with the spiral vessels running up the tentacles.

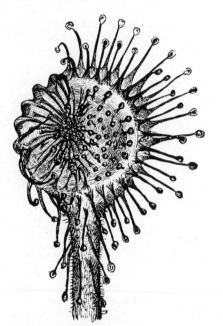

Drosera rotundifolia, the Round-leaved Sundew, showing a leaf with its tentacles inflected over a bit of meat

The way in which insects were actually captured was his first experiment. He placed a living fly on the glands in the centre of a leaf. A motor impulse was at once transmitted to the tentacles, the nearer ones answering first and slowly bending towards the centre, and then the farther ones, until they formed a cage imprisoning the fly. This took from one hour to four or five or more hours, depending on the size of the prey, whether it was soluble or inorganic, and on the age of the leaf. A living fly was more efficient than a dead one, as in struggling it pressed against the glands of many tentacles, but rapid results were obtained with drops of various fluids, for instance of saliva or of a solution of any salt of ammonia, when a tentacle would begin to bend in only ten seconds.

Sometimes large soluble objects caused the whole leaf to curve and even form a cup, as happened when Darwin placed bits of hard-boiled egg on three leaves.

The secretion from the glands was so glutinous that it could be drawn out into long threads. It was colourless but stained little balls of paper pale pink. The placing of an object of any kind on a gland caused it to secrete more freely. Particles of sugar had a marked effect; while immersion in solutions of chloride of gold and some other salts excited the glands to largely increased secretion, immersion in solutions of acids excited them so much that when he lifted out the leaves long ropes of the sticky glue were hanging from them.

After contact with animal matter the secretion not only increased in quantity but changed its nature and became acid. The leaf now began to digest its prey. But Darwin discovered that the secretion was not only a gastric juice but acted also as an antiseptic. During very warm weather he placed two equal-sized bits of raw meat, one on a *Drosera* leaf and the other surrounded by wet moss. After forty-eight hours the piece on the moss was swarming with infusoria and was much decayed, while the piece on the leaf was still perfectly fresh. The digestive process took several days, after which the tentacles began to resume their normal positions, their glands secreted less freely and finally became dry. This enabled the leaf to enjoy a spring-clean, for any breath of air puffed away undigested material such as the hard wing-cases of small beetles. The glands now quickly began to secrete again, and as soon as full-sized drops were formed the tentacles were ready for a new victim.

The absorption of animal matter from captured insects explains how the sundew can flourish in poor peaty soil, sometimes where nothing else but *Sphagnum* moss grows—and mosses, as Darwin pointed out, depend entirely on the atmosphere for their nourishment. So, he said, the sundew feeds like an animal. It drinks by means of its roots, and it must drink a great deal so as to manufacture the many drops of fluid round its glands, sometimes as many as 260, exposed during the whole day to a glaring sun.

There was another animal-like phenomenon Darwin discovered during his experiments on the *Drosera*. He found that the plant had muscles!

How this was proved is a story in itself. All living things are formed of protoplasm, usually a clear colourless jelly-like or fluid substance consisting of carbon, hydrogen, oxygen and other elements. One of its most distinctive properties is its ability to contract, and this happens when particles of protoplasm are so arranged that they act in concert, producing a cumulative effect. So when we look at a sundew leaf curling up to form a cup, what we are seeing is the leaf contracting its muscles.

This much Darwin had observed with his naked eye. But what was interesting him much more at this moment was what he saw under his compound microscope when he took a young but fully matured leaf that had never been excited or become inflected and examined its tentacles. He saw that the cells forming the hair-like stalks were filled, like the gland cells, with purple fluid. The walls were lined with a layer of colourless circulating protoplasm. He now excited the glands of several plants by repeatedly touching some of the tentacles, by placing different particles on them, and by drops of various fluids. After some hours the tentacles looked entirely different. The purple fluid had coagulated and become masses of purple matter suspended in almost colourless fluid. He called the process aggregation, and noted that it began within the gland and then travelled down the tentacles. Soon after the tentacles had re-expanded, the little masses dissolved, the purple fluid becoming as transparent as it was before. The process of redissolution travelled upwards from the bases of the tentacles to the glands, in the reverse direction from that of aggregation.

His microscope revealed that the masses of aggregated matter were of different shapes, round or oval, sometimes elongated, or irregular with 'thread- or necklace-like or club-like' projections. 'These little masses,' he wrote, 'incessantly change their forms and positions, being never at rest. A single mass will often separate into two, which afterwards re-unite. Their movements are slow, and resemble those of Amoeba or of the white corpuscles of the blood.'

He 'was able by varying the light and using a high power, to detect a connecting thread of extreme tenuity, which evidently served as a channel of communication between the two. On the other hand, such connecting threads are sometimes seen to break, and their extremities quickly become club-headed.' He saw that 'Small spheres of protoplasm, apparently quite free, are often driven by the current round the cells; and filaments attached to the central masses are swayed to and fro, as if struggling to escape. Altogether, one of these cells with the ever-changing central masses, and with the layer of protoplasm flowing round the walls, presents a wonderful scene of vital activity.'

What he was watching was the wonderful life going on inside a cell and the streaming movement by which food substances are distributed. He did not know this but was so excited by what he saw that he called in his scientific friends. 'Tentacles in an aggregated condition were shown to Prof. Huxley, Dr Hooker and Dr Burdon Sanderson,' he wrote, 'who observed the changes under the microscope, and were much struck with the whole phenomenon.'

John Burdon Sanderson, Professor of Botany at St Mary's Hospital, had been experimenting with the thermo-electric pile. He asked Darwin to see if heat rigor occurred when he immersed the leaves in hot water.

Darwin proceeded to find out, heating two leaves to 130°F., with the result that every tentacle became closely inflected. He then put one leaf in cold water and it re-expanded. The other leaf was heated to 145°F. and did not re-expand. 'Is not this latter case heat rigor?' he wanted to know.

By September Burdon Sanderson was thoroughly involved. Darwin sent him two plants of the Venus's Fly-trap 'with five goodish leaves' and detailed instructions as to their welfare, adding: 'If you get a positive result, I should think you ought to publish it separately, and I could quote it; or I should be most glad to introduce any note by you into my account.'

Burdon Sanderson eventually published separately, reading a paper on the subject at the British Association meeting of 1873.

In the latter part of the eighteenth century Madame Luigi Galvani, wife of the Italian physiologist, had observed that when the muscles of frogs were brought into contact with two metals, a convulsion resulted. Her husband went on to invent the galvanic battery. Using a galvanometer Burdon Sanderson was able to prove that a definite electric current existed in the leaves of carnivorous plants. He connected a leaf of the *Dionaea* to the delicate instrument and brushed the tentacles. The needle of the galvanometer registered the electromotive force as the leaf contracted.

Joseph Hooker, as president of the Royal Society, was now the crowned head of science. In his address to the British Association at Belfast in 1874 he said: 'All students of the vegetable side of organised nature were astonished to hear from Dr Sanderson that certain experiments, which, at the instigation of Mr Darwin, he had made, proved to demonstration that when a leaf of Dionaea contracts, the effects produced are precisely similar to those which occur when muscle contracts. Not merely then are the phenomena of digestion in this wonderful plant like those of animals; but the phenomena of contractility agree with those of animals also.'

Dionaea muscipula, the sinister Venus's Fly-trap, had a method of capturing its prey that was simple but deadly. The leaf was divided into two lobes with needle-like spikes along the outer edges. Studding the upper surface of each lobe were three tiny spines. These were the 'electric switches' that set the leaf in action. An insect had only to touch one of them with a delicate foot and the trap snapped shut—'with quite a loud flap,' as Darwin remarked. And it was useless for the insect to watch where it put its feet, for by the time it got well inside the leaf a smothering glue was exuding. In its struggles to free a leg or a wing it inevitably touched one of the switches.

As usual, Darwin made many experiments. When a leaf snapped shut it turned itself into a temporary stomach. The presence of an insect had stimulated the surface glands to pour forth their acid secretion, and this now acted like the gastric juices in the stomach of an animal. But the digestive process was not so easy to watch as it had been with the *Drosera*, for everything went on behind closed doors. To see what was happening Darwin had to pin open the lobes with a thin wedge driven between them, and he was astonished at the force with which the spikes resisted. In fact they would generally break rather than yield. Several days after a Fly-trap had caught an insect its body was surprisingly softened. It could even digest a beetle, though it left the hard chitinous coat untouched.

As usual, too, Darwin played tricks on his subject. The lobes would snap shut when touched by a blade of grass, or when anything was blown on to them by the wind. But *Dionaea* was not to be duped. The leaf instantly closed and as quickly opened again, a faculty important to the plant if the intruder was of no use to it, for as long as the leaf remained shut it could not of course capture an insect. 'Quickly', incidentally, was relative, for in rejecting such unwanted objects as bits of wood, cork, or balls of paper, the lobe did not reopen for 24 hours. It took 15 days to digest one fly and reopen, 24 days over a second fly, the same time over a wood-louse, and 36 days over a large Daddy-long-legs. When very small insects were caught the lobes reopened as if they had trapped nothing. Darwin supposed that this was because the victims were too small to be crushed or to have any flesh on them. Minute insects were sometimes able to escape between the spikes.

No such escape was offered by *Aldrovanda vesiculosa*, a miniature aquatic *Dionaea*. The leaves were arranged around the stem in a series of wheels, the seven leaf-stalks resembling spokes. At the end of each spoke was a tiny transparent bilobed leaf surrounded by six stiff bristles. When a larva, water-insect or crustacean swam into the trap it could not avoid touching the bristles, which immediately embraced their victim, swept it inside and closed the exit. Darwin observed the tiny creatures swimming about inside like fish in an aquarium. They were doomed to certain death.

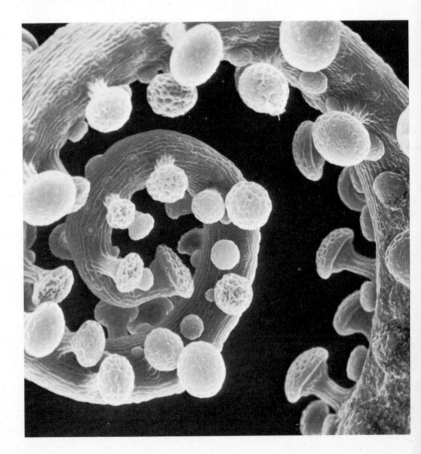

Drosophyllum lusitanicum with the stalked glands which capture its insect prey

Variations in
Antirrhinum,
including the highly
mutable (spotty) form
obtained by Darwin

Normal and peloric
forms of Antirrhinum
in a 3:1 ratio, which
Darwin also obtained

Three layers of cells in the sides of the lobes had the power of exuding a clear liquid to dissolve and digest their prey. The lobes had another power. So that the digestive fluid would not be too diluted they were able to squeeze out the intake of water.

One thing puzzled Darwin: the infolded rims of the leaf lobes. Further examination revealed that they acted as useful filters of the water being expelled. This contained diluted animal matter, and this the rims could absorb, so nothing was lost. It was a remarkable case of different parts of the same leaf serving very different purposes: one part for true digestion, and another for absorption.

'We can thus also understand,' wrote Darwin, 'how, by the gradual loss of either power, a plant must be gradually adapted from the one function to the exclusion of another.'

When he came to examine two other carnivorous plants—the Common Butterwort and species of the bladderwort—he found that this was exactly what had happened.

Meanwhile he turned his attention to *Drosophyllum lusitanicum* which is found only in Portugal and southern Spain. Living plants came from William Chaster Tait who with his brother Alfred lived at Oporto. Both were botanists, Alfred passionately interested in daffodils, William particularly in forestry: he was a pioneer in introducing the eucalyptus as a forest tree into Portugal.

Drosophyllum was supposed to be a rare plant, but William Tait told Darwin that it grew all over the slopes of the dry hills near Oporto, and that vast numbers of flies were to be seen adhering to its leaves. The villagers called it the 'Fly-catcher' and hung it up in their cottages for this purpose.

When the plants arrived Darwin put them in his hot-house. It was early April and the weather was cold and insects scarce. Yet the drosophyllums caught so many flies that he was astounded.

It was a pretty plant, about a foot tall with large yellow single flowers in a corymb at the top of a leafy stem. The leaves were curious: they not only unrolled like a fern but had their edges rolled inwards. They sprang from a thick woody stem and at a little distance looked like long blades of grass. A near view showed that they were beset with stalked glands topped by hollow caps of bright pink or purple, looking like rows of miniature toadstools. Unlike the tentacles of the sundew these had no power of movement except to exude the dewdrops of glue with which to ensnare their victims. *Drosophyllum* spread its net wide. Each flower-stalk and calyx had the same mushroom glands, each with its alluring dewdrop. There were other traps, glands without stalks which grew both on the upper and lower surfaces of the leaves. These were colourless and so small as to be scarcely visible to the naked eye. They were none the less useful to the plant. In fact, Darwin found that it was the secretion from these tiny glands which did the main work of dissolving and absorbing. When an insect alighted on a leaf the sticky drops from the stalked glands adhered to its wings, feet or body, and as it crawled onward in an effort to free itself, other drops stuck to it. At last, unable to move, it sank down and died, resting on the tiny surface glands which then began consuming their meal.

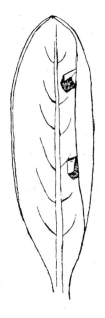

A leaf of *Pinguicula vulgaris*, the Common Butterwort, with the right margin inflected against two square bits of meat

Among the carnivorous plants Darwin investigated (and there was a surprising number, including the innocent-looking *Primula sinensis* and two species of saxifrage) was *Pinguicula vulgaris*, the Common Butterwort. On the morning of the 23rd of June 1874 he received a parcel of its leaves from Miss Amy Ruck of Pantlludw to whom his son Francis was engaged—they married in the following month and came to live in Downe village, Frank from now on to be his father's assistant.

Darwin was 'in that state in which I would sacrifice friend or foe,' as he wrote to William Thiselton Dyer who was now Assistant Director at Kew. 'I have ascertained that bits of certain leaves, for instance, spinach, excites much secretion in Pinguicula, and that the glands absorb matter from the leaves.' Those sent from Wales not only had a number of captured insects on them but a good many leaves blown on to them, and two seed-capsules, and he had found that the protoplasm in the glands beneath the little leaves had undergone aggregation. 'Therefore, absurd as it may sound, I am prepared to affirm that *Pinguicula* is not only insectivorous, but graminivorous, and granivorous!' It seemed like a joke, but it was all perfectly serious.

In July 1868 there appeared in the *Quarterly Magazine of the High Wycombe Natural History Society* an article by the botanist Robert Holland who had found that water insects are often imprisoned in the bladders of *Utricularia vulgaris*. He suspected that they were destined for the plant to feed on. Darwin procured living specimens from the New Forest and from

INSECTIVOROUS PLANTS:

(*left*) *Dionaea
muscipula*, Venus's
Fly-trap

(*right*) The lobe
showing the three
trigger-hairs which
spring the trap

The death-cell
aquarium of the
Bladderwort
Utricularia vulgaris

INSECTIVOROUS PLANTS:

(right) Nepenthes pervillei, the Peruvian Pitcher Plant, which drowns its victims in a pool of anaesthetising liquid

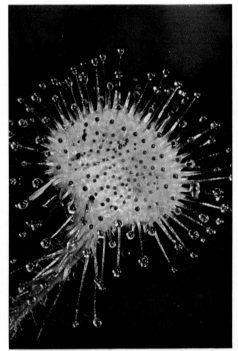

(above left) The glistening tentacles of a sundew, *Drosera rotundifolia*, which give the plant its name

(left) The Sundew captures an ant

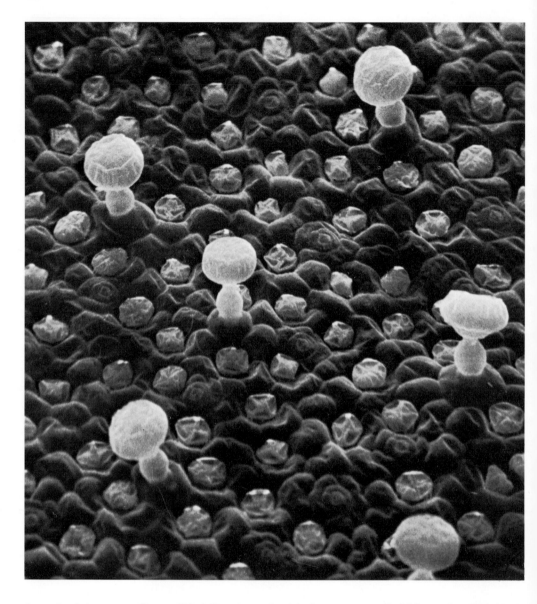

A scanning electron-micrograph of a Butterwort leaf surface, magnified 1420 times

Cornwall but they turned out to be a very rare British species, *Utricularia neglecta*. The plant had no roots. It consisted of a branch bearing small bladders mounted on footstalks, and it floated about like a pirate on the high seas in search of its quarry. The bladders were always filled with water and sometimes contained bubbles of air. Like the *Aldrovanda*, each had bristles at the entrance, but these were jointed and flexible and made incessant random movements due to microscopic particles suspended in a fluid. These particles, Darwin observed, slowly changed their position, travelling back and forth from one end of a bristle to the other. He called the bristles antennae: they were sensory appendages which picked up signals exactly like the antenna or aerial of a radio receiver. There was also a valve at the entrance, which at the slightest touch imprisoned its victim.

Another bladderwort lived in trees. Its native home was tropical South America and Darwin first examined dried specimens provided by Kew, and then live plants sent him by Lady Dorothy Nevill who had been so useful with orchids. On the rhizomes of the dried plants were particles of soil and grit. The live plants were growing in peaty soil and it was the rhizomes that produced the bladders. Here the trap-door of the valve was steeply tilted and it had a roof over the cavity to prevent earth from blocking it up. Again its bladders were filled with water and an occasional air-bubble. When Darwin and Frank opened some of them they found plenty of animal remains.

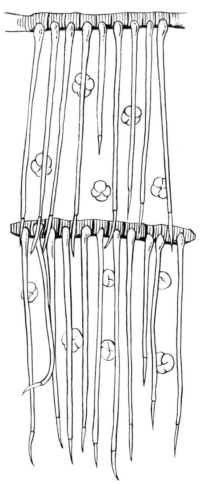

The grids of *Genlisia ornata* which Darwin likened to a paper of pins from which there was no escape

Utricularia nelumbifolia was perhaps the most remarkable of all the bladderworts. It grew in the Organ Mountains of Brazil, but despite this habitat it was an aquatic and was only to be found growing in the water which collected in the bottom of the leaves of a large *Tillandsia* inhabiting an arid part of the mountains about 5,000 feet above sea level. It propagated itself not only by seed but by runners which it threw out from the base of the flower-stem, and a runner was always to be found directing itself towards the nearest *Tillandsia*, where it inserted its point into the water, settled down to grow and in due time sent out its own runners.

As if all these horrors were not enough, *Genlisea ornata* had something fresh to show him. Nothing could escape from the trap it set, for beyond the entrance to the death chamber was a passage lined with grids of long thin transparent hairs with sharp points, each row pointing down to the row below it. Once inside, the victim could not retreat. Darwin likened its construction to an eel-trap made of an old-fashioned paper of pins. Even if the victim survived this long journey—of 10.583 millimetres—death by drowning in the bladder awaited it at the end.

How cruel was Nature, Darwin had once exclaimed to Joseph Hooker. He must have thought so now when, having finished *Insectivorous Plants*, he surveyed his rogues' gallery of carnivores.

But it was all part of variation, struggle, and survival, and as such part of evolution.

15

Darwin's Hero, rule and method

Darwin found that even a hermaphrodite or self-pollinating flower requires to be crossed occasionally, if the plant is to be kept fertile.

He did thousands of experiments, sometimes breeding a plant to the tenth generation in order to detect reverses from previous results.

These experiments, embodied in Cross- and Self-Fertilisation, *took him eleven years. In one he counted 20,000 seeds of the Purple Loosestrife to prove a point.*

The drawing shows the flowers of the Morning Glory (Ipomoea purpurea) *and the bigger form of Darwin's 'Hero'.*

WHEN DARWIN'S BOOK *The Effects of Cross- and Self-Fertilisation* was published in 1876 the *Gardeners' Chronicle* hailed it as a perfect mine of facts for seed-growers and hybridisers, who 'will find, as we have already pointed out, and as we shall have occasion to repeat again and again, that much that was mere haphazard and of a tentative nature in their practices has been by Mr Darwin reduced to rule and method.' The review of the book was continued in no less than seven issues. It was 'an abstract suitable for our purposes' and consisted of the practical applications resulting from Darwin's ;very numerous, protracted, and laborious experiments.' The reviewer was Henslow's son George who, like his father, combined pastoral duties with botany. A popular writer and lecturer he was to become honorary professor to the Royal Horticultural Society.

Protracted Darwin's experiments certainly were. They had been going on for eleven years, though his notes on fertilisation dated back farther. In 1866 when he was studying inheritance for his book on *Variation* he had raised two beds of *Linaria vulgaris*, the yellow-flowered Common Toadflax, one set being the offspring of cross- and the other of self-fertilisation, and he was astonished to see that the offspring of the selfed plants were clearly less vigorous than the others. It seemed to him incredible that this could be due to a single act of self-fertilisation, and it was only in the following year when he saw the same result in a similar experiment with the Wild Carnation, *Dianthus caryophyllus*, that his attention was thoroughly aroused and he decided to make a series of experiments on this new subject.

The date 1866 refers to the announcement to Asa Gray of the start of his experiments. But on page 96 of his first Transmutation Notebook, begun in July 1837, we find him asking: 'Do not plants, which have male and female organs together, yet receive influence from other plants—Does not Lyell give some argument about varieties being difficult to keep on account of pollen from other plants because this may be applied to show all plants do receive intermixture.'

This was the theme of his new book. In the *Fertilisation of Orchids* he had shown how perfect were the means for cross-fertilisation. Now he showed how important were the results.

He started with two plants which happened to be in flower in his greenhouse: *Mimulus luteus*, the Blotched Monkey Flower, and *Ipomoea purpurea*, the climber Morning Glory. Covering them with nets to protect them from insects, he pollinated some flowers on a single plant of both species with their own pollen, crossing others with pollen from the flowers of other mimulus and ipomoea plants. The crossed and selfed seeds from them were sown on opposite sides of the same pot and treated exactly alike, and he always took care that the seeds were thoroughly ripe when he gathered them. When the plants were fully grown he measured and compared them, finding that with both species, as with the linaria and dianthus, the crossed seedlings were conspicuously taller and in every way superior.

Darwin went about his experiments as scientifically as he knew how, his selfing plants always being protected by a net stretched on a frame and large enough to cover the plant (together with the pot, when one was used) without touching it. This was important, for if the flowers touched the net a bee could alight and cross-pollinate them, which he had known to happen, and if the net were wet the pollen could be injured. He used at first white cotton net of a fine mesh, and later net with a mesh of only one-tenth of an inch in diameter, finding that this effectually excluded all insects but the tiny Thrips, which apparently no net could exclude.

Both the crossed and selfed seeds were then placed in damp sand on opposite sides of a glass tumbler and covered by a glass plate, a partition having been put between the two lots. The tumbler was then put on the warm chimney-piece of his study where he could watch the seeds germinating. Sometimes a few would germinate on one side before any on the other, and these he always threw away. But as often as a pair germinated at the same time they were planted on opposite sides of a pot, always with the surface partitioning between them. He was even careful to see that the partition was placed endwise to the light, so that the plants on both sides shared the light equally. He went on until a score of more seedlings of exactly the same age were planted in several pots. Then, if one of the seedlings became sickly or was in any way injured it was pulled up and thrown away, together with its antagonist on the opposite side of the pot.

Having done everything he could to ensure equality, 'I do not believe it possible,' Darwin wrote, 'that two sets of plants could have been subjected to more closely similar conditions, than were my crossed and self-fertilised seedlings.' He made sure that the soil in which they were grown was thoroughly mixed, so as to be uniform. He watered the pairs equally and at the same time. And lest it be thought that the net-dodging Thrips had influenced the results obtained from self-pollination he pointed out that some crossed seedlings would therefore be included among them, and that this would tend to diminish and not to increase any superiority in the average height or fertility of the crossed over the selfed plants.

There was one last precaution. The botanists Gottlieb Kölreuter and Joseph Gärtner had declared that with some plants as many as fifty or sixty pollen-grains were necessary for the fertilisation of all the ovules in an ovary. In France Charles Victor Naudin had found that if only one or two of its very large pollen-grains were placed on the stigma of *Mirabilis* the progeny were dwarfed. Darwin was therefore careful to give an ample supply and generally covered the stigma with it. Then remembering that Gärtner thought an excess of pollen was perhaps injurious, he put it to the test by applying both small and large quantities. After many trials resulting in differences too slight to be of any significance, Darwin concluded that his experiments were not being affected by the amount of pollen, enough having been used in all cases.

In selecting plants for his main work he chose species belonging to widely different families native to different countries all over the world. He did not intend to raise crossed and selfed plants for more than a generation, but as soon as these were in flower he decided to raise one more generation. Then, finding himself more and more fascinated with some of the species, he continued the trials through ten successive generations.

One such experiment, and the first of the long series, was again with his Morning Glory. Ten flowers of this plant were fertilised with pollen from the same flower, and ten other flowers on the same plant crossed with pollen from another Morning Glory. The first part of the experiment was superfluous, because this convolvulus is self-fertile. Darwin knew this but did the trial in order to make the experiments correspond. He found that the average height of the crossed progeny was 86 inches, while that of the selfed progeny was only 65.66 inches, so that the crossed exceeded the selfed in a ratio of 100 to 76.

Flowers on the crossed plants of this generation were now crossed with pollen from different plants of the same generation; and flowers on the selfed plants fertilised with pollen from the same flower. Again every single crossed plant was taller than its competitor, with an average height of 84.16 inches to 66.33 inches, or a ratio of 100 to 79. (The ratios were always given, and at the end of each series of experiments the results were shown in detailed comparative tables. We remember that Darwin was no mathematician: it was his learned and statistically-minded cousin Francis Galton who tabulated the results.)

He went on to deal with third-generation plants, and again all the crossed plants were taller than their competitors. This continued to the tenth generation, the crossed plants being much taller, flowering earlier, and producing more capsules and more seeds in them.

There was an interesting exception in the sixth generation when a self-pollinated plant growing in Pot 2 exceeded the height (though only by half an inch) of its crossed opponent. The victory was won after a long struggle. At first the selfed plant was several inches taller, then when its opponent was four and a half feet high the two were equal. The crossed plant then grew a little taller than the selfed plant but was ultimately beaten by the victorious half-inch. Darwin was so much surprised by this result that he named the plant the Hero and saved its seeds for experiments on its descendants.

He wanted to see whether Hero would transmit his superior growth to the seedlings. Several of his flowers were therefore fertilised with his own pollen, and the seedlings thus raised were put into competition with ordinary selfed plants and with intercrossed plants, all of the same generation. Hero's children proved superior, winning the battle against the ordinary selfed plants by an average of 74.54 inches to 62.58, being likewise victorious over the intercrossed plants with an average height of 88.91 inches to 84.16. Hero's great-grandchildren upheld the family honour!

Yet this again was a side-issue. The main experiments continued to show that the crossed Morning Glories were superior in every way to the selfed plants, averaging 93.7 inches and the selfed only 50.4, with totals of 468.5 and 252.0 inches.

Darwin had raised his original Morning Glory plants from bought seeds, and found that the flowers of this first batch and those of the next few generations varied a great deal in the depth of the purple tint. Many were pink or pinkish, and occasionally there was a white variety. It was interesting that the crossed plants continued to vary in this way to the tenth generation, while all the selfed plants tended to the one colour, a rich dark purple, and by the seventh generation had stabilised this tint.

'My attention was first called to this fact,' Darwin recorded, 'by my gardener remarking that there was no occasion to label the self-fertilised plants, as they could always be known by their colour.' Darwin attributed this extraordinary uniformity of colour to inheritance not having been interfered with by crosses, combined with the fact that they had all been grown under the same conditions.

Selfing experiments over several generations with *Petunia violacea*, *Dianthus caryophyllus* and *Mimulus luteus* had the same result. So, wrote Darwin, 'florists may learn from the four cases which have been fully described, that they have the power of fixing each fleeting variety of colour, if they will fertilise the flowers of the desired kind with their own pollen for half-a-dozen generations, and grow the seedlings under the same conditions.'

For his next series of experiments he took six genera of the Scrophularia family: *Mimulus, Digitalis, Calceolaria, Linaria, Verbascum,* and *Vandellia*, a plant now lost to cultivation.

Though the species name of *Mimulus luteus* means yellow, the plants Darwin raised from seeds he bought varied so much in the colour of their flowers that hardly two plants had them quite alike. They were all shades of yellow blotched with purple, crimson, orange or coppery-brown. In disbelief he sent several specimens to Kew where Joseph Hooker confirmed that they were all *luteus*.

The flowers were plainly well adapted for pollination by insects, as were those of a closely allied species, *Mimulus rosea*, which Darwin was also growing. He watched bees entering the flowers and getting their backs well dusted with pollen, then entering another flower where the pollen was licked off their backs by the two-lipped stigma which closed like forceps on the pollen-grains. But if no pollen was enclosed between the lips, they opened again. A most ingenious mechanism! For if a bee entered a flower with no pollen on its back it touched the stigma which quickly closed. This prevented selfing. It retired dusted with pollen from the stamens, and as soon as it entered the next flower it left plenty of pollen on this stigma, and the flower was thus cross-pollinated. Yet despite this precaution, Darwin found that when he netted plants the flowers fertilised themselves perfectly well and produced plenty of seed!

He bred *Mimulus rosea* through three generations, finding as before that the crossed seedlings were superior in every way. Then among the fourth selfed generation appeared a variety bearing large peculiarly coloured flowers. It grew much taller than the others and proved more highly self-fertile. It was in fact another Hero. Darwin named it the White variety for its large crimson-blotched white flowers.

This fourth generation surprised him, for these plants showed a complete reversal, the crossed plants being inferior in height, though retaining the habit of flowering earlier. Many of the selfed plants were the tall White. Another change showed up in the fifth generation, the selfed plants having become more fertile than the crossed. Again the selfed were decidedly the tallest and finest plants, and most of them were Whites. The pattern was repeated in the sixth generation, and this time every one of the selfed plants bloomed white. This became uniform in the seventh, eighth and ninth generations, the flowers of the crossed plants differing greatly in colour.

Having proved by the seventh generation that the tall White transmitted its characters faithfully, Darwin began a new line of investigation—to see whether intercrossing two selfed plants of the sixth generation would give their offspring any advantage over the offspring of flowers on one of the same plants fertilised with their own pollen. Results proved no reversal, the average height of the intercrossed plants being 9.96 inches and the self-fertilised 10.96.

He now wondered what would be the result of a cross with a new stock, and for this he used plants raised from seed which had come from a garden in Chelsea. Taking their pollen he crossed plants of the eighth selfed generation, and put their seedlings into competition with seedlings of an intercross between selfed plants of the same generation (the eighth), and also with seedlings from selfed plants of the ninth generation. The results were astonishing. The Chelsea-crossed were well ahead of the others, with a ratio to the intercrossed of 100 to 56, and to the selfed of 100 to 52.

The Chelsea-crossed plants had begun to show their superiority when only one inch high. When fully grown they were much more branched, with larger leaves and larger flowers. They were also extraordinarily fertile, producing 272 capsules to the 24 of the intercrossed and the 17 of the selfed plants. The difference in the weights of seed was almost unbelievable, the Chelsea-crossed and intercrossed having a ratio of 100 to 4; the Chelsea-crossed and selfed seeds 100 to 3, compared with the 100 to 73 of the intercrossed and selfed.

He put the plants to the severest tests. Early in the autumn he bedded them out in open ground, a trial of survival for any plant kept in a warm greenhouse. All three lots suffered greatly, but the Chelsea-crossed much less. On the 3rd of October the Chelsea-crossed began to flower again and continued to flower for some time. Not a single flower appeared on the two other lots. Their stems wilted almost to ground-level and they seemed half dead. Early in December there was a sharp frost, this time bringing down the stems of the Chelsea-crossed. But on the 23rd of December they began to shoot up again from the roots, while the other two lots were quite dead. The fact that emerged from the Chelsea experiment was, of course, that the original stock had benefited by the introduction of 'new blood' from other plants grown under different conditions—in other words, as we know now, new genetic variability.

No one watching a bumble-bee crawl up into the flowers of a Foxglove would suspect what complicated mechanisms are prepared for it inside. It is important for the Foxglove to avoid self-pollination, and it avoids it in this way: the two upper and longer stamens (standing close to the stigma and therefore the most likely to pollinate it) shed their pollen on to the bee's back before the two lower and shorter stamens. The Foxglove has another precaution: the pollen is mature and mostly shed before the stigma of the same flower is ripe for pollination. The large pollen-producing anthers stand at first crosswise, and if they were to dehisce in this position (that is, open to discharge the pollen) they would uselessly smear it on the back and sides of the bumble-bee. But before they dehisce, the anthers twist round and place themselves longwise.

Yet another safety device is employed by the Foxglove. The lower and inner side of the entrance to the flower is thickly carpeted with hairs, and these collect so much pollen that bees Darwin watched were copiously dusted with it. But this pollen can never be applied to the stigma, since retreating bees do not turn their under-surfaces upwards. Darwin was at first puzzled as to whether these hairs were of any use. The explanation was that smaller kinds of bees are not fitted to pollinate Foxglove flowers, and if they were allowed to enter easily they would steal the nectar and fewer bumble-bees would visit the flowers. The big bumble-bee can crawl into the flowers with the greatest ease, and he uses the hairs as footholds while sucking the honey. Smaller bees are impeded by the hairs; and even if they succeed in struggling through them, when they reach the slippery precipice above the nectary they are completely baffled.

Darwin now tried to baffle the Foxglove. This was on a visit to Barmouth in June 1869 where he had taken his family for a holiday and to recuperate from a bad fall from his usually quiet cob Tommy. Caerdeon, the house where they stayed, was on the north shore of the beautiful Barmouth estuary, pleasantly close to wild hill country behind, and to the picturesque wooded hummocks between the steeper hills and the river. It was ideal terrain for Foxgloves. He found groves of them, and covering one plant with a net he dusted the stigmas of six flowers with their own pollen and six others with pollen from another plant growing a few feet away. Occasionally he shook the netted plant violently to simulate a breeze and facilitate self-pollination. This plant had ninety-two flowers (besides the dozen he had pollinated), and of these only twenty-four produced capsules, whereas almost all the flowers on the uncovered Foxgloves growing nearby were fruitful. Of his twenty-four selfed capsules only two were full of seed, six contained a moderate amount and the remaining sixteen very few seeds. Darwin thought that a little pollen adhering to the anthers after they had dehisced, and accidentally falling on to the stigmas when mature, must have partially self-pollinated the twenty-four flowers. He had good reason to think so, for the margins of Foxglove flowers in withering do not curl inwards, nor do the flowers in dropping off screw round on their axes—either of which would bring the pollen-covered hairs into contact with the stigma and so effect full self-pollination.

At the end of July when they returned to Down House he took with him ripe capsules from the crossed and netted plants, germinating the seeds on sand and planting the seedlings as usual in pairs on the opposite sides of five pots in his greenhouse. After a time they appeared to be starved, so without disturbing them he turned them out and planted them in open ground in two close parallel rows. Their leaves were then between five and eight inches long. Comparing the longest leaf on each side of each pot, those of the crossed plants were longer than the leaves of the selfed plants by an average of 0.4 of an inch. In the following summer the tallest flower-stem on each plant was measured when fully grown. There were seventeen crossed plants, of which one had not produced a flower-stem. Nine of the original selfed plants had died during the winter and spring, leaving only eight to be measured, but the average height of the flower-stem for the crossed plants was 51.33 inches, that of the eight selfed plants only 35.87.

'But,' said Darwin, 'this difference in height does not at all give a fair idea of the vast superiority of the crossed plants.' These had produced sixty-four flower-stems, an average of four to each plant, while the selfed

The Yucca, whose fantastic partnership with the Pronuba moth Darwin described in *Cross- and Self-Fertilisation*

plants produced only fifteen, with an average of 1.87 on each plant, and as usual they were less luxuriant.

After more experiments on the Foxglove had proved that even intercrossing flowers on the same plant produced better results than selfing, Darwin turned to other plants, finding again and again that the general law of crossing held good.

This possibility of cross-pollinating depended of course on the co-operation of insects, often of insects belonging to special groups, and sometimes of one particular special insect. There was the wonderful partnership between the yucca and the tiny Pronuba moth called *Tegiticula yuccasella*. No other insect, not even another kind of moth, can pollinate the yucca.

The story is a fantastic one of split-second timing. 'A wonderful instinct,' wrote Darwin, 'guides this moth to place pollen on the stigma, so that the ovaries may be developed on which the larvae feed.'

It is at the moment of the moth's emergence from its chrysalis in the ground that the creamy bells of the yucca flower begin to open. The little Pronuba now spreads its glistening white wings and flies about till it finds a mate. It is a cruel mating. Soon after, the male dies. The female catches the heavy fragrance of a now open flower and there it collects pollen from the anthers. With this it flies to another flower, crawls down and lays its eggs within the ovarium, runs up to the stigma and thrusts the pollen which it has chewed into a ball into the stigmatic opening, thus effecting cross-pollination. Inside the ovarium the hatched larvae feed on the swelling seeds, but they do not damage more than a few, twenty out of an average 200 in a seed-pod, small payment for the service which the moth has rendered to its partner. For apart from a nursery for its eggs and caterpillars and a moiety of sustenance for them the Pronuba seeks no other reward. It does not ask for honey: its mouth is not adapted for sucking but solely for shaping the pollen ball. In Britain where there are no Pronubas and where no other insect has learnt this complicated ritual, we must—if we would have flowers on our plants year after year—imitate the little Pronuba by transferring pollen from one flower to another. Otherwise we may well believe the legend that the yucca blooms only once in a span of years.

Most insects on a flight, as Darwin found, generally keep to the same species of a plant.

'Humble and hive-bees are good botanists,' he wrote, 'for they know that varieties may differ widely in the colour of their flowers and yet belong to the same species. I have repeatedly seen humble-bees flying straight from a plant of the ordinary red *Dictamnus fraxinella* [the Burning Bush] to a white variety; from one to another very differently coloured variety of *Delphinium consolida*, and of *Primula veris*; from a dark purple to a bright yellow variety of *Viola tricolor*; and with two species of *Papaver*, from one variety to another which differed much in colour; but in this latter case some of the bees flew indifferently to either species, although passing by other genera, and thus acted as if the two species were merely varieties.'

He had shown that cross-fertilisation takes place within the same species and that only rarely is crossing effected between two separate species. It is in this way that species have been kept constant.

If you look at the arum commonly called Lords-and-Ladies or Cuckoo-pint, you will see inside the hooded spathe a circle of filaments surrounding the spear-like spadix. It had been thought by botanists that flies entering the spathe were trapped by these hairs and that they never escaped. But as this was not a bee-flower, and as only minute dipteran flies seemed to be its sole visitors, how could pollen be carried from one arum to another if on a first visit the insects were sentenced to life-imprisonment?

As long ago as the spring of 1842 Darwin had examined several spathes, and finding from thirty to sixty Diptera of three species in some of them (and many flies lying dead at the bottom, which seemed to bear out what the botanists had said) he tied a fine muslin bag tightly round one of the spathes. Returning in an hour's time he found several little flies crawling about inside the bag.

'I then gathered a spathe and breathed hard into it,' he wrote. 'Several flies soon crawled out, and all without exception were dusted with arum pollen. These flies quickly flew away, and I distinctly saw three of them fly to another plant about a yard off. They alighted on the inner or concave surface of the spathe, and suddenly flew down into the flower. I then opened this flower, and although not a single anther had burst, several grains of pollen were lying at the bottom, which must have been brought by one of these flies or by some other insect. In another flower little flies were crawling about, and I saw them leave pollen on the stigma.'

But before the insects left the flower a great deal had been happening inside it. Most flowers give off a scent which attracts their own particular kind of insect visitor. The arum exudes a smell like rotting meat to attract carrion-feeding flies like the Diptera, and when the flower is ready to be pollinated it is the job of the fleshy spadix to heat up like an electric element and spread this odour. In answer to this seemingly juicy invitation a fly alights, and the moment it touches the inside of the deep cup it finds itself cascading to the bottom on a slide which tiny droplets of oil have made as slippery as a skating rink. Half-way down is the ring of stiff hairs to prevent bigger insects crawling down the column. If the little Diptera tried to escape now by this staircase, the hairs would form a barrier. As it is, the insect shoots through them and finds at the bottom of the cup a circle of female flowers exuding drops of nectar. Above is a circle of male flowers, and while the Diptera is feasting the stamens on these flowers open and release a shower of pollen. Soon after, the bristly hairs that have barred the insect's way to freedom begin to soften, making the way easy now for it to crawl up the middle part of the flower and fly away. It goes immediately to another arum flower, for it cannot resist that rotting smell—and finds itself in a new prison. But this time it leaves on

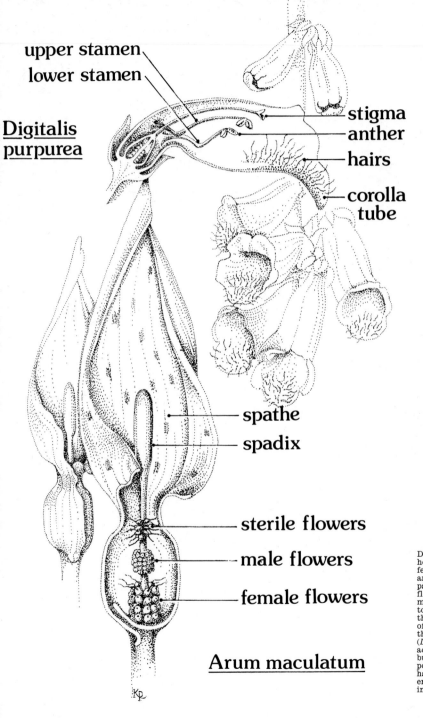

upper stamen

lower stamen

Digitalis purpurea

stigma

anther

hairs

corolla tube

spathe

spadix

sterile flowers

male flowers

female flowers

Arum maculatum

Darwin discovered how the arum is fertilised. It was yet another partnership—the flower imprisoning midges until they touched off the spring that releases a shower of pollen. And he found that the Foxglove (*Digitalis purpurea*) accepts only the bumble-bee as its pollinator, the rough hairs preventing the entrance of smaller insects the stigma the pollen it has been showered with by the first flower. The dead flies Darwin saw must have feasted so greedily that they were too exhausted to climb up the stairway to freedom.

Darwin was very fond of the little blue lobelia which was often used as a bedding plant in those days, and pondering the question of how bees recognise the flowers of the same species, and believing that the coloured corolla was their chief guide, he thought he would play a trick on them. On a fine day when honey-bees were continually visiting the little blue flowers he cut off all the petals of some, and only the lower striped petals of others, and waited to see what would happen. These flowers were not once again sucked by the bees, although some actually crawled over them. The removal of only the two small upper petals made no difference.

Quite the contrary happened to some large plants of the Dusky Cranesbill: flowers which had shed all their petals were still secreting an abundance of nectar and being visited by bumble-bees. Darwin thought that if the bees were attracted to these flowers, they might have learnt that the denuded lobelias were equally worth visiting.

Some that visited Darwin's kitchen garden were more cunning. In the summer of 1857 when he was working on the fertilisation of the Scarlet Runner Bean he saw bees busy sucking at the mouths of the flowers. 'But one day,' he tells us, 'I found several humble-bees employed in cutting holes in flower after flower; and on the next day every single hive-bee, without exception, instead of alighting on the left wing-petal and sucking the flower in the proper manner, flew straight without the least hesitation to the calyx, and sucked through the holes which had been made only the day before by the humble-bees; and they continued this habit for many following days.' Till the nectaries were sucked dry. A bee would often meet a dry nectary on its daily rounds. Darwin found by timing them that these short-cuts enabled it to visit twice as many flowers.

Other flowers such as *Stachys coccinea, Penstemon argutus* and *Salvia grahamii* were perforated by bees, as he observed in 1841 and 1842 on visits to Maer, but only where there were large beds of them. The booty was rich and attracted many bees; competition was fierce. Two plants of the *Stachys* growing separately were not perforated.

The extraordinary industry of bees and the number of flowers they visited in a short time, so that each flower was visited repeatedly by a succession of bees, must, Darwin saw, greatly increase the chance of each receiving pollen from a different plant. When the nectar was hidden they could not tell without inserting their long tongues whether it had been taken by other bees, and this forced them to visit many more flowers than they otherwise would. But they tried to lose as little time as possible. In flowers having several nectaries, if they found one dry, they did not try the others but, as he often observed, passed on to another flower. Even where hundreds of thousands of flowers grew together—in a group of heathers, for instance—every single flower was visited. So quickly they flew from flower to flower that Darwin decided to time them. He found that bumble-bees fly at the rate of ten miles an hour.

Investigating the Common Vetch (*Vicia sativa*), he examined the glands of its stipules (the outgrowths at the base of the leaf-stalk). These glands secrete minute drops of nectar beloved by sugar-loving ants, honey-bees and wasps, and Darwin repeatedly observed that as soon as the sun was hidden behind clouds the secretion ceased and the bees left the field, and that as soon as the sun broke out again they returned to their feast. He was therefore able to suggest that primarily the sugar in nectar was excreted as a waste product of chemical changes in the sap, dependent on the sun shining brightly. This view was disputed, but when Dr Maxwell Tylden Masters, a distinguished botanist and editor of the *Gardeners' Chronicle*, heard the subject discussed at a meeting of the Royal Horticultural Society, he told Darwin that he had no doubts about it.

It was in these ways, letting no detail escape his notice, even in a realm of what others might consider to be a side-issue, that the fact-finding Darwin accumulated the pile of evidence from which he deduced and proved. Truth to him was the whole, the sum of each why and wherefore. He was interested in the secretion of nectar, for when it happened within a flower it was used in the important work of cross-pollination.

There were, of course, exceptions; the hermaphrodites having both male and female organs in the same flower were therefore able to

(*below left*) *Limnanthes douglasii*, the 'Poached Egg Plant', which Darwin found was self-pollinating, despite the fact that it attracts bees. But when crossed by bees it flowers earlier

(*below*) Similarly with the Borage. Keeping a count of bees visiting *Borago officinalis*, Darwin found that more frequented it than almost any other plant he observed. But it was capable of self-pollination, should the bees fail, though yielding only a quarter of the number of seeds

pollinate themselves without the help of insects. Yet even they retained traces of having been formerly adapted for cross-pollination. Flowers bloomed in the ancient world before the advent of winged insects, ready when the wind blew to release a shower of pollen, some of which must land on the stigmas of other plants of the species. Conifers and cycads were among the earliest plants, Darwin recalled, and they had kept this habit, while the common rhubarb is in an intermediate state, visited by tiny Diptera flies which come away covered with pollen, yet the flower-head if gently shaken releasing a cloud of it.

Darwin never lost sight of the evolutionary process. To him it was the key to the plant's basic structure, and it was a general law of nature that flowers were adapted to be crossed by pollen from a different plant. He wrote: 'These exceptions need not make us doubt the truth of the above rule, any more than the existence of some few plants which produce flowers and yet never set seed should make us doubt that flowers are adapted for the production of seed and the propagation of the species.'

In his Introduction to *Cross- and Self-Fertilisation*, summing up what he hoped to prove, Darwin wrote: 'The most important conclusion at which I have arrived is that the mere act of crossing by itself does no good. The good depends on the individuals which are crossed differing slightly in constitution, owing to their progenitors having been subjected during several generations to slightly different conditions.'

The benefit derived from slight changes in the conditions of life stands, he pointed out, in the closest connection with life itself. 'It throws light on the origin of the two sexes and on their separation or union in the same individual, and lastly on the whole subject of hybridism, which is one of the greatest obstacles to the general acceptance and progress of the great principle of evolution.'

By the end of the book he had removed all doubt.

16

Legitimate and Illegitimate Marriages

Mars ♂ Ferrum

Mercury ☿ Argent. viv.

Venus ♀ Cuprum.

Medieval alchemists used the planetary symbols as a kind of shorthand to signify certain metals: ♂ for Mars and iron; ♀ for Venus and copper, and ☿ for Mercury and quicksilver.

Linnaeus transferred them to biology, in 1751 when discussing hybrid plants denoting the supposed male parent species by the sign of Mars, the female by that of Venus, and the hermaphrodite by Mercury.

It was Darwin who found that some plants of the same species having two or even three different forms—long-styled, short-styled, mid-styled—must have reciprocal marriages with plants having stamens of the same height, in order to obtain perfect fertility.

His work on this subject had a special bearing on the origin of species, and it threw a new light on hybridism.

'NO LITTLE DISCOVERY of mine,' Darwin wrote in his *Autobiography*, 'ever gave me so much pleasure as the making out the meaning of heterostyled flowers.'

He was speaking of his book *The Different Forms of Flowers on Plants of the Same Species* which was published in 1877. As usual it was the outcome of many years of observation and an almost world-wide correspondence to check facts he could not verify in England. So, in preparing the ground, went letters to Roland Trimen at Cape Town, the entomologist and botanist; to George Henry Kendrick Thwaites, director of the Peradeniya botanic garden in Ceylon, to John Traherne Moggridge at Menton, to William H. Leggett in New York, to his old faithfuls Asa Gray, Fritz Müller in Brazil and his brother Hermann at Lippstadt, and of course to Hooker.

Many things had been happening to Joseph Hooker. Since 1865 when his father died he had been in command at Kew, and one of his first thoughts was to bring the Gardens into line with the more modern botanical establishments in Europe. His friend Thomas Jodrell Phillips Jodrell donated a sum sufficient to build a properly equipped experimental laboratory. It was opened in 1876 and William Thiselton Dyer put in charge of it. In the same year Joseph Hooker remarried: Frances, tragically, had died two years before, leaving six motherless children to be looked after by twenty-year-old Harriet, and Darwin immediately offered to take in the whole family. Three years later Harriet married Thiselton Dyer. The two held the fort at Kew while the newly knighted Sir Joseph went off to America to join Asa Gray on a botanical expedition in the Rockies.

In 1875 Sir Charles Lyell had died, to Darwin as to Hooker a sad departing of a great and good friend. The two interchanged letters about a suitable memorial. 'When I think,' Darwin wrote, 'how Lyell revolutionised Geology, and aided in the progress of so many other branches of science, I wish that something could be done in his honour.' Lyell was buried in Westminster Abbey.

There were changes in the Darwin household, with a baby in the old nurseries, for Frank lost his wife in the autumn of 1876 after only two years of marriage and came to live again at Down House with his little son Bernard who was a delight to Emma as he was to his grandfather.

What a year was 1877! William was married to a charming American, Sara Sedgwick. In February on the occasion of his father's sixty-eighth birthday came a magnificent folio album bound in velvet and silver, containing the photographs of 154 men of science in Germany, and with it from Holland an album with the photographs of 217 distinguished professors and lovers of science. The handsome title-page of the German album bore the inscription: *Dem Reformator der Naturgeschichte,*

CHARLES DARWIN. The students at Edinburgh asked him to allow himself to be nominated as candidate for the Lord Rectorship. He declined for health reasons, but he was present at the University of Cambridge on the 17th of November when the honorary degree of LL.D. was conferred upon him. He was very touched by the honour, which Emma and Bessy, George, Horace and Lenny watched him receive. Emma, despite a 'baddish headache', enjoyed it all. 'I felt very grand walking about with my LL.D. in his silk gown,' she wrote to William.

It was fitting that 'the distinguished Master of Natural Science' should become a Doctor of Laws, he who had established so many of them.

Different Forms of Flowers was another of Darwin's *pièces justicatives*. In the *Origin* he had given a tantalisingly brief outline of the subject, following this up with five papers read at the Linnean Society between 1861 and 1868. The subject became so important to him that he greatly expanded the material, adding the results of observations on new cases of plants having two and three different kinds of flowers. The work had a special bearing on the problem of the origin of species. It threw a new light on hybridism.

Several plants belonging to distinct orders present two forms. A casual glance at their flowers reveals no difference, but close examination shows that one form has a long style with short stamens, the other a short style with long stamens. The two have differently sized pollen-grains. Darwin called the two-formed kind 'dimorphic'. There is a third form called trimorphic whose three forms likewise differ in the lengths of their styles and stamens and in the size and colour of the pollen-grains; and as in each of the three forms there are two sets of stamens, they possess between them six sets and three kinds of styles. These organs are so proportioned in length to each other that half the stamens in two of the forms stand on a level with the stigma of the third form.

So that these plants can obtain full fertility it is necessary that the stigma of the one form should be pollinated by pollen taken from the stamens of corresponding height in another form. With dimorphic species two unions are fully fertile, and these Darwin called legitimate marriages; and two, which he called illegitimate, are more or less infertile. With trimorphic species six marriages are legitimate and twelve are illegitimate.

It had long been known to botanists that the common cowslip had two forms, but the difference was regarded as mere variability. Florists who cultivated the primula and auricula were also aware of the two kinds of flowers, and they called the ones displaying the globular stigma at the mouth of the corolla 'pin-eyed', and those displaying the anthers 'thrum-eyed'—from the likeness to the weaver's thrums or ends of threads. 'Village children notice this difference,' Darwin wrote, 'as they can best make necklaces by threading and slipping the corollas of the long-styled flowers into one another.' This they did with cowslips or primroses, the first flowers Darwin examined, and he noted that the two forms never

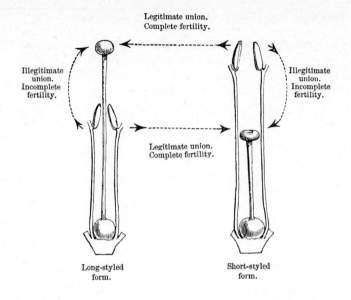

Legitimate union.
Complete fertility.

Illegitimate
union.
Incomplete
fertility.

Illegitimate
union.
Incomplete
fertility.

Legitimate union.
Complete fertility.

Long-styled
form.

Short-styled
form.

Legitimate and
illegitimate marriages
of the dimorphic
Primrose

appeared on the same plant but that each form was constant on its own plant year after year.

But what was the meaning of all this? The first thought that occurred to Darwin was (an idea he later dropped) that *Primula* was tending towards a dioecious condition, that is, having the male and female organs on different plants, for it seemed to him that the long-styled flower with its rougher stigma and smaller pollen-grains was more feminine in nature and would produce more seed; and that the short-styled flower with its longer stamens and larger pollen-grains was more masculine. He began his experiments in 1860, marking a few cowslips of both forms growing in his garden, some growing in an open field and others in a shady wood, and gathered and weighed the seed. In all the lots, contrary to his expectations, it was the short-styled or 'male' plants that produced far more flower-heads and yielded the most seed. By all standards of comparison therefore the short-styled form was more fertile.

In the following year he carried out another trial. During the previous autumn he had transplanted a number of the wild plants into a large bed in his garden. They were now growing in good soil instead of a shady wood or struggling with other plants in the open field, so it was not surprising that they produced more seed. Nevertheless the short-styled plants still produced more seed than the long-styled, and the same result held good in the next two years. Professor Daniel Oliver at Kew found himself co-opted into the experiments, and he discovered that the ovules in unexpanded long-styled flowers were considerably larger: hence the fewer seeds.

Darwin at the same time was doing other trials with a few selected cowslips, netting both forms. Artificially pollinated ones produced an abundance of seed. The others did not. The same experiment with ten pots of polyanthuses and cowslips of both forms in his greenhouse, all netted,

(*left*) The Primrose (*Primula vulgaris*)

(*below left*) The Oxlip (*Primula elatior*)

(*below right*) The Cowslip (*Primula veris*)

Botanists thought the Oxlip to be a cross between the Primrose and Cowslip. Darwin found that it bred true and was therefore a species on its own did not set one pod. This proved that the visits of insects were necessary for fertilisation.

In artificially pollinating the cowslips Darwin worked through all the permutations of legitimate and illegitimate marriages. The triumphs of the first over the second left not the least doubt of their superiority, 'and we have here a case to which no parallel exists in the vegetable or, indeed, in the animal kingdom,' he wrote. The primula was divided into two types which could not be called distinct sexes, for both were hermaphrodites. Yet each required reciprocal union for perfect fertility. As with the two sexes of quadruped animals, here were hermaphrodite plants related to each other like males and females! Darwin pointed out that there were many hermaphrodite animals such as snails and earthworms, and plants other than primulas, which could not fertilise themselves but must unite with another hermaphrodite.

He next experimented with the Oxlip, *Primula elatior*, a plant rare in Britain even in his day and still to be found only in some parts of East Anglia, though common in Europe. It was thought to be a variety and not a true species. Living plants were sent to him by Henry Doubleday who wrote about them in *The Phytologist* and first called attention to their existence in England. Darwin called it the 'Bardfield Oxlip of English Authors', so as not to confuse it with the one then called the 'Common Oxlip' but which Darwin found to be a hybrid between the primrose and cowslip. The various marriages of the Bardfield oxlip proved much more sterile than those of the cowslip, but, breeding true, it declared itself to be a true species.

In 1862 Darwin was corresponding with John Scott who was then head propagator at the Royal Botanic Garden, Edinburgh. The young man's letters showed that he was a wonderful observer, and Darwin thought of employing him to work out various problems connected with inter-crossing. This is why, when he was investigating the Common Primrose (*Primula vulgaris*), we read that 'Mr J. Scott examined 100 plants growing near Edinburgh.' They shared (by post) many experiments, on the primrose and on other plants, until 1864 when Scott who was unhappy under James McNab, the curator, went to Calcutta to take charge of its botanic garden. Joseph Hooker secured him the post and Darwin paid his passage. The Scott–Darwin experiments on the primrose proved once more that illegitimate marriages were not nearly so fertile.

Both the cowslip and the primrose depended for complete fertilisation on insects, the cowslip being visited during the day by the larger bumble-bees and at night by moths, the primrose depending almost entirely on moths. Their flowers had a different scent. Both forms of the primrose when legitimately and naturally pollinated yielded twice as much seed as the cowslip. When illegitimately pollinated they were likewise more fertile than the cowslip's two forms, and when netted the long-styled primrose yielded a large amount of seed and the long-styled cowslip not a single one. When Darwin inter-crossed them he experienced considerable difficulty (which is usually the case when species are crossed with species), and a high proportion were sterile. This was trustworthy evidence that the cowslip, primrose, and Bardfield oxlip were distinct species—not, as the botanists thought, varieties of the same species.

He was always ahead of the botanists, though this, far from displeasing them, commanded their deep respect. Word had only to go round that he was studying a new subject and information or parcels of plants arrived at Down House. In 1876 he was writing to Daniel Oliver: 'You must be a clair-voyant or something of that kind to have sent me such beautiful plants. Twenty-five years ago I described in my father's garden two forms of *Linum flavum* (thinking it a case of mere variation); from that day to this I have several times looked, but never saw the second form till it arrived from Kew.'

It was known that several species of *Linum* were dimorphic, but this had been overlooked in the *grandiflorum* species until Darwin discovered this and other interesting facts about it. His son William was always to be relied on for collecting, and from the Isle of Wight came 202 plants. Examining the beautiful rose-coloured flowers Darwin could easily distinguish the two forms. For although foliage, corolla, stamens and pollen-grains were alike, the five stigmas in the short-styled forms lay inside the tube of the corolla, while those in the long-styled stood nearly upright and level with the anthers of the stamens.

In 1861 he began growing them in his garden. Of eleven plants three were short-styled and eight long-styled, two of the latter growing in a bed a hundred yards from the others and separated from them by a screen of evergreens. Experimenting with these two he marked twelve flowers, placing on their stigmas a little pollen from the short-styled plants. The stigmas of the long-styled flowers were already blue with their own pollen and it was late in the season, the 15th of September. 'It seemed almost childish to expect any result,' wrote Darwin. He certainly did not expect the result he obtained, for the ovaries of the twelve flowers all swelled and six fine capsules were produced, the seed of which germinated the following year. In that summer the two original plants produced a vast number of flowers, but they all proved absolutely barren. Of the nine other plants, the short-styled were more fertile with their own pollen than were the long-styled.

After many more experiments carried out in his garden and in his greenhouse, a remarkable fact emerged. Although the pollen-grains of the two forms were indistinguishable even under the microscope and the stigmas differed materially only in length, yet they were widely dissimilar in their mutual reaction '—the stigmas of each form being almost powerless on their own pollen, but causing, apparently by simple contact (for I could detect no viscid secretion), the pollen-grains of the opposite form to protrude their tubes.' He went on: 'It may be said that the two pollens and the two stigmas mutually recognise each other by some means.'

This was yet another case of Darwin's insight into the physiology of plants. He could realise that here were two types of pollen, one of which was acceptable to the plant and the other unacceptable. It was the mysterious mutual recognition 'by some means' that, in 1943, led Ronald A. Fisher and Kenneth Mather (head of the Genetics Department at the John Innes Institute) to take on Darwin's experiments from that point. They found that pollen of a similar genotype would be successful in growing down the style and fertilising the ovules. This they called the Incompatibility system. The plant required 'pollen-grains of the opposite form', as Darwin had pointed out.

'Botanists,' he wrote, 'in speaking of the fertilisation of various flowers, often refer to the wind or to insects as if the alternative were indifferent. This view, according to my experience, is entirely erroneous.' He had noticed that when the wind was the pollen-carrier the flowers were markedly different from those that relied on insects. Conifers, rhubarb and spinach produced enormous quantities of fine dusty pollen: it blew from them in clouds, and in plants like grasses and docks the stigmas were downy or feathery so as to catch chance-blown grains. Their flowers did not secrete nectar and they were not brightly coloured to guide insects to them. Those which relied on insect partners were not only brightly coloured for day visitors, pale-coloured or white to attract night-flying moths, they had an endless number of adaptations to ensure the safe transport of their pollen by living workers. *Linum perenne* had a very special adaptation.

In the three linums he examined—*grandiflorum, flavum* and now *perenne*—both forms had their stigmatic surfaces facing the centre of the flower. But this was true of the long-styled *perenne* only while it was in bud. By the time the flower was fully open the five stigmas had twisted round to face the outside. Insects were attracted to all these linums by five drops of nectar at the base of the stamens, so that to reach the honey they had to insert their tongues outside the ring of filaments and between them and the petals. In all the short-styled forms (whose stigmas faced the centre of the flower), if the styles had retained their upright position the insects would not have been able to dust off any pollen. As it was, when the plant was ready for fertilisation, their styles bent downwards till the stigmas were lying within the tube of the corolla. Every insect seeking nectar would now brush off the pollen.

In the long-styled forms the anthers and stigmas of *grandiflorum* stood directly above the space leading to the nectar; in *flavum* the pistil was nearly twice as long as in the short-styled. So the flowers of both were easily pollinated.

'In the case of *Linum perenne*, affairs are arranged more perfectly,' Darwin saw, for the stamens in the two forms stood at different heights, so that pollen from the anthers of the longer stamens would adhere to one part of an insect's body, afterwards to be brushed off by the rough stigmas of the longer pistils; while pollen from the anthers of the shorter stamens would adhere to a different part of the insect's body, to be brushed off by the stigmas of the shorter pistils: and this was what was required for the legitimate marriage of both forms.

Thus Darwin taught the flower-breeder science. By showing how bees pollinated the different forms of these flowers, the breeder could emulate the process and procure a harvest of good seed, as of course, could the ordinary gardener.

After experimenting with many more dimorphic flowers Darwin turned to those with three forms. It was the Rev. M. Vaucher of Geneva who first observed trimorphism, and later the German botanist Wirtgen. 'But,' said Darwin, 'these botanists, not being guided by any theory or even suspicion of their functional differences, did not perceive some of the most curious points of difference in their structure.' He explained these with the help of the diagram reproduced here. It shows the three forms of *Lythrum salicaria*, the Purple Loosestrife, in their natural position but with the petals and calyx on the near side removed, and magnified six times.

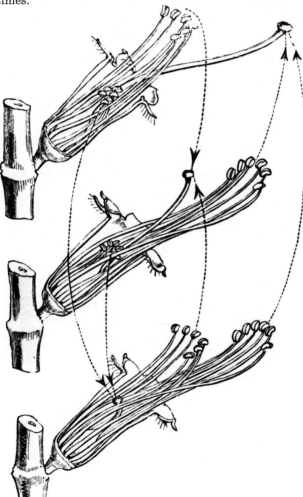

The trimorphic *Lythrum salicaria*, the Purple Loosestrife, showing its three forms of marriages, legitimate and illegitimate

'In their manner of fertilisation,' Darwin wrote, 'these plants offer a more remarkable case than can be found in any other plant or animal.'

Remarkable it certainly was. 'Nature,' wrote Darwin, 'has ordained a most complex marriage-arrangement, namely a triple union between three hermaphrodites,—each hermaphrodite being in its female organ quite distinct from the other two hermaphrodites and partially distinct in its male organs, and each furnished with two sets of males.'

In the diagram the dotted lines with the arrows show the directions in which pollen must be carried to each stigma to ensure full fertility. Darwin had to make eighteen different unions to ascertain the fertilising power of three forms, six legitimate and twelve illegitimate. Again he found that the permitted marriages won handsomely over the clandestine ones. From his experiments with the other plants, this could have been expected, but Darwin's tables of results also showed the remarkable fact that the sterility of the marriage increased as the distance between the stigma and the stamens increased. There was no exception to this rule.

This is a brief account of his experiments on lythrums. At the end he was able to write: 'We must look exclusively to functional differences in the sexual elements as the cause of the sterility of the species when first crossed and of their hybrid offspring. It was this consideration which led me to make the many observations recorded in this chapter, and which in my opinion make them worthy of publication.'

The purpose of heterostyly was to ensure cross-fertilisation. The same end was gained in other kinds of plants by different means—by the maturing of the pollen and stigma of the same flower at different periods, by the male and female organs being on different plants, by self-sterility, by the structure of the flower in relation to the visits of insects, and by what Darwin called the prepotency of pollen from another individual over a plant's own pollen—by this inferring that only foreign pollen was acceptable to a plant, and we have seen what Fisher and Mather developed out of his penetrating insight.

Plants well adapted for cross-pollination by insects were often those with an irregular flower like the antirrhinum, and Darwin pointed out that it would have been of little or no use to such plants to have become heterostyled. Lipped plants like peas and beans, salvia, Catmint and lavender, orchids and Foxgloves, all have irregular flowers (that is, irregular in shape), and of these not a single species is heterostyled, with the exception of the genus *Pontederia*, of which the aquatic Pickerel Weed is a member.

Why various plants should have different characteristics was always of interest to Darwin, for in them might be read their long lineage from the primordial past. Rudimentary and abortive organs are a strong clue to this. There are many groups of species and single species allied to hermaphrodites, the female flower showing rudiments of male organs and male flowers showing rudiments of female organs, and Darwin believed these to be descended from plants which formerly had the two sexes combined in the same flower. How and why had such hermaphrodites become bisexual? He answered his own Why easily. If stamens were to abort, females and hermaphrodites would be left, and there were many instances of this; and if the male organs of the hermaphrodite were then to abort, the result would be a dioecious plant. Similarly, if the female organs aborted, males and hermaphrodites would be left, and the hermaphrodites might then be converted into females. He could see that if a species were subjected to unfavourable conditions because of severe competition with other plants or from any other cause, the fact of a plant having to play the parts of both male and female and also mature the ovules might prove too great a strain for it. The separation of the sexes would then be highly beneficial.

In the United States several varieties of strawberry were trimorphic and tending to separate their sexes. The females, which had small flowers, produced a heavy crop of fruit. The hermaphrodites, which had mid-sized flowers, 'seldom produce other than a very scanty crop of inferior and imperfect berries', as Darwin knew from Leonard Wray, a grower; while the males, which had large flowers, produced no berries. But the females produced no runners, the two other forms many, and they consequently increased rapidly and tended to supplant the females. Darwin thought this strong tendency to sex-separation, much more common than in Europe, to be the result of the direct action of the climate on the reproductive organs. To get over the difficulty, skilful American growers planted seven rows of female plants, then one row of hermaphrodites, and so on throughout the field.

Darwin described various other hermaphrodite plants that were tending to become dioecious. From abroad his observers were sending him back information. At Harvard, Asa Gray was studying *Rhamnus lanceolatus* for him. John Traherne Moggridge was examining the Common Wild Thyme at Menton, Thwaites the brambles in Ceylon, and Fritz Müller a species of *Aegiphila*, one of the verbenas. At home he had quite a battalion, including William who was measuring the pollen of pulmonarias and 'by the help of drawing or rather making marks under Camera lucida' counting the grains.

In 1864 Darwin discovered a new kind of dimorphism. Asa Gray had written about dimorphic plantains, and because these had only female and hermaphrodite forms Darwin thought they must belong to some different class. 'How could the wind,' he asked, 'which is the agent of fertilisation with Plantago, fertilise "reciprocally dimorphic" flowers like Primula? Theory says this cannot be.' For the wind to place pollen from a long stamen neatly on to a long-styled stigma, or from a short stamen on to a short-styled stigma would be a much too chancy process. This required the manoeuvrings of a bumble-bee. He called the new class 'gyno-dioecious', and, since his theory proved correct, it was accepted by the botanists.

He began hunting round for other plants which might be gyno-dioecious. Mints, he found, were of this class, Hyssop, Catmint and Balm. Another was the Wild Thyme, the first of the class he had noticed. When he was on holiday near Torquay (the '8 weeks & a day' of 1861) he could, 'after a little practice, distinguish the two forms while walking quickly past them.' The females had fewer and rather smaller flowers and very long stamens. So added to Darwin's remarkable powers of observation was this remarkable acuity of vision. His brother Erasmus once referred to 'those telescopes you call eyes'. Finding a large hermaphrodite and a large female of nearly the same size, he gathered all the flower-heads when the seeds were ripe and took them home and grew them. The female seed weighed more than twice as much as the seed from the hermaphrodite, and this was the same with other gyno-dioecious species.

Few people perhaps realise that the violet produces two completely different kinds of flowers. When the one we all know is withered and over another is secretly developing amongst the foliage. It is not nearly so pretty, for its petals are only five minute scales. It is different in many other ways, having no spur-like nectary, its tiny anthers containing very little pollen and its style being hooked. It produces no honey and exudes no fragrance, for the reason that it does not need to attract bees or other insects: in any case they could not pollinate it, for the flower is cleistogamic, or closed, and is self-fertilising. However, it bears just as many seed-capsules as the showy flowers, as Darwin found from experiments with them. 'What queer little flowers they are,' he exclaimed to Hooker.

The capsules produced by the cleistogamic single white variety (*Viola odorata*) bury themselves in the soil to mature. Hermann Müller wrote to say that his brother Fritz had found a white-flowered species bearing subterranean cleistogamic flowers. John Scott sent him seed of the tiny *Viola nana* from the Sikkim Terai, from which, summer after summer Darwin raised many plants with an abundance of cleistogamic flowers but never a petalled one. Of all the violets and wild pansies only *Viola tricolor* did not have them.

He compiled a list of all the known genera and their species which produced these curious flowers, growing many of them and observing how their organs had been modified into rudiments. In some cases only a single anther was left, containing a mere few pollen-grains of diminished size. In other cases the stigma had disappeared, leaving a simple open passage into the ovarium. There was a complete loss of certain parts of service to the perfect flowers but of none to the cleistogamic. 'We here see, as throughout nature,' he wrote, 'that as soon as any part or character becomes superfluous it tends sooner or later to disappear.'

Of the fifty-five genera on his list, thirty-two had irregular perfect flowers. These included *Impatiens fulva*, the Orange Balsam, dead-nettles, vetches and peas, and being irregular these were flowers specially adapted for pollination by insects. Why then had they been provided with a second means of producing seeds? Darwin explained: 'Flowers thus constructed are liable during certain seasons to be imperfectly fertilised, namely, when the proper insects are scarce; and it is difficult to avoid the belief that the production of cleistogamic flowers, which ensures under all circumstances a full supply of seed, has been in part determined by the

The Violet has two completely different flowers, the familiar one blooming in spring and requiring cross-pollination by bees, the other a cleistogamic flower appearing in autumn and able to pollinate itself—nature's way of ensuring fertilisation if the spring is wet and no bees are about

Not all flowers are pollinated by bees. Those of the Spindle tree (*Euonymus europea*), whose fruits are shown, are visited by St Mark's Flies and Fever Flies

The Lungwort (*Pulmonaria officinalis*) is dimorphic. The short-styled form yields many more flowers and sets more fruits. But the fruits yield fewer seeds than long-styled plants. The Lungwort is visited by bees, but Darwin found that when long-styled plants were illegitimately pollinated they were highly fertile perfect flowers being liable to fail in their fertilisation.' He did not think this was the complete answer. Plants might produce their perfect flowers too early or too late in the seasons and so have no pollen; cold and darkness or too much heat and light might reduce the size of the plant's corolla. In these cases, he said, natural selection might well complete the work and render it strictly cleistogamic, such flowers being admirably fitted to yield a copious supply of seed at a wonderfully small cost to the plant.

Different Forms of Flowers was one of Darwin's most important books. His discovery of legitimate and illegitimate unions, whereby the first produced a normal crop of seed and the other resulted in impaired fertility, enabled hybridisers to work their plants successfully and avoid sterility. His theory of opposite-form pollen that resulted in an understanding of the incompatibility phenomenon, was of inestimable value, particularly to breeders of fruit trees.

The use or meaning of heterostyly had not until now been explained. No one had pointed out that its own pollen placed on the stigma of such a plant 'in full health and capable of bearing seed' would have no more effect than 'so much inorganic dust'.

17

The Sleep of Plants and other Matters

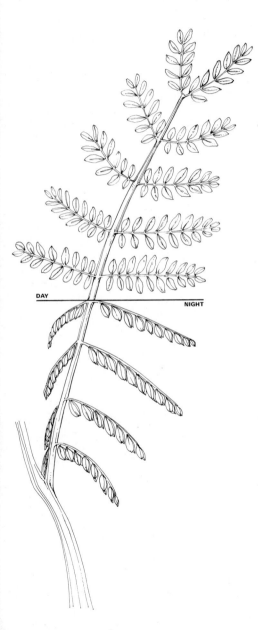

DAY   NIGHT

In a plant everything moves. It is the spiralling movement of the first root emerging from the seed that enables it to find moisture in the soil and avoid obstacles like stones; the same spiralling movement by which the first shoot finds its way upward through the soil.

Leaves describe a circle when they turn to the light. Climbing plants continually turn in their search for a prop.

At night the leaflets of a pinnate leaf like that of the acacia turn to each other and fold together.
    This, Darwin discovered, was their protection against radiation.

D ID YOU KNOW—when you look at a tree—that every one of its innumerable growing-shoots is describing small ellipses? That each leaf-stalk and leaf is doing the same thing? That if we could look below the ground we could see the tip of each rootlet endeavouring to sweep its own small ellipses and circles? Did you know that when night comes plants go to 'sleep'? That they suffer from 'insomnia' if deprived of enough light during the day?

These were some of the fascinating behaviour-patterns of plants Charles Darwin investigated for his book on *The Power of Movement in Plants*, published in 1880. It was destined to mark an era in biological science.

The reason for the book was the solving of a problem that had cropped up in the writing of *Climbing Plants*; for, as Darwin explained in his *Autobiography*: 'In accordance with the principles of evolution it was impossible to account for climbing plants having been developed in so many different groups, unless all kinds of plants possessed some slight power of movement of an analogous kind.' This he proved to be the case, and as his investigations proceeded he was led to a wider generalisation— that the great and important classes of movements, excited by such factors as light and the attraction of gravity, were all modified forms of what can loosely be called going round in circles. Darwin's name for it was circumnutation, a term which has become part of botanical language.

He began work in the summer of 1877. He was now seventy-one and because of recurring illness was feeling his age. But Frank had been his secretary and research assistant for three years now, and in July after the publication of *Different Forms of Flowers* they were free to tackle *Movement in Plants*. 'This was a tough piece of work,' Darwin recalled, but by autumn he was exclaiming to William Thiselton Dyer: 'I am all on fire at the work.' Thiselton Dyer was particularly interested in what Darwin was doing, because of the work going on at the Jodrell Laboratory.

At Down House Darwin had his own 'Jodrell', his experiments taking place in his hot-house, his greenhouse, his study and, when laid low by illness, in his bedroom. A list of the materials he used in his work on radicles (the root-tip and 'brain' of a plant) must have amazed Thiselton Dyer. Today, those who work in a modern laboratory where the equipment is even more sophisticated than in Dyer's Jodrell, would use the term 'string and sealing-wax' to describe Darwin's primitive methods. For all that, his materials served a serious purpose and did the job required of them. They included the following: shellac, dry caustic (nitrate of silver), a razor, scissors, little squares (about 1/20th of an inch) of sanded paper as stiff as thin card (between 0.15 and 0.20 mm in thickness), ordinary card, little fragments of very thin glass, loops of thread, a little square of goldbeater's skin ('which is excessively thin'), thick bristles (0.33 mm in one diameter, 0.20 mm in the other) cut into lengths of about 1/20th of an inch, squares of extremely thin paper ('too thin for writing on'), blotting paper, water, Indian ink, cork lids from which he suspended sprouting beans so that their radicles dipped into water, tinfoil, paraffin lamps, tapers, twigs, needles and pins, flower pots, black grease, glass tubes, sand, quills, friable peat, rough cinders, olive oil, lamp-black, black paper, and sieves filled with damp sawdust.

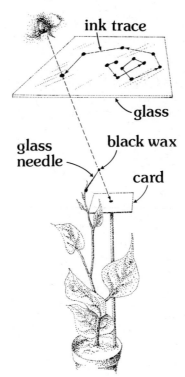

ink trace

glass

glass needle

black wax

card

Darwin's great discovery of circumnutation or the spiralling movements of roots, shoots and leaves in their search for soil, light or shade. The apparatus he used for studying circumnutation was simple but effective. A glass needle was attached to the shoot of a plant with glue. A blob of sealing wax was put on the end of this needle. A black dot was marked on a white card and this was fixed in position just beneath the needle. A glass sheet was placed some way above the plant to give a magnification factor. An ink spot was placed on the glass so that it, the dot on the card and the black wax blob were all in line. Later another dot was recorded, and so on, producing a magnified record of the movement of the shoot tip

He started with a seed of the common cabbage, and the moment its root-tip was 0.05 of an inch long he fastened the seed with shellac to a little plate of zinc, so that the radicle stood up vertically. A fine glass filament was then fixed next to the seed, and a minute bead cemented to it. Morsels of damp sponge were then put round the seed to provide the moisture it would normally find in the earth, and finally it was laid on a piece of glass smoked with lamp-black, so that the movements of the bead could be traced. For the next sixty hours Darwin and his son kept watch as the radicle grew from 0.05 to 0.11 of an inch and the glass bead to which it was fixed circumnutated across the smoked surface. It was an irregular zigzag course it traced, though more or less spiral, and at the end of their observations the tip, instead of standing vertically upwards, had become so bowed downwards by the pull of gravity that it almost touched the zinc plate. Measurements made with compasses on other cabbage seeds showed that the tip alone—and for a length of only 2/100ths to 3/100ths of

By means of smoked
pieces of glass Darwin
watched the radicle of
a bean, its first root,
trace a downward path
in search of the soil

an inch—was affected by gravity, while the tracing showed that it had
continued to circumnutate the entire time. Another seed was treated and
observed, this time with the radicle protruding 0.1 of an inch and fastened
so that it projected upwards not quite vertically. They found that,
because of this, geotropism (plant-growth in relation to gravity) came
into play at once; but again an irregular zigzag course was traced, which
showed that there was sometimes growth on one side and sometimes on
the other. The bead occasionally remaining stationary for about an hour,
growth was occurring on the side opposite to that which caused the
geotropic curvature. When seeds were upside down so that the radicle
protruded obliquely, the tracings were only slightly zigzag, but the
movement was always in the same general downward direction. The
summing up of this first experiment proved what is an important fact to
any gardener with a packet of seeds to sow—that, whether or not the seed
is planted upside-down, the root-tip, the first part of the plant to grow,
will always find its way down into the earth, aided by gravity and the
seeking movement of circumnutation, to find the moisture it primarily
needs, and an anchorage.

The next experiment was when the seeds were a stage further forward
in their development and had sprouted their cotyledons. It was with the
stem of these primary leaves that Darwin now concerned himself. The
botanists of his time called it the 'hypocotyledonous stem'; he shortened
it to hypocotyl and observed that it pushed its way upward in a
circumnutating fashion as much as the pressure of the surrounding soil
would allow, and it was interesting that it rapidly turned itself into an

arch like an inverted U. Nothing better could have been invented to break
the crust of the earth: it was like a shoulder heaving, and we have all seen
how a weed can burst its way through even a layer of tarmacadam. The
arching does more, as Darwin found: it protects the tender cotyledons,
and then as soon as the arch is above ground its inner or concave surface
grows more quickly than the upper or convex one. This separates the two
legs of the arch and draws the cotyledons out of the buried seed-coats. You
may ask where was the pull of gravity in all this. Another force was in
play—apogeotropism, the pull away from the earth, now called phototropism, the pull towards the light.

Darwin did many experiments on hypocotyls. He poured drops of
water on the earth and tamped it firmly down: the hypocotyls beat him at
this game. Then with fine silk he tied the two legs of an arch together, to
see how long it would continue to grow, and also whether the movement
when not disturbed by the straightening process indicated circumnutation. The hypocotyl did its best: it zigzagged in one direction and then
in the opposite one. After two days the summit of its arch had grown broad
and almost flat: a glass filament which had been fixed to the basal leg was
now removed and placed on top of the arch. It still continued to zigzag,
proving what Darwin had set out to prove—that whatever he did to it a
growing plant always burrows its way up to the sunshine.

To find out to what degree light affected them, young seedlings were
kept in complete darkness except for a minute or two during each
observation, when they were illuminated by a wax taper Frank held above
them. During the first day one hypocotyl changed its course thirteen
times, in long axes that criss-crossed each other at right angles or nearly
right angles. An older seedling which had formed a true leaf traced a very
complex figure, though the movement was not so great.

The seedlings went on growing in the dark. Under the microscope
Darwin measured the rate at which they oscillated back and forth across
micrometer divisions measuring 1/500ths of an inch. He was amazed at the
rapidity with which they moved, and the distance they travelled. In 6
minutes 40 seconds one tip grown in darkness passed across 10 divisions, a length of 1/50th of an inch. And these oscillations, Darwin noted,
were quite different from the trembling caused when, for instance, Frank
made a noise in the room or a door was shut in a distant part of the house.

Darwin experimented with seedlings of all sorts, representing the
entire plant kingdom except for non-flowering plants like mosses and
lichens: Corn Cockles from one of his own fields, a tropical Tree Cotton
from his hot-house, the little Tom Thumb nasturtium from his garden,
four species of oxalis, peas, the Horse Chestnut, the Scarlet Runner Bean
and the common bean, to mention only a handful. To complete the
representatives of the genera, seeds came from Kew (those of a seaside
plant, *Cassia tora*) and others from Fritz Müller in South America, and
from Lady Dorothy Nevill some 'very curious monsters' of seeds of a plant
he thought was allied to the *Medicago* or medicks.

At the end of this series of experiments, which went on for months day and night (as when he recorded that the stem of their fourth Tree Cotton seedling, lashed to a little stick, rose greatly at 3.45 a.m. and by 6.20 had fallen a little), Darwin said of them in a letter to Alphonse de Candolle: 'It always pleases me to exalt plants in the organic scale, and if you will take the trouble to read my last chapter when my book (which will be sadly too big) is published and sent to you, I hope and think that you also will admire some of the beautiful adaptations by which seedling plants are enabled to perform their proper functions.' In his *Autobiography* Darwin expressed the same pleasure in plants, but instead of the words 'organic scale' he wrote of them as being in the scale of 'organised beings'.

He now wanted to find out how the radicles of seedlings behaved when they met stones and other obstacles in the soil. Taking some germinating Broad Beans he put them in damp sand, then buried small sheets of glass upright in their path. The delicate root-tip was at first nonplussed when it met this obstacle, then tried so hard to force its way through that it became flattened. But after a while it solved the problem by climbing and gliding over the top. Now a thin slip of wood was cemented across the glass, at right angles to the radicle which was gliding down it. Encountering this the root-tip again flattened itself, after an interval of five hours solving the new puzzle by making another right-angled bend, growing alongside the slip of wood until it came to the end, when again it bent at right angles. It had one more hurdle. Coming to the edge of the glass plate, it bent once more, and this time descended straight down into the damp sand. A comforting thought for those who garden on stony ground.

Following these experiments Darwin decided to test the sensitiveness of a root-tip. The German botanist Julius Sachs had failed to detect that it was sensitive at all. Darwin after hundreds of experiments ('Little squares of tin-foil fixed with gum to one side of apex of a young and short radicle', 'A narrow chip of a quill was fixed'—and there were forty-four of these kinds of objects) proved that it was, and this was to be valuable information for future gardeners and propagators. He had noticed that the radicles grew much more quickly when subjected to considerable heat, and at first imagined that heat would also increase their sensitiveness. Jars with germinating beans suspended in damp air were placed on a chimney-piece where they were subjected to a temperature of 69°F and 72°F. Others were put in the hot-house where the temperature was even higher. Between five and six dozen trials were made to measure deflection from the little cards and splinters attached to them. To his astonishment only one radicle out of the total number showed moderately distinct deflection, while five others showed only slight and doubtful deflection. This was not the answer he expected, and he concluded that somewhere in the experiment he had made an inexplicable mistake. Then it occurred to him that radicles growing naturally in the earth would not be subjected to a temperature nearly so high as 70°F. He decided to make one last trial, now allowing the radicles of twelve beans to grow in a temperature of 55°F and 60°F. The result was that in each case the radicle was deflected in the course of a few hours. This proved that while a high temperature abnormally accelerated growth, 70°F destroyed the sensitiveness of the root-tip. Sachs had expressly stated that his beans were kept at a high temperature but had failed to make this last experiment, whereas it was typical of Darwin that he tested over and over again until the picture was complete. So, in the case of the bean or of any other seed, if you were to overheat them—they would grow. But if you then transplanted them, the 'brain' of the plant being destroyed, the root-tip would not know what to do. The seedling would wither and die.

He was not finished yet with the radicles. He next asked himself: taking it that the tip of the root is what can be regarded as the brain-machine of a plant, what happens if you injure part of it—does it still respond to gravity? His subjects were fifty-nine Scarlet Runner Beans, and some he rubbed on one side with dry caustic and other irritants, and from some he cut off one side with a razor. He found that the uninjured side took on the burden of the whole: it continued to grow and bend downward. So geotropism was ultimately victorious—even with two sliced radicles he sent almost crazy, looping and twisting before they, too, bent downward.

Another series of experiments was made to discover if the radicle found its way to moisture, and he established that it did.

Summing up his work on the marvellous root-tip, Darwin pointed out that 'a radicle may be compared with a burrowing animal such as a mole, which wishes to penetrate perpendicularly down into the ground. By continually moving his head from side to side, or circumnutating, he will feel any stone or other obstacle, as well as any difference in the hardness of the soil, and he will turn from that side; if the earth is damper on one side than on the other side he will turn thitherward as a better hunting-ground. Nevertheless, after each interruption, guided by the sense of gravity, he will be able to recover his downward course and burrow to a greater depth.'

He now went on to consider the circumnutation of mature plants, their stems and their leaves. A root has to burrow its way into the ground: the first leaf-complex has to burrow its way out of it. But why do leaves and stems describe these circles?

This was the question Darwin was asking.

To start off with, he thought it unlikely that plants should change their way of growing with advancing age, and that probably the various organs of all plants at all ages, as long as they continued to grow, would be found to circumnutate. To prove whether this was so, he began careful observations on about a score of genera belonging to widely distinct plant families inhabiting different countries. So far, nobody had ever thought to ask if stems and leaves moved in this way.

Some of the plants were woody-stemmed shrubs, which he thought might be least likely to circumnutate. Some were climbing plants, some fruit bushes, some perennials, some annuals. Others were conifers and bulbous plants, and one was the brilliant-flowered *Verbena melindres* he had seen growing in sheets of scarlet at Montevideo. A Mexican cactus was included, another unlikely subject, and an aquatic plant.

He began with their stems, and as with the seedlings he placed them on smoked glass to see what patterns, if any, they would trace. Each completed a series of zigzags, which would, of course, had the seedlings not been limited to a flat surface, have been circular. Even the woody stems of *Deutzia gracilis* changed their course greatly eleven times in a period of 10 hours 30 minutes, while the stem of a wild raspberry in a single morning almost completed a full circle and then moved far to the right to begin a fresh one. The cactus *Cereus speciosissimus*, though its branches looked utterly rigid, was also found to circumnutate, if to the much smaller degree of less than 1/20th of an inch. The verbena completed each of its circles in four hours. It was surprising that the candytuft *Iberis umbellata*, which was an annual, described only a single large ellipse in 24 hours.

Similarly he found that stolons circumnutated, those flexible runners in plants like strawberries, which travel along the surface of the ground and form new roots at a distance from the parent plant. These, like the root-tips of his beans, surmounted the obstacles he placed in their way, climbing over stones and pieces of glass. But one stolon had a fight for it. It swept up the glass in a curved line, could not make it, recoiled, and had another try. Ultimately it solved the difficulty by passing round one side of the glass instead of over it. Darwin drove a crowd of long pins into the ground. The stolons easily wound their way through them, but again there was the exception, one thick stolon being much delayed in its travels. At one place it was forced to turn at right angles, at another where it could not pass through the pins its hinder part became bowed. Then for some time it did nothing, and one would have sworn that it was pausing to take stock of the situation. Perhaps it was, for it then curved upwards and moved across the heads of the pins until it found an opening between two or three grouped farther apart. Now it happily descended and was able to pick up the path the others had followed. In another experiment the stolons found their way through a maze of other plants' stems, still by the wonderful exploring talent of circumnutation. Had the stolons not been able to circle hither and thither in search of exits, they would have gone round and round in the one spot until all their energy was spent, and got no farther.

Having seen that flower-stems obeyed the same useful law, Darwin now made exhaustive experiments with the stalks of leaves, and on finding that they, too, circumnutated he went on to investigate the curious phenomenon of the sleep of plants.

Have you ever wandered in your garden at night when the moonlight is bright enough to let you see? If so, have you noticed what a great number of plants have entirely altered their outlines? Go and look at a clump of lupins. You will be appalled, for its pretty fans of leaves instead of standing stiffly out will be hanging downwards. No doubt your reaction will be one of guilt and you will make a mental note to water it first thing

Darwin measures the angles of leaves awake and asleep

tomorrow morning, thinking that the plant must be drooping with thirst. But what of the Cucumber Tree whose network of roots must spread far down into the ground, finding moisture somewhere?—for its pinnate leaflets are hanging down like washing pinned on a line. Another surprise greets you in the kitchen garden where the blades of lettuces and radishes are standing up vertically and clinging close together. You will notice, too, that every little oxalis in the garden has furled its leaves as neatly as a City man's umbrella. If you did not know that plants go to sleep at night, this is what you are seeing, though Darwin did not imagine that plants sleep as we do. 'Hardly any one supposes that there is really any analogy between the sleep of animals and that of plants,' he wrote, 'whether of leaves or flowers.' He therefore invented the term 'nycti-tropism' and 'nyctitropic' to describe the night-turning or 'so-called sleep-movement of plants'. And having explained this he recalled that as long ago as the first century A.D., Pliny the Elder had observed this 'sleep of plants', and that in the eighteenth century the great Linnaeus, famous for his work on the classification of plants, had suggested in his *Somnus Plantarum* that leaves fold up or hang down to protect the young stems and buds from cold winds.

Darwin did not think this was the whole of the picture, and in a long series of complicated experiments he revealed some astonishing facts. First, it was true that leaves fold up or hang down in order to protect themselves from the cold, but what they were doing was turning their upper surfaces away from the zenith or direction of radiation. He took precious plants out of the greenhouse at night, and placing them on the lawn tied their leaves so that they could not move to close together or hang. A long list of 'killed, blackened and shrivelled' leaves was the

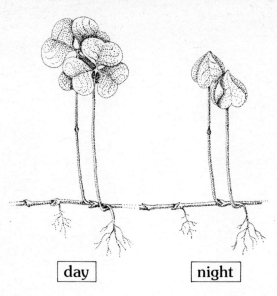

The leaves of *Oxalis
acetosella*, the Wood
Sorrel, awake and
asleep—folding up like
an umbrella

day   **night**

result, and he wrote to Hooker: 'I think we have *proved* that the sleep of plants is to lessen the injury to the leaves from radiation. This has interested me much, as it has been a problem since the time of Linnaeus. But,' he added remorsefully, 'we have killed or badly injured a multitude of plants: N.B.—*Oxalis carnosa* was most valuable, but last night was killed.'

How the three Down House gardeners must have wept to see pot after greenhouse pot of beauty thus slaughtered! The odd thing was that plants of *Lotus jacobaeus*, inhabitants of the tropical Cape Verde Islands, suffered not at all. They were exposed one night to a clear sky with the temperature of the surrounding grass 35°F, and on a second night for thirty minutes with the temperature of the grass between 37°F and 39°F, and not a single leaf, either the pinned-out or free ones, was in the least injured. Neither were those of the tropical *Cassia laevigata* nor *C. calliantha*. The explanation was that in the tropics the temperature can drop dramatically at night.

In the course of making this original discovery as to why plants sleep, Darwin discovered something else that was remarkable—that it was not only the influence of darkness and the threat of radiation which made them fold up but that they needed a sufficient amount of light during the day. Some very fine plants of *Abutilon darwinii* (again discovered by John Tweedie, to whom Darwin was a hero) stood in his large hall, which was lighted by day only from the glass cupola above. These plants never got to sleep, and neither did some nasturtiums he kept in pots in a sitting-room with a north-east window. But when he took them out of the house and exposed them to full daylight, they slept that night soundly.

Incidentally, during these investigations, he accidentally discovered that plants shaken violently will be deprived of sleep. Having put out on

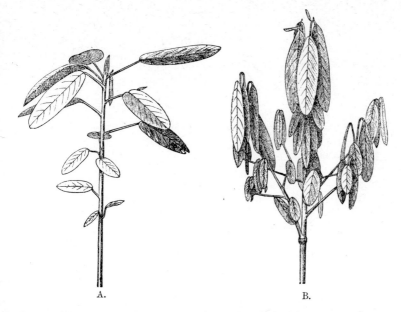

A.   B.

Desmodium gyrans, the Telegraph Plant, whose leaves during the day turn to the sun, at night droop downward, away from radiation

the lawn a *Maranta arundinacea*, which previously had led a quiet life in the hot-house, a wind unexpectedly blew up and roughly shook its leaves. The *Maranta* did not sleep for the next two nights, which did it no good. So if you go into your own glasshouse and think your plants look dusty, don't be tempted to shake them, but sponge their leaves—gently.

There are, however, as Darwin found, plants that appear to sleep during the day. Now you may hurry for your watering can, for this time the plant's folded or hung leaves indicate their mechanism to check evaporation, and that it is indeed thirsty.

It may be asked here how the leaves achieve their hanging or upright positions. At the base of the petiole or leaf-stalk is a group of small cells collectively called a pulvinus. Darwin explained that if a leaf is to rise, the pulvinus becomes inflated with water on the lower side where the cell walls are thinner, and the leaf is thus levered upwards. Conversely, if the leaf is to sink, these cell walls deflate.

We have seen how the leaves of plants turn to the light, that is to the sun, or, if the plant is indoors, towards the window. They do not all do this. The leaves of some plants when exposed to an intense light turn away from it, by rising or sinking or twisting.

To prove heliotropism or bending towards the light Darwin exposed several seedlings before his north-east window which he had masked with a linen blind, two layers of muslin curtain and a towel. So little light entered that a pencil cast no perceptible shadow on a white card, and the growing-tip did not bend at all towards the window. The towel was removed and two other muslin curtains hung up so that the light now passed through four of them and the linen blind. The pencil cast a shadow which could only just be distinguished, yet this very slight increase of light was enough to cause the seedlings to begin bending immediately in zigzag lines towards the window. Scores of experiments were done with all kinds of seedlings, all of which proved their sensitiveness to light. Some put in a box, darkened except for a pin-point where only the tiniest ray could struggle through, reached towards it in an almost straight line.

As to the plants which behaved in exactly the opposite way, Darwin tested this apheliotropism or antipathy to light with two plants, one of which was *Doxantha capreolata* (*Bignonia capreolata*), a climbing ever-green shrub. This, he found, depended on apheliotropism to guide it to the trunk of a tree, whose shadow would make a darkness. The other plant was the dainty little *Cyclamen persicum*. Gardeners who are alpine enthusiasts know how, as soon as its seed-capsules begin to swell, the flower-stalks increase in length and then slowly curve downwards until the capsules reach the ground where, if they can, they bury themselves. Darwin first attributed the movement to geotropism, the pull of gravity, but a pot which had lain horizontally with the capsules all pointing to the ground was rolled round until they pointed directly upwards. It was now placed, still horizontal, in a dark cupboard, and after four days and nights the capsules still pointed upwards. The pot was next brought back into the light, and after two days there was some bending downwards. A few days later the capsules were all pointing downwards. After more experiments with cyclamens of the same species Darwin came to the conclusion that the downward movement was indeed due to apheliotropism and not to gravity.

It was common belief at the time that heliotropism and apheliotropism were movements quite distinct from circumnutation. Darwin found that both were an extension of it. Tests proved that seedling plants were strongly heliotropic, a factor that was of great advantage to them in their struggle to expose their cotyledons to the light as quickly and fully as possible. Apogeotropism by itself would only blindly guide them upward. Meeting any over-lying obstacle they would circumnutate wildly. Only with the heliotropic-circumnutating combination could the seedling successfully dodge obstacles and be guided to the light.

This was one of heliotropism's side-effects or, as Darwin called them, transmitted effects, and this he now investigated. 'No one can look at a plant growing on a bank or on the borders of a thick wood, and doubt that the young stems and leaves place themselves so that the leaves may be well illuminated. They are thus enabled to decompose carbonic acid.' Insectivorous plants like the sundew and butterwort were not helio-tropic, he found. There was no need for them to be: as they did not live chiefly by decomposing carbonic acid from the air it was more important to them that their leaves should occupy the best position for trapping insects, whatever that position might be.

A tremendous fact now emerged that was to open up a completely new field in plant biology. It happened when Darwin was experimenting with the cotyledons of *Phalaris canariensis*. Watching their progress towards the light of a small lamp, he realised that the upper part determined the direction of the lower part. Experiments with cotyledons as tiny as 0.1 of an inch showed that the upper part reversed its curvature and then straightened itself. Had they not done this the growing-tip would ultimately have pointed to the ground instead of to the light. This in effect helped the seedling to find the shortest path from the buried seed to the light, 'on nearly the same principle,' Darwin wrote, 'that the eyes of most of the lower crawling animals are seated at the anterior ends of their bodies'—he often drew similes between plants and animals. Experiments showed that even when a bright light was shone on the lower half they did not bend in the least towards it; while a faint light shone on a narrow strip, even on one side of the upper part, induced curving. The transmission of any effect from light was a new physiological fact—and a break-through.

'These results,' Darwin concluded, 'seem to imply the presence of some matter in the upper part which is acted on by light, and which transmits its effects to the lower part. It has been shown that this transmission is independent of the bending of the upper sensitive part.'

Darwin was right. In fact, had he but known it, he was anticipating the line of research that was to lead to the discovery of hormones—half a century later. The 'matter' he thought must exist was a growth-regulator called auxin whose business it is to travel around in the plant and issue such instructions as 'Grow!' or 'Don't grow!' Auxin, the first plant hormone to be isolated, was not discovered until 1928 when Fritz W. Went, botany professor at Utrecht, was following Darwin's experiment on the same plant, *Phalaris canariensis*. But it required the contributions of many workers and the medium of twentieth-century biochemistry before the discovery was arrived at, and auxin isolated and eventually identified. Darwin's Canary Grass was a perfect subject, for grasses have a special organ, the coleoptile or modified first leaf, which produces auxin in its tip. The bean has it, too. So when Darwin sliced off one side of a bean radicle's tip with his razor and found that the radicle always bent (that is, continued to grow) on the uninjured side, this was the runner bean's auxin at work.

Was the tip permanently injured by this maltreatment? Darwin in another series of experiments discovered that such was the regenerative power of nature that after amputation the tip was completely recovered in three days, and to make up for the delay its growth-rate was often doubled in a day. Completely decapitated, new growing-points were formed after four days, and these new root-tips took up the behaviour-pattern of bending down to the earth. It was fortunate for the seedling that he did not amputate it farther than he did. The exact lengths he destroyed were 'from less than 1 to 1.5 mm (about 1/20th of an inch)': it is now known that 1.5 mm is the limit of its sensitivity.

After three years of experiments *The Power of Movement in Plants* was published on the 6th of November 1880. In it Darwin showed how, like animals, plants move, like animals they are sensitive, they need food and drink. Animals have memories: plants have a mechanism that serves the same purpose.

Some of these facts were known but only as isolated phenomena. It was Darwin who connected them together and gave them a fresh significance. As William Thiselton Dyer said in *Charles Darwin*, published in 1882: 'Whether this masterly conception of the unity of what has hitherto seemed a chaos of unrelated phenomena will be sustained, time alone will show. But no one can doubt the importance of what Mr Darwin has done, in showing that for the future the phenomena of plant movement can and indeed must be studied from a single point of view.'

His hopes for *The Power of Movement in Plants* have been fulfilled. Time has shown in full measure.

18
Memorial

Whether one glance back and compare his performance with the efforts of his predecessors, or look forward along the course which modern research is disclosing, we shall honour most in him not the rounded merit of finite accomplishment, but the creative power by which he inaugurated a line of discovery in variety and extension.

Through all ages he should be remembered as the first who showed clearly that the problems of Heredity and Variation are soluble by observation, and laid down the course by which we must proceed to their solution. . . . Evolution is a process of Variation and Heredity. The older writers, though they had some vague idea that it must be so, did not study Variation and Heredity. Darwin did, and so begat not a theory, but a science.

His elaborate speculations as to the genetic meaning of cytological appearances have led to a minute investigation of the visible phenomena occurring in those cell-divisions by which germ-cells arise.

William Bateson

In the conception of 'gemmules' as the carriers of inheritance he was not so distant from the twentieth-century conception of the gene.

Among botanists, Darwin's later work on plant movements had a special impact upon plant physiologists whose concern with his evolutionary arguments was slight indeed. Through a distinguished line of personalities Darwin's work on plant movements may be linked with present-day activity in the field of tropic responses, one outcome of which has been the discovery of the plant growth hormones.

John Heslop-Harrison

IN JULY 1881 DARWIN WROTE to Alfred Russel Wallace, the man Joseph Hooker called 'Darwin's true knight': 'What I shall do with my few remaining years of life I can hardly tell. I have everything to make me happy and contented, but life has become very wearisome to me.'

The early part of the year was occupied in completing his last book, begun in the previous autumn, *The Formation of Vegetable Mould through the Action of Worms*, in which he likened their work to that of 'a gardener who prepares fine soil for his choicest plants.' He had a great respect for worms. 'When we behold a wide, turf-covered expanse, we should remember that its smoothness, on which so much of its beauty depends, is mainly due to all the inequalities having been slowly levelled by worms. It is a marvellous reflection that the whole of the superficial mould over any such expanse has passed, and will again pass, every few years through the bodies of worms.' He spoke of the plough, one of the most ancient and most valuable of man's inventions, and of how, long before man existed, the land was in fact regularly ploughed, and still continues to be ploughed by earthworms. 'It may,' he said, 'be doubted whether there are many other animals which have played so important a part in the history of the world as have these lowly organised creatures.'

Sir Leslie Stephen was one who read the book, and in his biography of Jonathan Swift he compared the Dean's inspection of the manners and customs of servants to Darwin's observations on earthworms. 'The difference is,' he wrote, 'that Darwin had none but kindly feelings for worms.'

He had kindly feelings for all that lived. He brought up his children with a reverent love for living things: to them in all their after-lives nothing seemed 'common or unclean'. Whether an insect crawled or flew, were its colours bright or dull, its life above or below ground, the young Darwins all appeared to be, so to speak, on respectful terms with it, because it lived. Whenever he saw a spark of the naturalist's soul in a young person, Darwin knew how to fan it into life. On his walks he sometimes met young John Lubbock, the son of his neighbour, examining some living creature he had found. He encouraged these boyish researches, and the boy became one of the world's great scientific observers.

To all men, whoever they were, he was courteous. Francis Darwin recalled that his father had many letters from foolish, unscrupulous people, and how all of them received replies. 'He used to say that if he did not answer them, he had it on his conscience afterwards, and no doubt it was in great measure the courtesy with which he answered every one which produced the universal and widespread sense of his kindness of nature, which was so evident on his death.' Latterly he dictated his letters to Frank, and if it was to a European naturalist he hardly ever

(*opposite*) The portrait of Charles Darwin taken in 1868 in the Isle of Wight when he was fifty-nine—he had grown a beard two years before

Darwin wrote under this photograph: 'I like this photograph very much better than any other which has been taken of me.'

failed to say to him, 'You'd better try and write well, as it's to a foreigner.' This courtesy he extended to his fiercest and meanest opponents, hastening to excuse them. To his friends he was the very heart of kindness. When Huxley was overworked and urgently needed a long rest he instituted a fund that raised £2,100 to send him on holiday. When Wallace, home again in England, was facing financial defeat, Darwin backed him for the Civil List pension that relieved his anxieties. Anyone sick or in trouble in the village had his ready help. For thirty years he was treasurer of the Down[e] Friendly Club, keeping its accounts with minute and scrupulous exactness. Every Whit Monday the members of the Club used to march round with band and banner, and parade on the lawn in front of the house. There he met them, and explained to them their financial position. That it was prosperous gave him pleasure, since it meant a sure standby for its members.

The local schoolchildren had an annual frolic on the Down House lawn. Two toys, relics of the young Darwins' childhood, were always brought out to amuse them. One was the planed board, about fourteen inches wide and eight or nine feet long with a rim fixed to either side, which their father had made for them to slide down the stairs! The other was a rocking-boat.

He was a giant among men, yet nothing was too small for his notice. Mrs L. A. Nash, an American who with her husband lived in Downe village for four years, recalled this in the 'Memories of Charles Darwin' she wrote for the *Overland Monthly* in 1890, ten years after her return to San Francisco. 'There was such a sweet, childlike simplicity about that great man,' she said, 'that one forgot he was great, because he was always interested in the littles that make up so much of life.' She remembered the day she went to Down House with a young niece who chanced to have a bunch of some wild berries in her hand. 'We had left, and got as far as the gate into the road, when Mr Darwin came running after us. "You will think me crazy, but after you had left I thought I should like your niece to let me have some of those berries; you see the bloom is still on them." Just then he was studying the wherefore of the bloom on fruit. When his mind was at work on any subject nothing cognate to it ever escaped his notice.'

She last saw him in 1880. In October of the following year he was investigating the action of carbonate of ammonia on the roots of certain plants, the subject of two papers read to the Linnean Society by Frank on the 6th and 16th of March 1882.

They were his last published papers. But not his last act for science. In his 1880 report on the work at Kew, Hooker had appealed for clerical help in keeping the inter-leaved copy of Steudel's *Nomenclator* up to date. This was a catalogue of all known plants up to the year 1840. The same work had had to be done on Pritzel's *Index Iconum*, a catalogue of all published figures of plants up to 1866. Hooker now felt that a new and complete catalogue was necessary, a want which Darwin appreciated. He told Joseph that he would supply the necessary funds, and in January 1882 sent him a first £250 to launch the monumental task, leaving a letter 'To my

Executors & other children' that he had promised to pay £250 annually 'for 4 or 5 years', and that in the event of his death before the completion of the work 'I desire that my children may combine & arrange for the annual payment of the above sum.' The *Index Kewensis* is his lasting memorial, for it is continually in use and continually supplemented to keep it up to date.

On the 7th of March 1882 he was on his daily perambulation of the Sand Walk when he had a seizure. It was his last visit to the 'thinking walk' where so many of his revolutionary ideas had been turned over, in a mind that was never still. He recovered, and on the 17th of April in Frank's temporary absence recorded for him the progress of an experiment he was doing. But on the following night he had a severe attack and passed into a faint from which he regained consciousness knowing his end was near. 'I am not the least afraid to die,' he said. About half-past three on the afternoon of Wednesday the 19th of April in the seventy-fourth year of his age he left the world to which he had given so much.

Emma and his children wanted him to be buried at Downe, where other Darwins lay, one his brother Erasmus whom he had outlived by eight months. But the nation claimed him—he was laid to rest in Westminster Abbey among his illustrious brother scientists: Newton, Faraday, Herschel, and his friend Sir Charles Lyell. The pall-bearers were Sir Joseph Hooker, Sir John Lubbock and Alfred Russel Wallace, with the Dukes of Devonshire and Argyll, the Earl of Derby, the American Minister J. Russell Lowell, William Spottiswoode the president of the Royal Society, and Frederic William Farrar, canon of Westminster. He was followed by his five sons and two daughters.

The Abbey was crowded. Every scientific society was represented; Lord Spencer, president of the Council, representing Queen Victoria, with members of Parliament and universities and foreign embassies as well as a great gathering of the public who had come to honour him. J. Frederick Bridge, the distinguished organist, composed an anthem for the occasion, sung to the words from the Book of Proverbs, 'Happy is the man that findeth wisdom, and getteth understanding.'

The tributes that filled columns and pages of every newspaper and journal were not only from leaders of thought everywhere. At least one was written by a man whose name means nothing to us. But in the *Gardeners' Chronicle* of April 29th this is what he said: 'Please permit a practical gardener to lay a wreathe on the bier of our great teacher. It is impossible for me to express my full indebtedness to him.' He tried. He spoke of the extent of Darwin's researches and the extreme value of his labours. 'But you cannot so well estimate what mental quickening and pleasure Darwin brought to hundreds—probably thousands—of all but unknown practical horticulturists. His facts, so carefully collated, so powerfully marshalled, brought new light, fresh inspirations, a higher intellectual life to myriads of plodding workers in the field of horticulture.' He had been amazed at the reverent timidity of Darwin in heaping up fact upon fact 'until they seemed piled to a mountain height', and then, instead of dogmatic conclusion or assumption, 'merely a mild supposition, or "may be so".' He added:

It will not be until the major portion of Darwin's facts are known and thoroughly understood, and their far-reaching results grasped by practical horticulturists, that the richer fruits of his labours will begin to appear. For not only are Darwin's facts valuable in themselves, but most of them are like living seeds and will originate species, genera, varieties of similar or correlated facts; and this horticultural knowledge will root deeper, spread wider and rise higher, and its practice also be improved through all coming ages by the labour and example of Charles Darwin, for not only the work done but the manner of doing it are invaluable.

One particular trait in his character had endeared him to the writer and to many like him. 'No practical man, however humble his station, that had a fact to record was considered unworthy of his notice or a note of thanks.'

He ended:

In losing Darwin few among us but feel we have lost a friend as well as a great teacher. No man has done more to raise horticulture than he who has been laid in his right place in the Great Abbey. His friends may be assured that he lives in the hearts of many among us and being dead he will yet speak to us through his marvellous works, which have done so much and will yet do more to raise the science and practice of horticulture, as well as botany, to a higher, nobler level than either of them has ever reached heretofore.

He signed himself *D. T. Fish.*

The editorial in the *Gardeners' Chronicle* hailed Darwin as 'the physiologist who has done most in our time to advance the science of horticulture.' The memorial by Asa Gray, written on the day of his funeral, spoke of the amount of work he had done for the pure love of it, and said that as a philosopher and scientific investigator, 'what Galileo was to physical science in his time, Darwin is to biological science in ours.'

Huxley, in his most magnificent bulldog and battling vein, wrote the obituary for *Nature.* He began with a formal approach:

Very few, even among those who have taken the keenest interest in the progress of the revolution in natural knowledge set afoot by the publication of the 'Origin of Species'; and who have watched, not without astonishment, the rapid and complete change which has been effected both inside and outside the boundaries of the scientific world in the attitude of men's minds towards the doctrines which are expounded in that great work, can have been prepared for the extraordinary manifestation of affectionate regard for the man, and of profound reverence for the philosopher, which followed the announcement on Thursday last, of the death of Mr Darwin.

Now in more bulldog vein he went on:

Not only in these islands, where so many have felt the fascination of personal contact with an intellect which had no superior, and with a character which was even nobler than the intellect; but, in all parts of the civilised world, it would seem that those whose business it is to feel the pulse of nations and to know what interests the masses of mankind, were well aware that thousands of their readers would think the world the poorer for Darwin's death, and would dwell with eager interest upon every incident of his history. In France, in Germany, in Austro-Hungary, in Italy, in the United States, writers of all shades of opinion, for once unanimous, have paid a willing tribute to the worth of our great countryman, ignored in life by the official representatives of the kingdom, but laid in death among his peers in Westminster Abbey by the will of the intelligence of the nation.

He spoke of his 'intense and almost passionate honesty by which all his thoughts were irradiated, as by a central fire.' It was, he said, this rarest and greatest of endowments which kept his vivid imagination and great speculative powers within due bounds; which compelled him to undertake the prodigious labours of original investigation and of reading, upon which his published works were based; which led him to allow neither himself nor others to be deceived by phrases, and to spare neither time nor pains in order to obtain clear and distinct ideas upon every topic with which he occupied himself.

He ended with these words: 'He found a great truth, trodden underfoot, reviled by bigots, and ridiculed by all the world; he lived long enough to see it, chiefly by his own efforts, irrefragably established in science, inseparably incorporated with the common thoughts of men.'

By mistake, Hooker as well as Huxley had been asked to write the *Nature* obituary. Thankfully he was spared the task, for Darwin's death had struck him terribly. He was 'utterly unhinged and unfit for work,' as he wrote to Huxley. He was also very unwell.

But long before, in 1868, in his presidential address at the British Association meeting at Norwich, Joseph Hooker had spoken of the great advances made in botany during the previous few years. The greatest discoveries had been physiological, and, he said, 'I here allude to the series of papers on the Fertilisation of Plants which we owe to Mr Darwin.' Referring to his book on orchids he declared that the work 'has thrown more light upon the structure and functions of the floral organs of this immense and anomalous family of plants, than had been shed by the labours of all previous botanical writers. It has further opened up entirely new fields of research, and discovered new and important principles that apply to the whole vegetable kingdom.' As to Darwin's investigations of the primrose and cowslip the results had taken the botanists by surprise, the plants being so familiar, their two forms of flowers so well known to every intelligent observer, and his explanation so simple. 'For my own part I felt that my botanical knowledge of these homely plants had been but little deeper than Peter Bell's, to whom

A primrose by a river's brim
A yellow primrose was to him,
And it was nothing more.'

Thus spoke the man who was the greatest botanist of his time, whose knowledge of plants was encyclopaedic, whose massive Floras remain standard works of reference. And Darwin had yet to publish *Insectivorous Plants, Cross- and Self-Fertilisation*, his expanded *Different Forms of Flowers*, and *The Power of Movement in Plants*, each to be hailed as a masterpiece.

A Darwin Medal was struck, a Darwin Prize had already been instituted, a magnificent statue of him was placed in the Natural History Museum in London's South Kensington. In Holland the Darwin Tulip was named for him. J. C. Lenglart, a tulip fancier of Lille, had in his possession a bright red tulip called 'Princesse Aldobrandini' which was bought by Messrs Krelage of Haarlem. From it a new strain was raised, and in April 1889 Dr E. H. Krelage wrote to Francis Darwin asking if he might name the new tulip after his illustrious father. The Darwins were introduced into Britain that same year. A vigorous race, they have become the most popular of all tulips.

Since then the tributes have gone on accumulating. The centenary of his birth year, coinciding with the jubilee of the publication of the *Origin of Species*, was the occasion of a great international congress at Cambridge, as was the conference on the centenary of the *Origin*, arranged by the Botanical Society of the British Isles.

To commemorate the centenary and jubilee of 1909 a collection of essays by the leading scientists was published by the Cambridge University Press under the title *Darwin and Modern Science*. On the eve of the *Origin*'s centenary two volumes of essays were published: *Darwin's Biological Work*, and *A Century of Darwin* in which Professor John Heslop-Harrison who is, as I write, the Director of Kew, contributed a masterly analysis which brought Darwin right into the field of modern research: pointing out, for instance, that in his chapters on geographical distribution in the *Origin* Darwin had worked out the flow and counterflow of plants through the Ice Age, a theory that failed to gain general acceptance until comparatively recently when he was vindicated by the discovery of evidence in sub-fossil remains that plants had indeed migrated and immigrated during that period; evidence, too, of the isolated plant colonies of which he had written. He drew attention to Darwin's work on plant movements, linked to present-day work in the field of tropic responses, one outcome of which has been the discovery of plant hormones. Of Darwin's detailed powers of observation, he said it was doubtful whether any biologist had ever surpassed him. He warned the systematist or morphologist who might be blind to the implications of evolution when dealing with the lower levels of organic variation, and the analytically-minded physiologist who might scorn phylo-genetical speculation, that there was a lesson for both in the spirit—cautious yet enterprising, speculative yet reasoned—in which Darwin used the evolutionary hypothesis, the most important generalisation of biology, in his own botanical researches.

It may still be asked why it was Charles Darwin, no professed botanist, who came to discover so much about plants.

In working out his theory of Natural Selection, 'I test it in plants,' he said. They were convenient subjects for studying organic phenomena, and amenable. He could grow them quickly and breed them in isolation. From every country and climate, from mountain and valley stations across the whole world, he experimented with plants of every kind. And in plants he could trace the evolutionary factor back to a single cell.

Working with them in his greenhouse, or in one of his trial plots, or on his study table, he treated them almost as human beings, praising them if they did what was expected of them, scolding them if they behaved differently. Frank remembered how he spoke in a half-provoked, half-admiring way of the ingenuity of a mimosa leaf in screwing itself out of a basin of water in which he had tried to fix it.

There was something more. Even added to his towering genius and the labours and infinite patience he expended, as William Thiselton Dyer wrote of him: 'Mr Darwin—if one may venture on language which will strike no one who had conversed with him as over-strained—seemed by gentle persuasion to have penetrated that reserve of nature which baffles smaller men.'

Illustration Acknowledgements

For permission to use original Darwin drawings, photographs, and excerpts from Darwin manuscripts I wish to thank the Syndics of the Cambridge University Library, and personally E. B. Ceadel, the Librarian, and Peter J. Gautrey for their co-operation and help. I am indebted to the Botany School, the University of Cambridge, for photographing the *Scalesia* herbarium sheet; to Lady Barlow for the kind loan of family photographs; to George Darwin for permission to use the three Darwin chalk drawings, which have been photographed by Stephen Moreton Prichard; and to Professor R. D. Keynes for the loan of the Julia Margaret Cameron portrait. For the use of the Leonard Darwin photographs of Down House and the miniature of Robert Waring Darwin I acknowledge the kind permission of the President and Council of the Royal College of Surgeons of England, and I most gratefully thank Messrs William Clowes and Sons Ltd. for their kindness in reproducing ten engravings in Charles Darwin's books. I also thank the Ipswich Museum for permission to use the portrait of J. S. Henslow, the Linnean Society of London for that of T. H. Huxley, the National Maritime Museum of London for the *Beagle* painting, the National Portrait Gallery for the portrait of Erasmus Darwin, the Wedgwood Society for that of Josiah Wedgwood, the Darwin Museum, Moscow, for the group of Darwin, Huxley and Lyell, as well as for the picture of A. R. Wallace; and Stanford University Press, California, for permission to reproduce two drawings from the *Flora of the Galápagos Islands*.

To Professor R. Markham, Director of the John Innes Institute, I am grateful for providing the beautiful antirrhinum pictures, and to Professor Uno Eliasson, Department of Systematic Botany, University of Göteborg, for supplying pictures of Galapagos plants. I also thank Dr E. A. Ellis and Clare Williams for the loan of slides of British wild orchids, and George Hurn for his of *Catasetum macrocarpum*.

My sincere thanks go to Gavin Wakley and Richard Robins of the Electron Microscope Unit, the Botany School, the University of Oxford, for taking the scanning electron-micrographs specially on my behalf, and to Dr B. E. Juniper for making this possible; and I thank Oxford Scientific Films for their co-operation over other exciting pictures of insectivorous plants. Finally I am grateful to Dr Keith Roberts and Brian Hughes for the endless trouble they took in working with me to interpret Darwin's ideas in their drawings, which they did with skill and imagination; and for their excellent work in reproducing some of the older photographs and portraits I thank my friends Grace Woodbridge and Ford Jenkins. Not least, I express my thanks to the Trustees of the Botanical Research Fund for financial help towards defraying the expenses of travelling and of photographic work for many of the illustrations.

Glossary of Botanical Terms

Anther. *See* Stamen.

Apheliotropism. The movements of plants away from the light.

Apogeotropism. The movements of plants towards the light. Now called phototropism.

Auxin. A plant hormone regulating growth.

Bract. Leaf or scale below the calyx. Usually green but sometimes brightly coloured as in the poinsettia whose 'flower' is composed of bracts.

Bud-variation. Darwin's term for what gardeners call a sport and scientists a mutation: an accidental change in the genetical make-up of a plant.

Calyx. The ring of sepals on the outside of a flower.

Carpel. A component part of the pistil forming a closed receptacle for the ovules.

Caudicle. The slender stalk carrying the pollen-masses in orchidaceous plants.

Circumnutation. The spiralling movements described by the roots and growing-shoots.

Cleistogamic. A permanently closed flower. It is always self-fertilising and sometimes lacks many parts of the perfect flower. Some plants, the violet for one, produce both cleistogamic and perfect flowers.

Clinandr[i]um. In orchids a cavity in the apex of the column, often containing the stigmatic surface.

Coleoptile. The protective casing outside the shoot of a grass or cereal seedling. It produces auxin in its tip.

Column. In all common orchids the female pistils and male stamens are united and together form the column.

Corolla. The petals of a flower, which may be separate or united.

Corymb. A flat-topped flower-cluster whose flower-stalks spring from the main stem at different points, unlike an umbel in which they radiate from a single point like the spokes of an umbrella.

Cotyledon. A seed-leaf, usually different in appearance from the adult leaf.

Crucifer. Any plant belonging to the family Cruciferae, the flowers of which have four petals forming a cross.

Cryptogam. A plant having no flowers, e.g. lichens, mosses, ferns.

Cultivar. *See* Variety.

Cyme. A term covering types of flower-heads on which the growing points end in a flower, the central flowers opening first.

Cytology. The study of cell structure.

Dehisce. The act of opening by an anther to discharge pollen; by a pollen grain to discharge its tube, and the opening of a fruit when ripe.

Dicotyledon. A flowering plant having two seed-leaves.

Dimorphism. Two forms of flowers or leaves on plants of the same species.

Dioecious. Having male and female flowers on separate plants of the same species.

Edentate. An animal without incisor and canine teeth.

Exserting. The thrusting out of the tube from a pollen grain.

Family. (Formerly called Natural Order), Genus (plural Genera), and Species: the three main categories of biological classification. A family (e.g. Orchidaceae) contains one or more genera (e.g. *Orchis, Habenaria*), and a genus contains one or more species (e.g. *Orchis mascula, O. militaris*, etc.) *See also* Variety, Tribe.

Filament. *See* Stamen.

Flora. The totality of plants growing naturally in a country or district. Also applied to a book describing them.

Gastropod. A mollusc with its locomotive organ placed ventrally.

Gemmules. Darwin's term for the carriers of inheritance, which he believed were 'thrown off' from each different part of the body or plant, later to be contained within each bud, ovule and pollen grain.

Gene. A 'particle' in the nucleus which determines the presence or absence of certain characteristics in organisms.

Genotype. The genetic constitution of an individual, i.e. all the different genes present, whether expressed or not.

Genus. *See* Family.

Geotropism. The pull of gravity exerted on a root-tip.

Gynodioecious. Having female and hermaphrodite flowers on separate plants.

Heliotropism. *See* Phototropism.

Hemiptera. Plant bugs (aphids, froghoppers, etc.).

Hermaphrodite. Containing both male and female organs.

Heterostyled, Heterostyly. Having styles of different lengths in plants of the same species. *See* Pistil.

Homologous. Having the same relation or relative position, corresponding.

Homomorphic. Having perfect flowers of only one type or kind.

Hypocotyl. The stem of a cotyledon or seed-leaf.

Internode. The part of the stem between two nodes.

Monocotyledon. A flowering plant having only one cotyledon or seed-leaf.

Monoecious. Having separate male and female flowers, but both on the same plant.

Monster. An abnormal flower, that is one departing from the normal form of the species.

Mutation. A change in a gene or chromosome. The result is popularly called a sport. It is a spontaneous variation from an original type.

Natural Order. *See* Family.

Node. A joint, the point on an adult plant stem at which leaves or growth-buds appear.

Nyctitropism. The so-called sleep of plants when leaves turn their surfaces away from the direction of radiation, leaflets folding together in pairs, leaves standing vertically close together or hanging downwards.

Ovary, Ovarium. *See* Pistil.

Ovules. The unripened seeds within the ovary.

Pedicel, Peduncle, Petiole. Various kinds of stalks. The pedicel is the stalk of a single flower; peduncle is applied to the main stalk of a cluster, petiole to the stalk of a leaf.

Peloric. A regular or symmetrical flower on a plant normally producing flowers irregular in shape, such as the antirrhinum.

Perfect flower. One complete in all its parts, as opposed, for instance, to a cleistogamic flower.

Phenotype. The observable characteristics of an organism.

Phototropism. The movement of plants towards the light.

Phyllode. A flat expanded petiole replacing the blade of a foliage leaf. PHYLLATE. Of that form.

Phylogeny. The evolutionary history of plants and animals.

Pin-eyed. With the mouth of the corolla displaying the globular stigma, in such flowers as the primrose and auricula.

Pinnate. A compound leaf bearing separate leaflets on each side of the midrib. Acacia and rowan leaves are examples.

Pistil. The female organ, comprising (1) the STIGMA at the top, which receives the pollen; (2) the STYLE, the tube down which the pollen grain grows to reach (3) the OVARY at the bottom and so fertilise the ovules.

Plumule. The minute bud between the first leaves of a seedling.

Pollen. *See* Stamen.

Pollinium. Plural pollinia. The pollen-masses in orchids, each pollinium being borne on a stalk called the caudicle and fixed by a sticky disc.

Prepotency. Darwin's term for the inheritance of dominant characters.

Pulvinus. A group of small cells at the base of a petiole which inflates with water to lever a leaf upwards and deflates to make the leaf droop.

Race. A loose term used for a variant within a species.

Radicle. The embryo root of a seed.

Rhizome. An underground stem growing horizontally: in Solomon's Seal and bearded irises thick and fleshy, in Couch Grass white and stringy.

Rostellum. In orchids the cup containing the pollinia.

Sepals. The leaf-like structures forming a whorl (calyx) under the petals and protecting the flower in bud.

Spadix. A thick column in such plants as the arum, formed by the male and female flowers growing together.

Spathe. A bract enclosing one or several flowers, in the arum forming a hood over the column (spadix), in daffodils the sheath surrounding the flower bud.

Species. *See* Family.

Stamen. The male organ of a flower, usually composed of a thin stalk (FILAMENT), and a head (ANTHER) which produces and releases the POLLEN containing the male sex cells. Darwin's plural for the word was stamina.

Stigma. *See* Pistil.

Stipule. Outgrowth at the base of a leaf-stalk in such plants as the pea.

Stolon. A flexible runner in plants like the strawberry which forms new roots at a distance from the parent plant.

Style. *See* Pistil.

Taxonomy. The principles and practice of classification.

Thrum-eyed. With the mouth of the corolla displaying the anthers, in such flowers as the primrose and auricula. The name derives from the weaver's ends of threads called thrums.

Tribe. A category of classification between a family and a genus.

Trimorphic. Having three forms of flowers on plants of the same species.

Tropic responses. Changes in the direction of growth in response to light, darkness, or to other stimuli.

Variety. A variant within a species. Strictly applicable to variants originating in the wild, but popularly applied to variants selected or bred by man, the technical name for which is cultivar.

Bibliography

Apart from Charles Darwin's original material in manuscript at the University Library, Cambridge, Darwin correspondence and family letters there, at Kew and elsewhere, the following are the main sources of reference:

The Life and Letters of Charles Darwin, ed. Francis Darwin. 3 vols. Murray, 1888

More Letters of Charles Darwin, ed. Francis Darwin and A. C. Seward. 2 vols. Murray, 1903

Emma Darwin: a Century of Family Letters, ed. Henrietta Litchfield. 2 vols. Murray, 1915

Extracts from Letters addressed to Professor Henslow by C. Darwin, Esq. (Beagle letters). CUP, 1960

'Some unpublished letters of Charles Darwin', ed. Sir Gavin de Beer. *Notes and Records of the Royal Society of London*, Vol. 14, 1959

'Further unpublished letters of Charles Darwin', ed. Sir Gavin de Beer. *Annals of Science*, Vol. 14, June 1958, No. 2. Published 1960

'The Darwin letters at Shrewsbury School', Sir Gavin de Beer. *Notes and Records of the Royal Society of London*, Vol. 23, No. 1, June 1968

Darwin and Henslow: the Growth of an Idea (Letters 1831–1860), ed. Nora Barlow. Murray, 1967

Life and Letters of Sir Joseph Dalton Hooker, ed. Leonard Huxley. Murray, 1918

'Darwin's Journal', ed. Sir Gavin de Beer. *Bulletin of The British Museum (Natural History)*, Vol. 2, No. 1, Historical Series, Nov. 1959

The Autobiography of Charles Darwin, ed. Nora Barlow. Collins, 1958

Charles Darwin and the Voyage of the Beagle, ed. Nora Barlow. Pilot Press, 1945

Charles Darwin's Diary of the Voyage of H.M.S. 'Beagle', ed. Nora Barlow. CUP, 1933

'Darwin's Notebooks on Transmutation of Species', ed. Sir Gavin de Beer. *Bulletins of the British Museum (Natural History)*, Historical Series, 1960–8

Foundations of the Origin of Species (the Essays of 1842 and 1844), ed. Francis Darwin. CUP, 1909

Charles Darwin's Natural Selection, ed. R. C. Stauffer. CUP, 1975

The Works of Charles Darwin, R. B. Freeman. Dawsons of Pall Mall, 1965

Appleman, Philip *Darwin*. Norton Critical Editions, 1970

Ashworth, J. H. 'Charles Darwin as a student in Edinburgh', *Proc. Roy. Soc. Edin.*, Vol. 65, 1935

Barnett, S. A. (ed.) *A Century of Darwin*. Heinemann, 1958

Barrett, Paul H. 'The Sedgwick-Darwin Geologic Tour of North Wales'. *Proc. Amer. Phil. Soc.*; Vol. 118, No. 2, April 1974

Beddall, Barbara G. 'Wallace, Darwin, and the Theory of Natural Selection'. *Journal of the History of Biology*, Fall 1968, Vol. 1, No. 2

Bell, P. R. (ed.) *Darwin's Biological Work: Some Aspects Reconsidered.* CUP, 1959

Briggs, D. and S. M. Walters *Plant Variation and Evolution.* Weidenfeld & Nicolson, 1969

Chambers, Robert *Vestiges of the Natural History of Creation.* Churchill, 1951

Darwin, Charles (*In the following, dates prefacing titles denote year of first publication*)

1839   *Journal of researches into the geology and natural history of the various countries visited during the voyage of H.M.S. Beagle round the world.* Henry Colburn, London, 1839; and Ward Lock, Minerva Library No. 2 (Darwin's second edition text) with Introduction by G. T. Bettany

1840–   *The zoology of the voyage of H.M.S. Beagle under the command of*
1843   *Captain Fitzroy, R.N., during the years 1832 to 1836.* . . . edited and superintended by Charles Darwin. Smith Elder & Co., London, 1840–43

1842   *The structure and distribution of coral reefs, being the first part of the geology of the voyage of the Beagle.* Smith Elder & Co., London, 1842; and *Geological observations on coral reefs, volcanic islands, and on South America* etc. 1851

1851,   *A monograph of the sub-class Cirripedia* . . . The Ray Society,
1854   London, 1851, 1854 [Living Cirripedia]

1851,   *A monograph of the fossil Lepadidae* . . . The Palaeontographical
1854   Society, London, 1851, 1854 [Fossil Cirripedia]

1858   'On the tendency of species to form varieties, and on the perpetuation of varieties and species by natural means of selection.' Charles Darwin and Alfred Wallace. *Journal of the Proceedings of the Linnean Society of London, Zoology.* Vol. III, No. 9, pp. 45–62, 1859

1859   *On the origin of species by means of natural selection, or the preservation of favoured races in the struggle for life.* John Murray, London. 1st edition, 1859; 2nd edition, 5th and 6th

1862   *The various contrivances by which orchids are fertilised by insects.* John Murray, London, 2nd edition, 1877

1865   *The movements and habits of CLIMBING PLANTS.* John Murray, London. 2nd edition, 1875

1868   *The variation of animals and plants under domestication.* 2 vols. John Murray, London. 2nd edition, 1875

1875   *Insectivorous plants.* John Murray, London. Third thousand, 1875

1876   *The Effects of Cross- and Self-Fertilisation in the Vegetable Kingdom.* John Murray, London. 1st edition, 1876

1877   *The different forms of flowers on plants of the same species.* John Murray, London. 1st edition, 1877

1880 *The power of movement in plants.* John Murray, London. 1st edition, 1880

1881 *The formation of vegetable mould, through the action of worms, with observations on their habits.* John Murray, London. Thirteenth thousand, 1904

(For lists of Darwin's scientific, including botanical, papers see *Life and Letters*, vol. III, Appendix II, and, in typescript, 'New Bibliography of Darwin Papers', Paul H. Barrett.)

Darwin, Erasmus *The Botanic Garden.* Johnson, 1791; *Zoonomia.* Johnson, 1794–6; *The Temple of Nature.* Johnson, 1803

Darwin, Francis 'The Botanical Work of Darwin'. *Annals of Botany*, 1899; *Rustic Sounds and Other Studies.* Murray, 1917

de Beer, Sir Gavin *Charles Darwin.* Nelson, 1963

Galston, Arthur W. and Davies, Peter J. *Control Mechanisms in Plant Development.* Prentice-Hall, 1970

Geikie, Sir Archibald 'Charles Darwin as Geologist'. Rede Lecture, CUP, 1909

Ghiselin, Michael T. *The Triumph of the Darwinian Method.* University of California Press, 1969

Gray, Asa 'Charles Darwin'. *Proc. Amer. Acad. of Arts and Sciences*, Vol. XVII, May 1882

Greene, John C. *The Death of Adam: Evolution and its Impact on Western Thought.* Iowa State University Press, 1959

Gustafsson, Åke 'The Life of Gregor Johann Mendel—Tragic or Not?' *Hereditas*, Vol. 62, 1969

Henslow, George 'Darwin as Ecologist'. *Journal of the Royal Hort. Soc.*, 1912–13

Herbert, Sandra 'Darwin, Malthus, and Selection'. *Journal of the History of Biology.* Spring 1971, Vol. 4, No. 1

Keith, Sir Arthur *Darwin Revalued.* Watts, 1955

King-Hele, Desmond *Erasmus Darwin.* Macmillan, 1963; 'The Lunar Society of Birmingham'. *Nature.* Oct. 15, 1966

Krause, Ernst *Life of Erasmus Darwin, with a preliminary notice by Charles Darwin.* Murray, 1879

Lyell, Sir Charles *Principles of Geology*, 5th, 6th eds. Murray, 1837, 1840

Malthus, Thomas *An Essay on the Principle of Population.* Chivers-Penguin, 1970

Marchant, James *Alfred Russel Wallace: Letters and Reminiscences.* Cassell, 1916

Meteyard, Eliza *A Group of Englishmen.* Longmans Green, 1871

Olby, Robert and Gautrey, Peter 'Eleven References to Mendel before 1900'. *Annals of Science.* Vol. 24, No. 1, March 1968

Pfeffer, W. 'Geotropic Sensitiveness of the Root-tip'. *Annals of Botany.* Sept. 1894

Raverat, Gwen *Period Piece.* Faber & Faber, 1960

Ridley, H. N. *The Dispersal of Plants Throughout the World.* Reeve, 1930

Romer, Alfred Sherwood *Vertebrate Palaeontology.* University of Chicago Press

Seward, A. C. (ed.) *Darwin and Modern Science*. CUP, 1909

Sheppard, P. M. *Natural Selection and Heredity*. Hutchinson, 1971

Steward, F. C. *Growth and Organisation in Plants*. Addison-Wesley, 1968

Vorzimmer, Peter J. *Charles Darwin: The Years of Controversy*. Temple University Press, Philadelphia, 1970; 'Darwin and Mendel: the Historical Connection'. *Isis*, 1965

Wallace, Alfred Russel *Darwinism: an Exposition of the Theory of Natural Selection*. Macmillan, 1889

Wiggins, Ira L. and Porter, Duncan M. *Flora of the Galápagos Islands*. Stanford University Press, 1971

Williams-Ellis, Amabel *Darwin's Moon*. Blackie, 1966

Wilson, J. Tuzo (ed.) *Continents Adrift*. Scientific American, 1972

Index

Illustrations are denoted by **bold** type

Abrolhos 62; plants 62, 66
Abutilon darwinii 286
Acacia pycnantha 187; *A. tortuosa* 97
Acaena magellanica 70, **69**
Adiantum henslovianum 97
Adventure, H.M.S. 56–7, 111–12;
  schooner 73, 77, 79, 86–7
Aegiphila sp. 273
Agassiz, Louis 192
Ageratums 61
Ainsworth, William Francis 33–4
Albemarle Island 95, 98; plants
  95, 144
Aldrovanda vesiculosa 240, 242
Allen, Elizabeth 128
——, John Hensleigh 128
Alsophila quadripinnata 83
America 58, 99, 153, 165, 192, 233,
  273, 297; flora 144, 154, 169, 182
America, North, flora 182, Asa
  Gray's 153–4
America, South 49, 70, 72, 75, 77,
  79, 81, 90, 108, 112, 247; flora 90,
  98, 134–6, 153; fossils 67, 74, 119,
  165; land expeditions 62–3, 66–7,
  73–7, 81–2, 86–9, 92, **64**, **80**; plants
  112, 115, 134–8, 281
Amygdalis persica (*Persica
  vulgaris* and *P. laevis*) 225
Andes 81, 86, 88–9, 224;
  Cordilleras of 72, 81–2, 89, 91, 99,
  108, **84**; flora 91, 134, 137, 177
Andira inermis 101–2
Anemone decapetala 79; *A.
  triternata* 85
Angraecum sesquipedale 202, **204**
Animals 81, 98; comparison with
  plants *see* Natural Selection;
  pre-historic 67, 74, 76
*Animals and Plants under
  Domestication, The Variation of*
  188, 193–6, 209–10, 212, 222, 231,
  250; variation corner stone of
  evolutionary theory 222; man
  unable to cause variability,
  initial variation caused by
  nature 222; causes of
  variability 223; relates topic
  to Natural Selection 223;
  uncertainty of wild prototypes
  223; botanists neglect
  cultivated varieties 223;
  variations by graft and bud 223;
  opposes de Candolle's evidence
  223; savages and cultivation of

vegetables and cereals 223–4;
  increase in varieties of cabbage
  224; varieties of peas keep true
  because self-pollinating 224;
  origin of peach and nectarine,
  bud variations, sports, and
  classification 224–5;
  gooseberry improved by
  continued selection 226;
  hedgerow and forest trees 226;
  300 varieties of Burnet Rose
  226–7; development of
  Heartsease to pansy 227, **221**;
  all varieties of dahlia
  descended from single species
  227; flowers do not remain
  fixed under cultivation,
  intermediate stages lost to us
  228; Natural Selection acts by
  accumulating slight
  modifications 228; sudden
  leap, mutation 229;
  metamorphoses deceive
  botanists 228; reversion of
  suckers 228; factors of
  inheritance, reversion of
  cultivated plants and
  vegetables 228–9; irregular
  flowers show reversion 229;
  fifth stamen in peloric
  antirrhinum 229, **241**;
  progenitors leave impression
  capable of redevelopment
  229–30; prepotency of
  transmission 230; difference
  between embryo and adult 230;
  gemmules carriers of
  inheritance 230; theory of
  Pangenesis abused 230; obtains
  Mendelian ratios 230–1, **231**;
  Mendel ratio and Darwin
  prepotency 231; crossing a vast
  field, so subject of *Cross- and
  Self-Fertilisation* 231
*Annals and Magazine of Natural
  History* 114, 155; paper by A. R.
  Wallace 156, 160, 162
Anthoxanthum odoratum 154
Antirrhinum majus 187, 229–30, **241**
Apheliotropism 288
Argentine 56, 70, 73
Armadillo 67–8, 74, 119, 165
Arum maculatum 258, 260, **259**
Ascension Island 104; flora 183–4;
  geology 122

Asplenium obtusatum 83
Astelia pumila 88
Asterina darwinii **83**
Athenaeum Club 26; The
  Athenaeum 208
Atriplex sp. 153
Audubon, John James 36
Australia 100–1, 112, 125; flora 179
Autobiography of Charles Darwin
  18, 21, 27, 31, 33, 45, 48, 52, 70,
  119, 124–5, 128, 148, 173, 194, 208,
  210, 212, 264, 278, 282

Babington, Charles Cardale 112,
  153, 169
Baccharis darwinii 78
Bahia 61–2, 104; plants 61–2
Bahia Blanca 66–7, 73–4; plants 67
Balfour, John Hutton 192
Balm 273
Balsam, Orange 274
Banda Oriental 76–7
Baobab tree 60
Barbados Pride 60
Barnacles (Cirripedia)
Bateman, James 202
Bates, H. W. 155, 160
Bateson, William 291
Beagle, H.M.S. 15–16, 37, 43, 53–4,
  56–7, 107–8, 118, 134, 136, 147, 150,
  57; voyage of 58–105; Journal
  (*Diary* ed. N. Barlow) 67, 91,
  115–16; *Narrative*, vols. I & II 16,
  111, 121–2, vol. III (*Journal of
  Researches* by Charles Darwin)
  66, 91–2, 98, 108, 111, 113, 115,
  118–20, 134–7, 156; *Zoology* of
  voyage 109, 111, 118, 122
Beans 272, 281, **280**; Broad 282–3;
  Scarlet Runner 260, 272, 281,
  283, 289
Beaufort, Captain Francis, 50–1,
  53, 58–9
Bee Orchid (*Ophrys apifera*)
Beeches, Southern (*Nothofagus*)
Bees 171, 196–8, 257–8, 260–1, 270–1;
  flight of *Bombus hortorum* 150–1
Bell, Professor Thomas 109, 143
Bell Mountain 86, **80**
Bentham, George 163, 169, 208
Berberis 228; *B. buxifolia* 70; *B.
  darwinii* 82–3; *B. ilicifolia* 70
Berkeley, Rev. Miles Joseph 68,
  70, 114, 135, 138, 152–3, 208;
  article on fungi 114

Berkeley Sound 82
Berkenhout, John: *Botanical Lexicon* 21
Bignonia capreolata 218, 288; *B. venusta* 218
Biochemistry 289
Biology 15, 19, 188, 263
Birds 36, 58, 61–2, 75, 81, 102, 109, 122; finches 97–8
Bladder Cherry (*Physalis alkekengi*)
Bladderwort (*Utricularia*)
Borago officinalis **261**
Botanical Miscellany 85, 91
Botanical Society of the British Isles 298
Botany 20–2, 32, 39, 42–5, 58, 76–8, 86, 89–90, 98, 106, 109–10, 113–16, 127–8, 137, 153, 169, 183, 209, 223, 228, 231, 269–71, 273, 278, 280, 291, 296–9
Botofogo 62
Boyle, Robert 165
Bramble 273, **213**
Brazil 56, 61, 63, 66, 68, 75, 83, 95, 103, 116, 183
Britain, flora 78–9, 88, 93, 103, 153, 169, 177–8, 182–3
British Association 189, 239, 297
British Museum (Natural History) 89, 109, 112, 114, 298
Brown, Robert 89, 112–14, 130, 143, 199
——, —— (nurseryman) 226
Browne, William Alexander 34
Brunn Natural History Society 231
Bryonia dioica 219–20, **213**, **219**
Buckman, James 229
Buenos Aires 66, 73–6; flora 84
Butler, Dr Samuel 28–9
Buttercups (*Ranunculus*)
Butterfly Orchid (*Platanthera chlorantha*)
Butterton, George Ash 39
Butterwort (*Pinguicula*)
Bynoe, Benjamin 95

Cabbages 224, 230
Cabbage tree (*Andira inermis*)
Caesalpinia pulcherrima 60
Calceolaria 78, 253; *C. darwinii* 79, **65**, **82**
Cambridge 57, 107, 110–14, 118, 298; University of 30–1, 43–7, 53, 112, 148, 196, 234, 265; botany at 40–2, 44, 112; Botanic Garden 42; Christ's College 38–40; Darwin herbarium 61, 70, 78, 84–6, 91, 114–15, 134–5, 137, **106**; Library 15, 127–8, 141; University Press 298
Cambridge Philosophical Society 41–2, 104
Cambridge Ray Club 112
Cameron, Mrs Julia Margaret

234
Campions (*Silene*)
Canary Islands 46–7, 59
Candytuft (*Iberis umbellata*)
Cape of Good Hope 103
—— Negro 79
—— Tres Montes 87–8, 136
—— Verde Islands 59–60, 108
Cardamine geranifolia 70; *C. glacialis* 70
Cardoon (*Cynara cardunculus*)
Carex darwinii 87
Carnation, Wild (*Dianthus caryophyllus*)
Cassia caliantha 286; *C. laevigata* 286; *C. tora* 281
Catasetum macrocarpum 205, **65**; *C. saccatum*, 203, 205, **203**
Catmint 272–3
Celastrus orbiculatus **213**
Centaurea nigra 130
Century of Darwin, A 230, 298
Century of Family Letters, A 124, 150
Cephalanthera 195; *C. grandiflora* 200
Cerastium arvense 79
Cereus speciosissimus 284
Ceropegia gardneri 215–16
Chaffers, Edward Main 77
Chambers, Robert: *Vestiges of Creation* 141
Charles Island 94–5; flora 95, 98
Charlock 154
Chatham Island 93–5, 98; flora 93–4
Chepones 87
Chile 56, 88, 91, 93, 95, 111; flora 72, 88–9
Chiliotrichum darwinii 78
Chiloe, Island of 82–3, 87, 101; plants 86–9
Chlorea magellanica **69**
Chonos Islands 87–8; plants 88, 91, 111, 113, 136, 138, 224
Circumnutation 211–12, 214–17, 220, 277–81, 283–4, 288–9
Cirripedia 147–9
Cladonia 93
Classification 22–4, 41, 147, 154, 169–70, 189
Cleistogamic flowers 274–6
Clematis **212**
Clerodendron molle **96**
Clianthus formosus 128
Climbing Plants, The Movements and Habits of 63, 67, 212, 235, 278, **207**; why climbers climb 207; Asa Gray paper on 'Coiling of Tendrils' 210–11; their irritability 211; beautiful modifications 211; spontaneous movement of internodes 212; refutes Henslow's explanation of twiners 212; Darwin's book standard work 212;

modification of organs as aids, another proof of Theory 212; internodes on Hops, change of surface 214; semi-circle described by *Hoya carnosa* 214–15; *Cerropegia* and most twiners move anti-clockwise 215; English and tropical twiners 215; leaf-climbers intermediate between twiners and tendril-bearers 216–17; filaments rudimentary leaves 217, **216**; internodes sensitive to touch 217; climbing mechanisms 217–18, 220, **212**, **213**, **219**; tendrils modified leaves, flower-stalks modified branches 218; network of flexible fibres ensure support 218–19; *Echinocystis* tendrils form acute angles, power of stiffening 219; elasticity of White Bryony 219–20; leaf-climbers and tendril-bearers originally twiners 220; plants acquire and display movement as do animals, tendril-bearers high in scale of organisation 220; circumnutation 278
Clover 171
Cobaea scandens 218–19
Coke, Thomas William 173
Coldstream, John 34
Colletia longispina 67
Companion to the Botanical Magazine 85
Conifers 262
Coral Reefs 60; theory on 102, 108–9, 125–6; paper on 118–19
Cordia orientalis 102–3
Cordilleras *see* Andes
Corfield, Richard 86–7
Corncockles 281
Correns, Carl Erich 231
Corydalis 229
Covington, Syms 73–6, 94–7, 111, 118, 125, 132, 147
Cowslip (*Primula veris*)
Cross- and Self-Fertilisation of Plants, Effects of 18, 122, 231, 250, 262, 298, **261**; papers on 297. Aid of insects in cross-pollination 128; gives rule and method to hybridisers 250; offspring of self-pollinated plants not as vigorous 250; 11 years of experiments 250; experiments for germination, height 250–2; amount of pollen for fertilisation 251–2; trials through 10 successive generations 252–3; crossed Morning Glory Superior, Darwin's 'Hero' the exception 252–3, **249**; florists can select colours 253; varied colours in

flowers from purchased plants 253; adaptation of *Mimulus* and Foxglove to resist self-pollination 253–5; constant when bred to 7th generation 253–4; original stock benefits when crossed with new stock 254–5; again proves crossed superior 255–7; cross-pollination dependent on co-operation of insects 257; yucca and Pronuba moth 257, **256**; bees select plants of same species 257–8; crossing keeps species constant 258; Diptera flies pollinate arum 258, 260, **259**; cunning of bees in obtaining free nectar 260–1, **146**; sugar in nectar excreted as waste product 261; nectar important in cross-pollination 261; hermaphrodite flowers retain traces of pollination mechanism 262; good crossing depends on individuals slightly differing 262; light thrown on whole subject of hybridism 262

Cruckshanks, Alexander 81
Cruckshanksia glacialis 81–2, **72**
Cruger, Dr Hermann 205
Cuckoo-pint (*Arum maculatum*)
Cuming, Hugh 114, 143–4
Cunningham, Allan 112
Curtis, Rev. Dr 235
Cuvier, Georges 74, 109, 165
Cycads 262
Cyclamen persicum 288
Cycnoches ventricosum 206, **204**
Cynara cardunculus 75–6, 111
Cytology 188, 291
Cyttaria darwinii 68, 135

Daedalea erubescens 114
Dahl, Andreas Gustav 227
Dahlia 227
Darwin, Charles Robert **27**, **127**, **143**, **156**, **293**
  Ancestry and birth 22, 24–7, 298
  Career 22, 30–1, 35–6, 38, 42–3, 46–9, 55–9, 63, 67, 72, 104, 136–9, 147–9, 208; schools 28–31; Edinburgh University 31–3, 38, 44; Cambridge University 38–40, 42–7, 53, 148; *Beagle* voyage 43, 48–54, 56–105, 115–16, 134, 147; as botanist 20–2, 32, 39, 42–5, 58, 76–8, 86, 89, 98, 109–10, 113–16, 127–8, 137, 183, 209, 291, 299; as entomologist 30, 40, 58, 109; as geologist 30, 36, 47–8, 54, 58, 61, 72, 81, 88–9, 104, 107–8, 116, 118–19, 174–5; as naturalist 30, 32, 34–6, 42–3, 46, 49–50, 58, 70, 118, 137, 188;

as ornithologist 30, 81, 102, 109; his finches 97–8; as zoologist 36, 39, 58, 145
Character 18, 28, 30–1, 36–40, 42–3, 45–6, 48–52, 54, 59, 63, 107, 116, 124–5, 136, 143, 148–50, 178–9, 209, 292, 294–7
Childhood 21, 27–9
Contemporaries 16, 19–20, 30, 36, 41, 45–9, 53, 59, 61, 68, 74, 103, 107–9, 111–12, 120, 123–4, 130, 135, 138, 153, 155–60, 162–3, 166, 169, 180, 186, 188–90, 192, 199, 202, 205–6, 208, 224–31, 234, 238–9, 242–3, 261, 264, 266, 268–9, 271, 273–4, 278, 281, 295
Correspondence 16, 49–51, 67, 82–3, 104–5, 119–20, 228, 233, 235, 264, 268, 282, 292, 294, 296; with family 26, 29–31, 33, 38, 40, 51–4, 59–60, 103–4, 124, 131, 148, 150; with FitzRoy 57, 147; with W. D. Fox 45–7, 53–4, 108, 111, 118, 122, 130, 149, 233; with Asa Gray 19, 153–4, 178, 182, 192, 210–11, 235, 264; 1857 letter on Theory 160, 162–3; with Henslow 47–51, 53–4, 56, 60, 63–4, 66, 68, 70, 77, 82–6, 91, 97, 107–14, 118, 121–2, 134; with Hooker 16, 19, 71, 134–9, 147, 154, 159–60, 162–4, 177–8, 193–4, 209–12, 233, 264, 274, 285–6; publication of Theory 157–8, 162–3, **133**; with Huxley 149, 193, 233; with Lyell 123, publication of Theory 157–8, 162–3
Death 295, 297
Diary (personal) 130, 159, 163–4, 196, 209. *See also Beagle*
Down House 15, 130–2, 139–40, 142, 149–50, 163–4, 178–9, 192, 195, 212, 233–5, 256, 264; 269, 278, 294, **139**, **166**, **167**; greenhouse 209–10, 278, **210**; Sand Walk 140, 149, 179, 295, **140**
Downe village 131, 243, 294–5, Friendly Club 294
Evolution, his development of 17, 24, 119–20, 165–6, 189, 222, 233–4, 248, 262, 278, 291, 298–9; trees illustrating 120, **161**, **174**. *See* Natural Selection, main index
Experiments and Observations 32, 45, 78, 127–8, 130–1, 142–5, 147, 159–60, 171–3, 179–80, 183, 188, 193–200, 202–3, 205–6, 209–12, 214–20, 224, 226, 228–31, 235–40, 242–3, 246–58, 260–1, 264–6, 268–74, 278–89, 294–6, 299; power of observation 209, 298; acuity of vision 274; equipment 278–9,

microscopes 43, 47, 54, 147, 210, 237–8; **78**, **152**, **201**, **203**, **279**, **280**, **285**
Family 15, 18, 21, 24–31, 33, 37–40, 42, 49–53, 57–8, 104–5, 107–8, 118, 120–1, 124–5, 128, 130–1, 139–40, 142–3, 147–51, 163–4, 193–6, 202, 210, 228, 234–5, 243, 247, 264–5, 269, 273–4, 278–9, 281, 292, 294–5, **195**
Friends 28, 33–7, 39–40, 58, 86–7, 104–5, 107, 109, 142–3, 152–3, 189, 202, 208–9, 233, 294–5; W. D. Fox 40, 42, 53, 158; Captain FitzRoy 53–4, 56–60, 63, 66–8, 73–4, 76, 79, 81, 87, 107, 111, 190; Asa Gray 58, 153, 186, 208, 210–12, 219, 250, 273, 296, *Origin* 192; Henslow 16, 30, 40–51, 53–4, 57–61, 63, 77–8, 97, 102, 104, 106–8, 111–15, 118, 120, 127–8, 134–7, 142, 153, 177, 189, 192, 194, 208, 212; J. D. Hooker 16, 19–20, 58, 134, 136–9, 142–3, 147–8, 151, 177–9, 195–6, 202, 208–9, 238–9, 248, 253, 264, 268, his *Flora Antarctica* 16, 134–7, Galapagos flora 16, 95–6, 102, 114–15, 133–5, 137, 143–5, Geographical Distribution 71, 134–8, 144–5, 151, 177–9, 183–4, Natural Selection 138, 141–2, 154, 157–60, 162–6, accepts Theory 164, 179, partnership in work 177–9, *Origin* 189–90, 192, advice on future work 193, *Index Kewensis* 294–5; Huxley 147, 149, 192, 233, 238, 294, 296–7, *Origin* 189–90, 192–3; Lyell 60, 67–8, 107–11, 119, 122, 136–7, 142, 180, 193–6, 233, 250, 264, 295, Natural Selection 156–60, 162–5, accepts Theory 233
Galapagos Islands 16, 95, 98–9, 119, 134–8, 143–5, 177, **90**, **133**. *See also* main index
Geographical Distribution *see* main index
Geological Society 107–9, 122, 142
Honours 36, 41, 79, 95, 233, 265; plants named for 61, 67, 81, 87–8, 95, 100, 286, 298, **65**, **78**, **82**, **83**, **96**, **106**, **110**
Illness 58, 62–3, 75–6, 87, 120, 125, 130, 142, 148–9, 151, 159–60, 192, 210, 214, 221, 278
Life and Letters 15, 39, 136–7, 142–3, 163, 178–9; *More Letters* 15
Marriage 121–2, 124–5, 148, 150
Natural Selection, Theory of and *Origin of Species see* main index
Plants, interest in 15 20, 26,

28–9, 32, 45, 78, 104, 115–16, 127–8, 130, 195, 209–10, 228, 282, 299. *See also* main index and lists of illustrations
Plinian Society 32, 34–5
Servants 73–6, 94–7, 111, 118, 125, 132, 149–50, 154, 164, 285
Societies, membership of 34, 36, 107, 157
Transmutation Notebooks 117, 119, 122–4, 128, 130, 160, 172, 250, **161**
Travel 37, 46–7, 107, 115–16; *Beagle* 48–9, 53–4, voyage 58–105, land expeditions 62–3, 66–7, 73–7, 81–2, 86–7, 88–9, 92; holidays 164, 194, 196, 234–5, 255–6, 273–4; Wales 30, 37, 47–8
Tributes 209, 264, 291, 294–9; Medal 298, Prize 298, statue 298
Writings: 55, 60, 103, 193; *Autobiography* 18, 21, 27, 31, 33, 45, 48, 52, 70, 119, 124–5, 128, 148, 173, 194, 208, 210, 212, 264, 278, 282; *Beagle, Narrative of the voyages of H.M. Ships Adventure and* 16, 111, 121–2, *Journal of Researches* 66, 91–2, 98, 108, 111, 113, 115, 118–20, 134–6, 156, *Diary of Charles Darwin* (ed. N. Barlow) 67, 91, 115–16, *Coral Reefs* 60, 102, 108, 125–6, *Geological Observations on South America* 60, 122, 126, *Volcanic Islands* 60, 98, 131–2, 141, *Zoology of the voyage* 109, 111, 118, 122; *Animals and Plants* 188, 193–6, 209–10, 222–31, 250, 221, 225, 241 *and* main index; *Cirripedia* 147–9; *Climbing Plants* 63, 67, 207, 210–12, 214–20, 235, 278, **212, 213, 216, 219,** *and* main index; *Cross- and Self-fertilisation of Plants, Effects of* 18, 122, 128, 231, 249–58, 260–2, 297–8, **259,** *and* main index; *Descent of Man* 16, 234; *Different Forms of Flowers* 193–4, 210, 264–6, 268–76, 278, 297–8, 266, 267, 271, **263,** *and* main index; *Expression of the Emotions in Man and Animals* 234–5; *Fertilisation of Orchids* 19, 195–200, 202–3, 205–6, 208–9, 250, 297, **43, 197, 201, 203, 204,** *and* main index; *Insectivorous Plants* 70, 194, 235–40, 242–3, 246–8, 298, **236, 240, 243, 244, 245, 246, 247,** *and* main index; *Movements of Plants* 128, 154, 220, 278–90, 298, **277, 279, 280, 285, 286, 287,** *and* main index; *Origin of Species* 16–17, 19, 98, 104, 130,

136, 141, 147, 156–60, 163–6, 172, 188–90, 192–3, 212, 222, 228, 231, 233, 296, 298, *and* main index; *Vegetable Mould and Earthworms* 120, 131, 292. Publications in journals: *Annals and Magazine of Natural History* 155; *Gardeners' Chronicle* 19, 154–5, 'Nectar-secreting organs of Plants' 146, 154, 'Seedling Fruit Trees' 155, note on Bee Orchid 196; *Geological Society Journal* 'Elevation on coast of Chile' 111, 'Coral Islands' and 'Extinct Mammals' 118–19, 'Vegetable Mould' 120, 'Erratic Boulders' and 'Earthquakes' 126; *Linnean Society Journal* 194, 'Seeds in Salt Water' 157, paper on Natural Selection 163–4, 179, '*Linum*' 210, '*Lythrum salicaria*' 210, 'Climbing Plants' 212; 'Different Forms of Flowers' (5 papers) 265, 'Carbonate of Ammonia' 294; *Nature* 155; contribution to *Memoir* of Henslow 208
Darwin family
  Amy (*née* Ruck) 243, 264
  Anne Elizabeth 131, 149
  Bernard 264
  Caroline 37, 107, 120; correspondence 26, 33, 103
  Catherine 28, 124, 234, **27**; correspondence 29, 103, 131
  Charles Waring 163–4
  Elizabeth 149, 234, 265
  Emma (*née* Wedgwood) 37, 108, 124–5, 128, 130–1, 149–51, 163, 194–5, 234, 264–5, 295, 126, **195**; publication of 1844 Sketch 142
  Erasmus (1731–1802) 21–4, 27, 38, 165, 189, 235, **23**; *Botanic Garden* 22; *Phytologia* 22; *Temple of Nature* 23–4; *Zoonomia* 22–4
  Erasmus Alvey (1804–1881) 21, 28–31, 33, 39–40, 42, 53, 58, 107, 118, 125, 131, 142, 193, 210, 234, 274, 295; correspondence with 30–1
  Francis (Frank) 18, 28, 149, 151, 196, 234, 243, 247, 264, 278–9, 281, 292, 294–5, 298–9, **195**; illustrations for books 235; *Life and Letters* 15, 39, 136–7, 142–3, 163, 178–9; *More Letters* 15
  George 139–40, 149–51, 196, 202, 234, 265, **143**; illustrations for books 235
  Henrietta (Etty) *see* Litchfield
  Horace 149, 196, 234, 265, **195**
  Leonard 149, 163, 196, 234, 265, **195**

  Mary Eleanor 131
  Robert (of Elston) 40
  Robert Waring (1724–1816) 24
  Robert Waring (1766–1848) 21, 24–7, 30–1, 37–8, 49–53, 124–5, 148, 228, 234, **25**; correspondence with 33, 60, 131
  Sara (*née* Sedgwick) 264
  Susan 37, 131, 148, 234; correspondence with 53–4, 59, 148, 150
  Susannah (*née* Wedgwood) 26–7
  William Erasmus 131, 149, 196, 234–5, 264–5, 269, 273, **143**; illustrations for books 235
Darwin and Modern Science 298
Darwin's Biological Work 298
Darwiniothamnus 95; *D. lancifolius* **97**
Daubeny, Dr Charles Giles Bridle 189
Dawes, Richard (Dean of Hereford) 46
Dead-nettle 274
De Candolle, Alphonse 169, 224, 282; *Géographie Botanique Raisonnée* 223
De Jussieu, Antoine 41
De Vries, Hugo 231
Derby, Lord 118
Descartes, René 165
Descent of Man 16, 234
Desmodium 61; *D. gyrans* 154, **287**
Deutzia gracilis 284
Dianthus caryophyllus 250–1, 253
Diary (Darwin's personal) 130, 159, 163–4, 196, 209. *See also Beagle*
Different Forms of Flowers 210, 264, 278, 297–8; male and female flowers of cowslips 193–4; marriages of flowers with short and long styles 263, **275**; meaning of heterostyly 264, 276; important as bearing on origin and hybridism, 265; dimorphic and trimorphic, two and three forms in same species 265; fertility of legitimate and illegitimate marriages 265; pin-eyed and thrum-eyed primroses 265–6, **266**; short-styled cowslips more fertile 266; insects necessary for pollination 266, 268–9; hermaphrodites, comparison with animals 268; oxlip marriages more sterile than in cowslip 268; cowslip, primrose and oxlip all distinct species 268–9, **267**; acceptance of opposite-form pollen in *Linum grandiflorum* 269; two types of pollen-acceptance 269–70; Incompatibility theory

274; different adaptations for wind- or insect-pollination 270; teaches breeders to procure good seed harvest 270–1; curious structure in *Lythrum salicaria* 271–2, **271**; trimorphic hermaphrodites 271–2; different flower-structures and maturing date of pollen 272; irregular flowers adapted for cross-pollination by insects 272, 274, 276; rudimentary and abortive organs clue to ancestry 272–3; sex-separation planting of strawberries in America 273; hermaphrodite flowers tending to become dioecious 273, **275**; new class of gyno-dioecious plants and seed 271, 273; cleistogamic flowers, organs modified into rudiments 274, 276, **275**; legitimate and illegitimate marriages enable hybridisers to avoid sterility 276
Digitalis purpurea 253, 255–7, 272, **259**
Dimorphism 128, 193–4, 263–6, 268–71, 273–4, 276, 297, **266**
Dionaea muscipula 235, 238–40, **244**
Dolichos lablab 60
Don, Professor David 109–10
Donatia magellanica 88
Doubleday, Henry 268
Down House 15, 130–2, 139–40, 142, 149–50, 153, 178–9, 192, 195, 209–10, 212, 233–5, 256, 264, 269, 274, 294–5, **139**, **140**, **166**, **167**, **210**
Downe village 131, 243, 294–5; Friendly Club, 294
Downing, Charles 172, 226
Draba patagonica 67
Dracaena draco 46
Draper, Dr 189
Drimys winteri 83
Drosera 70, 194, 196; *D. rotundifolia* 235–8, **232**, **236**, **245**
Drosophyllum lusitanicum 242–3, **240**
Duncan, Dr Andrew 33
Dusky Cranesbill 260
Dyer, Harriet (*née* Hooker) 264
——, William Thiselton 243, 264, 278, 290–1, 299

Earle, Augustus 62
Early Purple Orchid (*Orchis mascula*)
Echinocystis lobata 211, 219
Edinburgh 36, 192, 268; *Journal of Science* 35; *New Philosophical Journal* 122; *Review* 192; Royal Medical Society of 36; Royal Society of 36; University of 31–40, 44, 58, 143, 265
Eleusina indica 62

Elizabeth Island 69, 79
Ellis, John 235
Embryos 160, 187–8, 230, **187**
Entomology 30, 40, 58, 62, 98, 109
Epilobium tetragonum 153
Epipactis 195
Erebus, H.M.S. 134, 136–7, 177
Erigeron lancifolium 95; *E. tenuifolium*, 95
Euonymus europaea **275**
Euphorbias 62, 94; *E. recurva* 94
Europe: flora 75, 77–8, 134–5, 153–4, 177, 182, 268, 273
Evelyn, John 227
Evening Primrose 78
Evolution 17, 24, 119–20, 165–6, 189, 222, 229, 233–5, 248, 262, 278, 291, 298–9; Darwin trees 120, 161, 175, **161**, **174**. *See* Natural Selection
Experiments and Observations *see* Darwin, Charles
Expression of the Emotions in Man and Animals 234–5

Fagus antarctica; *F. forsteri* (*Nothofagus*)
Falconer, Dr Hugh 142–3, 148, 164, 170, 233
Falkland Islands 70–1, 80–1, 85; flora 71, 81, 177
Fennel 75
Fernando Noronha 61, 137; plants 61, 111
Ferns 78, 83, 96–7, 99–100
Fertilisation of Orchids 19, 196, 208, 212, 250, 297; crossing keeps species constant, proves with orchids 195; co-adaptation 195; note on Bee Orchid 196, **129**; most modified forms in vegetable kingdom 196; special terms for mechanisms 196; pollination of *Orchis mascula* 196–9, **191**, **197**; perfect adaptation of Pyramidal Orchid for pollination by moth 198–9; *Neotina intacta* self-pollinating 199; proper season for pollination 199; Sprengel and sham-nectar-producers 199; discovers nectar in membranes 199–200, **129**; insects attracted by solid food 210; mechanism in *Pterostylis* 200; bee and rostellum of *Spiranthes spiralis* 200, **201**; insect partnership with Twayblade 202; mystery of pollination of *Cryptophoranthes* 202; Sphinx moth pollinator of long-nectaried orchid 202, **204**; *Catasetum* 3 forms on one plant 205, **65**, **203**; twisted mechanisms of *Mormodes ignea* 205–6; Vanda orchids, many

adaptations 206; orchids finest test-cases for theory of adaptive evolution 206; authority on orchids 208–9
Fisher, Ronald A. 269–70
FitzRoy, Captain Robert 49, 51, 53–4, 56–60, 63, 66–8, 73–4, 76, 79, 81, 87, 107, 111, 190, **50**; correspondence 57, 147
Flax (*Linum flavum*)
Fletcher, Mr 141
Floras: comparison of *see* Geographical Distribution; individual *see under* separate countries; by J. D. Hooker *see* Hooker
Flustra carbasea 35
Fly Orchid (*Ophrys insectifera*)
Forbes, Professor Edward 142–3
Formation of Vegetable Mould 120
Fossil trees 89, 112–13, 184–5
Fox, William Darwin 40, 42, 53, 158; correspondence 45–7, 53–4, 108, 111, 118, 122, 130, 149, 233
Foxglove (*Digitalis purpurea*)
Fraser, John 227
Fruit 60, 100, 122, 224–6, 283–4, 294
Fucus lorius 92
Fuegia Basket 68, 80
Fungi 68, 70, 78, 113–14, 138, **69**; article by Berkeley 114
Fyfe, George 34

Galapagoa darwinii 95; *G. fusca* 95
Galapagos Islands 92–9, 102, 119, 123, 156; currents 93, 98–9, **90**; flora 16, 90, 95–9, 104, 137–8, 143–5, **133**; geology 122; Hooker's Linnean Society paper 16, 95–6, 102, 114–15, 133, 143–5; plants 93–8, 108, 111, 113–15, 134–6, 143–5, 177; plant collectors 144
Galvani, Madame Luigi 239
Gardeners' Chronicle 19, 155, 209, 250, 261, 295–6; Darwin's letters and notes to 154–5; *Origin* reviews, and article by Patrick Matthews 192–3. *See* Darwin, Charles: Writings
Gärtner, Joseph 251
Genetics 18, 188, 228, 230–1, 255, 291
Genlisia ornata 248, **247**
Geographical Distribution, climatic influences 85–6, 91–2; floras, comparisons 69–70, 74–5, 77–9, 88–9, 93, 98–9, 102, 104, 134–5, 137–8, 153–4, 166, 177–9, 298, **90**, limits of 83, 87, 136; Cordilleras 72, 81–2, 89; Galapagos 137, 143–5, **133**; fossil evidence 72, 184–5; 'Keystone of creation' 71, 166, 177; methods of distribution 180, **181**; seeds: sea-water

germination 151–3, 155, 157, 180, **152**; viability 153; simultaneous creation 179–80, 184
Geological Observations on South America 60, 122
Geological Society 107–9, 122, 142
Geology 30, 36, 41, 46–8, 54, 58, 71, 76, 81, 86, 88–9, 106, 116, 118–19, 122, 264; volcanic islands 61, 98; *Beagle* specimens 60, 62, 64, 104, 107–8, 111, 118; fossil evidence in geographical distribution of plants 184–5
Geotropism 279–80, 282–3
Geraniums 45, 67, 78
Gerard, John 228
Geum rivale 153
Gloriosa plantii 217
Gonzales, Mariano 86–9, 91
Good Success Bay 69–70
Gooseberry Grower's Register 226
Gorse 103, **187**
Graham, James G. 165
——, Professor Robert 143
Grant, Professor Robert Edmund 34–6, 58, 107, 109
Grasses 62, 70, 78, 154, 177; Canary Grass (*Phalaris canariensis*); Sweet Grass (*Vulpia tenella*)
Gray, Asa 58, 153, 186, 208, 210–11, 219, 250, 264, 273, 296, **158**; 'Coiling of tendrils' paper 211–12; *Manual of Botany of Northern American States* 153–4; correspondence 19, 210–11, 235, 264, Geographical Distribution 153–4, 178, 182, *Origin* 160, 162–3, 192
——, John Edward: *Natural Arrangement of British Plants* 41
Greenwich 107–8
Green-winged Orchid (*Orchis morio*)
Guettardia speciosa 102
Gully, Dr James 149
Gunnera chilensis 87; *G. magellanica* 70; *G. manicata* 70
Gymnadenia conopsea 200

Häckel, Ernst 234
Hamadryas tomentosa **82**
Harling, Gunnar D. 95
Harris, Mr (schooners) 66, 73
Hawthorn 226
Heartsease (*Viola tricolor*)
Hebe darwiniana 100
Hedysarum (*Desmodium gyrans*)
Heliotropism 218, 220, 287–8
Helleborines (*Cephalanthera* and *Epipactis*)
Henslow, Frances see Hooker
——, George 250
——, Harriet (Mrs J. S.) 45, 53, 111
——, John Stevens 16, 30, 40–51, 53–4, 57–61, 63, 77–8, 97, 102, 104,

106–8, 111–15, 118, 120, 127–8, 134–5, 137, 153, 177, 194, 208, 212, **41**; 1844 Sketch 142; *Origin* 189, 192; correspondence 47–51, 53–4, 56, 60, 63–4, 66, 68, 70, 77, 82–6, 91, 97, 107–14, 118, 121–2, 134; *Memoir* of 208
Herbert, John Maurice 40
Hermaphrodites 188, 205, 249, 261–2, 268, 271–4
Herminium monorchis 195
Herschel, Sir John 46, 103, 108–9, 295
Heslop-Harrison, Professor John 230, 298
Hibiscus tiliaceus 103
Hitcham 112, 153
Hodgson, Brian 209
Holland, Robert: paper on Bladderwort 243
Honeysuckle 215
Hooker, Frances (*née* Henslow) 148, 208, 264
——, Harriet see Dyer
——, Sir Joseph Dalton 16, 19–20, 44, 58, 115, 134, 136–9, 142–3, 147–8, 151, 154, 177, 195–6, 202, 208, 238, 248, 253, 264, 268, 292, 295, **156**; Darwin's botanical work, praise for 208–9, 239, 297–8; *Flora Antarctica* 16, 134–7, 177; flora of the Galapagos 16, 95–6, 102, 114–15, 133–5, 137, 177, **133**, Linnean Society papers on 143–5; *Flora of New Zealand* 178–9; *Flora of Tasmania* 179; Geographical Distribution 71, 134–8, 144–5, 151, 153, 177–9, 183–4; *Himalayan Journals* 148; *Index Kewensis* 294–5; Natural Selection 138, 141–2, 157–60, 162–6, accepts Theory 164, 179; *Origin of Species*, defends 189–90, 192, future work on 193; partnership with Darwin 177–9; correspondence 16, 19, 71, 134–9, 147–8, 154, 157–60, 162–4, 177–8, 193–4, 209–12, 233, 264, 274, 285–6, **133**
——, Sir William Jackson 41–2, 81, 83, 85, 91, 95–6, 112, 114, 134, 136, 142
Hop (*Humulus lupulus*)
Hope, Dr Thomas Charles 33
Horner, Leonard 36
Horse Chestnut 281
Horticulturists 19, 141, 193, 221, 226–8, 250, 253, 265, 270–1, 273, 276, 280, 282, 295–6, 298
Hoya carnosa 214–15
Hubbersty, Nathan 37
Hubble, Dr Douglas 148–9
Humboldt, Alexander von 130, 135; *Personal Narrative* 46, 59, 61
Humulus lupulus 212, 214
Hutton, James 165

Huxley, Henrietta 233
——, Thomas Henry 147, 149, 192, 238, 294, 296–7, **158**; *Origin* 189–90, 192–3; plans after *Origin* 193; correspondence 149, 233; *Man's Place in Nature* 233
Hyacinth 228
Hybridisation 18–19, 122–3, 141, 172, 221, 224–8, 231, 250, 253, 262–3, 265, 270–1, 276
Hyssop 273

Iberis umbellata 284
Impatiens fulva 274
Inheritance 185, 228–31
Insectivorous Plants 70, 194, 235, 298; *Drosera* traps insects for special purpose 194, **232**; plant has nervous matter analogous to animal nerves 235; tentacles on glands, their functions 235–6, **236**, **245**; motor impulse of tentacles 236; digestive fluid 236–7; *Drosera* feeds like animal 237; leaf contracts by muscles 237–8; aggregated matter in tentacles resembles Amoeba 238; cell life, streaming of protoplasm 238; heat rigor 238; work of Burdon Sanderson on electromotive force 239; animal-like phenomena 239; capture of prey by Venus's Fly-trap 239–40, **244**; *Aldrovanda* sweeps prey into aquarium, dual purpose of leaves 240, 242; loss of one power, plants adapt functions 242; *Drosophyllum* tentacles have no power of movement, mushroom glands catch, dissolve and absorb flies 242–3, **240**; *Pinguicula* graminivorous and granivorous 243, **243**, **246**; sensory appendages of *Utricularia* 246–7, **244**; animal remains in bladders on rhizomes 247; *Tillandsia* host for *Utricularia nelumbifolia* 248; death-chamber grids of *Genlisia* 248, **247**; carnivores part of variation and struggle for existence 248
Insects see Entomology
Ipomoea purpurea 215, 250–3, **249**
Iresine 62
Ivy **213**

James Island 95–9; flora 95–7, 144
Jameson, Robert 34, 36
Jemmy Button 68, 79–80
Jenyns, Leonard 45–6, 53, 111
Jodrell, Thomas Jodrell Phillips 264

Journal of Botany 85, 112;
Darwin's journals of voyage
see Beagle
Judd, John Wesley 108

Kay, William 34
Keeling Islands (Cocos) 101–3;
flora 101–3; plants 111, 113–15;
article by Henslow 114
Kemp, Mr 153
Kerguelen Island flora 138, 177
Kew, Royal Botanic Gardens 58,
68, 96–7, 112, 134, 142, 148, 151,
153–4, 196, 202, 224, 243, 247, 264,
281, 294, 298; *Index Kewensis*
294–5; Jodrell Laboratory 264,
278
King, Philip Gidley 57, 112
——, Captain Philip Parker 57,
111–12
Kölreuter, Gottlieb 186, 251
Krelage, Messrs 298

Lablab or Hyacinth Bean 60
Lady's Tresses Orchid
(*Spiranthes spiralis*)
Lamarck, Jean Baptiste de 16,
68, 74, 120, 138, 165–6
Lantana fucata 62
Lathyrus maritimus 88; *L.
tomentosa* 67
Lavender 272
Lawson, Mr 94
Lawson's of Edinburgh 226
Leclerc, Georges Louis 165
Lee, James 227
Leggett, William H. 264
Leighton, Rev. William Allport
28
Lennon, Patrick 62
Lichfield Botanical Society 21
Life and Letters of Charles Darwin,
ed. F. Darwin 15, 39, 136–7, 142–3,
163, 178–9; *More Letters* 15
Lima 93, 112; flora 93
Limnanthes douglasii **261**
Linaria 229; *L. vulgaris* 250–1, 253
Lindley, John 153, 202, 205
Linnaeus, Carl 22–4, 30, 41, 147,
170, 223, **21**, **263**; *Somnus
Plantarum* 285–6
Linnean Society 109, 118, 157;
publications 133, 143, 157, 163–4,
179, 210, 212, 265, 294. *See*
Darwin, Charles: Writings
Linum flavum 128, 269–70; *L.
grandiflorum* 269–70; *L. perenne*
270–1
Listera ovata 195, 202
Litchfield, Henrietta (Etty) (*née*
Darwin) 31, 149, 151, 163–4, 194,
196, 234, **195**; *Emma Darwin:
A Century of Family Letters*
(ed.) 124, 150
——, Richard Buckley 234

Loasa eurantiaca 215
Lobelia 18, 260; *L. fulgens* 171
Loddiges, Conrad 123
Lomatia ferruginea 88
London 57, 107, 111, 118, 120, 124–5,
130–1, 142, 150–1, 193, 210, 234
*London Catalogue of British
Plants* 169
London Review 208
Lonsdale, William 109
Lotus jacobaeus 286
Loudon, J. C.: *Arboretum et
Fruticetum Britannicum* 225–6
Lowe, Robert (Viscount
Sherbrooke) 48
Lubbock, Sir John 147, 189, 209,
233, 295
——, John (son) 282
Lumb, Edward 75, 77
Lunar Society 22
Lychnis (*Silene*)
Lyell, Charles (of Kinnordy) 136
——, Sir Charles 60, 107–11, 119,
122, 136–7, 180, 194–6, 233, 250, 264,
295, **156**. Re Natural Selection
Theory 142, 156–60, 162–5;
books after *Origin* 193;
confession of faith to *Origin* 233;
correspondence 123; on Natural
Selection 157–8, 162–3;
Antiquities of Man 233;
Principles of Geology 60, 67–8, 108
——, Lady 194
Lythrum salicaria 210, 249, 271, **271**

Maccormick, Robert 57, 60, 136
MacGillivray, Professor William
34
Maer Hall 37–8, 48, 51, 120–1,
124–5, 128, 130, 139
Magazine of Zoology and Botany
78
Maldonado 71
Malthus, Rev. Thomas Robert:
The Principle of Population
123–4, 160
Maranta arundinacea 286–7; *M.
porteana* 61
Margyricarpus setosus 67
Marriages, legitimate and
illegitimate 18, 263, 265–6,
268–9, 272, 276
Marsh Orchid 199
Martyn, Professor Thomas 41
Masdevall, José 202
Masdevallia fenestrata 202
Masters, Dr Maxwell Tylden 261
Mather, Kenneth 270
Matthew, Patrick: *Naval Timber
and Arboriculture* 192–3
Maurandia lophospermum 217
Mauritius 103–4
Medicago 281
Melica papilionum 67
Mendel, Gregor 230–1
Mendelian ratio **231**

Mendoza 88–9, 91, 108
Microscopes 43, 47, 54, 147, 210,
237–8
Miller, Professor William
Hallowes 111
Mimosa 128, 299; Australian
mimosa (*Acacia pycnantha*)
Mimulus luteus 250–3
M. rosea 253–5
Mints 273
Mitraria coccinea 88
Modification and descent *see*
Natural Selection
Moggridge, John Treherne 199,
264, 273
Monke, Lady (Lady Mary
Bennet) 227
Monkey Flower (*Mimulus luteus*)*
Monocanthus viridis (*Catasetum
macrocarpum*)
Monro, Dr Alexander 33
Montevideo 63, 66–8, 71, 75–6, 85
Mormodes histrio **204**; *M. ignea*
205–6; *M. luxata* 206
Morning Glory (*Ipomoea*)
Mosses 93, 281
Mount, The 26, 37, 47, 130, 234
Mount Tarn 79
Movements of Plants 128, 154, 278,
289–90, 298; power of movement
of an analogous kind 278;
circumnutation 277–8;
equipment **279**; measures
radicle movement 279–80;
geotropism 280, **280**; arching of
cotyledon, power of the
hypocotyl 280–1; plants grow
to light 281; growth of
seedlings in dark 281; day and
night recordings of movement
282; exalts plants in organic
scale 282; radicles avoid
obstacles 282; sensitiveness of
root-tip and effect of heat
282–3; injured root-tip 283;
movement to moisture 283;
radicle compared to burrowing
animal 283; circumnutation
283–4; plants alter outline at
night 284–5, **277**, **285**, **286**; sleep
of plants and animals,
nyctitropism 285; protection
from radiation 285–6, **287**;
tropical plants accustomed to
drop in temperature 286; effect
of lack of daylight 286; plants
shaken violently deprived of
sleep 286–7; mechanism to
check evaporation 287;
turning to light (heliotropism)
and apheliotropism 287–8;
bending of cyclamen to bury
seed 288; heliotropism and
apheliotropism extension of
circumnutation 288;
movement to light and

insectivorous plants 288; 'brain' in upper part of cotyledon likened to anterior end of lower crawling animal 288–9; anticipates discovery of hormones, auxin 289; amputation of hypocotyl 289; plants' resemblances to animals 289–90; phenomenon of plant movement must be studied from single point of view 290

Muhlenbergia rariflora **83**

Müller, Fritz 202, 264, 273–4, 281

——, Hermann 264, 274

Murray, John 189, 192–3, 208, 212, 235

Musk Orchid (*Herminium monorchis*)

Mutation 226–8

Myanthus barbatus (Catasetum macrocarpum)

Myrtus luma 87

Myzodendron brachystachum 68, **69**

Nash, Mrs L. A. 294

Nasturtium (*Tropaeolum*)

Natural History 34–6, 42–3, 49–50, 106, 109, 136–7, 169, 188; *Beagle* specimens 59, 63–4, 66, 70, 84, 106. *See* Darwin, Charles: Career

Natural Selection, Charles Darwin's Theory of 58–60, 68, 71, 74, 91–2, 99, 119, 123–4, 138–9, 149, 151, 154–5, 192–3

Development of: species not immutable 138–9; 1842 Sketch 130, 162; 1844 Sketch 141–2, 162, 173; Wallace's theory and sketch 156–60; 20 years' work 160; Evolutionary Tree 120, 161, **161**; joint paper with Wallace's read at Linnean Society 163, publication of 179; publication offered of extended paper but commences book 164; Hooker accepts Theory 164, 179; meaning of term Natural Selection 170; extension of Theory 173–5; genealogical tree **174**, as model for natural classification 188–90; future fields of inquiry 188; Lyell accepts Theory 233; progress of acceptance 233; modern scientists should learn from Darwin 298–9; plants trace evolutionary factor back to single cell 299

Collecting of facts 154–5, 264, 269, 273

Comparison with animals 122–4; rejects Lamarck's theory 68, 120, 166; adaptation 119–20;

gradation of species 122; variation of animals ('test it in plants') 154, 166, 299; animals' effect on vegetation 159–60, 171; embryonic seedlings and animals 160, 187–8, **187**; wars in nature 171–2; man cannot produce variability—in wild Natural Selection preserves beneficial variation 172; vigour and fertility 172; rarity precursor to extinction 172–3; ecological explanation for divergence 173; tree model for natural classification 175; descent from single prototype 184, 186, 299; queries fossil evidence 184–5; plants acquire movement 220; 'nervous matter' analogous to animal nerves 235; both have muscles 237–8; feed and drink 237, 289; have movement, sensitivity and memory 290; fertility of hermaphrodites 268; curiosity of triple union in *Lythrum* 271–2, **271**; radicle compared to burrowing animal 283; analogy of sleep 285; similarity between cotyledon and crawling animal 289

Modification and descent: resistance to conversion 117, 130; atavism (reversion) 122, 228–30; flower structure 127–8, conversion of organs element of success 130; similarity of seedlings, unlike adults 160; superfluous organs 173; belief in single prototype 184; rudimentary organs reveal scheme 185–8, **186**; hermaphrodites 188; adaptive evolution shown in orchids 206; climbing plants 217; metamorphosed leaves in tendrils 218; intermediate stages lost in cultivation 228; mutation 227; embryos throw flood of light 230; 'gemmules' factor of inheritance 230; Mendelian ratios obtained 231; plants adapt functions 242; hermaphrodites and cross-pollination 261–2; crucifers and cycads wind-pollinated 262; conditions for good crossing 262; origin of the two sexes 262; reciprocal marriages special bearing on origin of species 263;

legitimate and illegitimate marriages 265–6, 268–9; Oxlip true species 268–9; opposite-form pollen 269–70, 276; rudimentary and aborted organs clues to ancestry 272–3; heterostyly 276; circumnutation 278

Partnership with insects: sees relationship 122; *Opuntia*, simulates pollination 78, 128, **78**; introduced plants 122; mimosa stamens and Glory Pea 128; cross-pollination keeps species constant 128, 258; dimorphism in Flax flowers 128; night-pollination 128; Sprengel's fertilisation theory 130; abortive anthers in Thyme 130; pea flower self-pollinating, nectar-secreting 155, **146**; co-adaptation 169–70, 195; correct insects 171, 257, 268–9; development of nectar-secreting flowers 188; safety device of cleistogamic flowers 274, 276

Plants in relation to 17, 67, 74, 119, 122, 137, 170, 193

Survival: favourable variations 124, 170; viability of seeds 154; leaf-folding 154, 285–6; cross-breeding 155, 172, 257; inter-dependency 170; high productivity of introduced plants 170; climatic conditions 171; power of movement 220; mosses dependent on atmosphere 237; mechanisms of insectivorous plants 248; cleistogamic flowers 274, 276

Variation: hybridisation and preservation of species 122–3; variability 124, 127–8, 170; conditions for 141–2; clue to natural classification 147, 169; supplied by inheritance 155; accumulated effects of selection 167–8; plants in wild 168; variety and species 168–9; different opinions 169; tables of gradations 169; species to genera 170; corner-stone of Theory 222; theories explaining facts 223; breeders can fix flower colours 253

Nature 155, 297

Naudin, Charles Victor 251

Neotina intacta 199

Nepenthes 218; *N. pervillei* **245**

Nevill, Lady Dorothy 202, 247, 281

New Zealand 100, 138; flora

177–8, 183–4
Newman, Col. H. W. 171
Newton, Professor Alfred 180
——, Sir Isaac 165, 295; *Principia Mathematica* 24
Nicholson, George: *Dictionary of Gardening* 208–9, 224
Nothofagus antarctica 136;
   N. betuloides 68–9, 83, 113, 135–6;
   N. forsteri 136

Ochrosia parviflora 102
Odontoglossum grande **204**
Oenothera dentata 78
Oliver, Daniel 208, 211, 266, 269
Ombu tree 75
Ophrys apifera 196, **129**
Opuntia darwinii 78, 128, **78**;
   O. galapageia **96**
Orchids 171, 191, 195–6, 202, 206, 272
Orchis maculata 200; *O. mascula* 196–9, **191**, **197**; *O. morio* 43, 45, 199–200, **43**, **129**; *O. pyramidalis* 195, 198–9
Oreopolis glacialis 81–2
Origin of Species 16–17, 19, 98, 104, 130, 136, 141, 147, 163–6, 172, 188–90, 192–3, 212, 222, 228, 231, 233, 296, 298; 1842 Sketch 130, 162; 1844 Sketch 141–2, 162, 173; Wallace's theory, publication of paper 156–60; Darwin's dilemma 162; joint paper at Linnean Society 163; publication 179; Society offers extended publication in *Journal* but Darwin begins book 164; Abstract of 'big book' published as *Origin* 164; 'Why the *Origin*?' 165–6; how plants have changed, will tell complete story 166; tree model for all classification 161, 175, **161**, **174**; conclusions, publication and reception 189–90; reviews 192–3; 'Book of the Day' 192; Matthews claims priority 193; second edition and work on 'big book', altered to series of books, space limited in *Origin* 193; foreign editions 233; rapid and complete change effected in men's minds towards doctrine 296; established in science 297. *See also* Natural Selection
Owen, Richard 107–9, 189–90, 192
——, Sarah 37
Oxalis 281, 285, *A. acetosella* **286**;
   A. carnosa 286
Oxlip (*Primula elatior*)

Palm, Ludwig H.: *Uber das Winden der Pflanzen* 212
Pampas lands 71, 76, 111

Panagyrum darwinii 78
Pangenesis 230
Pansy 227
Parana river 75–6
Paris Museum of Natural History 109
Paronychia gorgonocoma 60
Parthenon, The 208
Passion Flower 219
Patagones 73, 78
Patagonia 100; flora 69, 72, 79, 81, 89; plants 85, 113, 135, 177
Paul, William 226, 228
Pavonias 62
Peach or Nectarine tree (*Amygdalus persica*)
Peacock, Rev. George 49–51, 58
Peas 188, 224, 230, 272, 274, 281, **146**, **225**, **231**
Pemphis acidula 102
Penstemon argutus 260
Pernettya pumila 70
Persoon, Christian Hendrick 178
Peru 92–3, 95; plants 93
Petunia violacea 253
Phalaris 67: *P. canariensis* 288–9
Phoradendron henslovii **97**
Phototropism 281, 288–9
Physalis alkekengi 60
Phytolacca dioica 75
Pickering, Sir George 148
Pictet, François Jules 192
Pinguicula 70, 235, 242–3, **243**, **246**
Pisonia darwinii 61
Pitcher Plants (*Nepenthes*)
Plantago 273
Plants 15–20, 26, 28–9, 32, 45, 130, 209–10, 228, 232; of *Beagle* 55, 58–63, 66–7, 70, 73, 75, 77–89, 91–7, 100–3, 109–10, 144, 177, **65**, **69**, **78**, **82**, **83**, **96**, **97**, **110**; fungi 68, 70, 78, 113–14, 138, **69**, seeds 75, 77, 84–7, 91, 98, 102, 111; herbarium 70, 72, 78, 84–7, 91, 93–4, 96, 98, 104, 108, 112–13, 115, 134–5, 137, **106**, identification of 68, 70, 85, 89, 91, 111–15, 135, 144; geographical distribution 69–72, 74–5, 77–9, 81–3, 85–9, 91–3, 98–9, 104, 134–8, 143–5, 151–4, 177–80, 182–5, **72**, **176**; Natural Selection 17, 67, 74, 78, 91–2, 119–20, 122–4, 127–8, 130, 137, 141, 146–7, 154, 185–9, 193, 298–9, **146**, **187**; nomenclature 147, 154; Writings: *Climbing Plants* 211–20, **207**, **212**, **213**, **216**, **219**; *Cross- and Self-Fertilisation* 250–8, 260–1, **249**, **256**, **259**, **261**; *Different Forms of Flowers* 193–4, 265–6, 268–71, 272–4, 276, **266**, **267**, **271**, **275**; *Fertilisation of Orchids* 195–200, 202–3, 205–6, 208–9, **65**, **129**, **191**, **197**, **201**, **203**, **204**; *Insectivorous Plants* 235–40, 242–3, 246–8, **232**, **236**, **240**, **243**,

244, **245**, **246**, **247**; *Movements of Plants* 277–89, **277**, **280**, **285**, **286**, **287**; *Variation of Animals and Plants* 221–31, **221**, **225**, **241**. *See also* Darwin, Charles
Platanthera chlorantha 200
Plinian Society 32, 34–5
Pliny the Elder 285
Pollen 18, 122, 197–200, 202, 205–6, 230, 251–60, 262, 265–6, 269–70, 276, **43**; pollination 78, 128, 155, 176, 195, 250–3, 255–8, 261–2, 265, 268–70, 272, **78**, **146**, **191**, **197**, **201**, **203**
Polygala paniculata 62
Polypodium ignominia 83;
   P. pleiosoros 96
Polystichum adiantiforme 78
Pontederia sp. 272
Pontobdella muricata 32, 35, **32**
Port Desire 77, 79; flora 85
—— Famine 79
—— Louis 70
—— St Julien 77, 79; plants 79
Porto Praya 59–60
Potatoes (*Solanum*)
'Prepotency' 230–1
Primrose (*Primula vulgaris*)
Primula elatior (Oxlip) 268–9, **267**;
   P. sinensis 243; *P. veris* (Cowslip) 193–4, 265–6, 268–9, 297, **267**; *P. vulgaris* (Primrose) 177, 265–6, 268–9, 273, 297–8, **266**, **267**
Pronuba moth 257
Pterostylis longifolia 200, **201**;
   P. trullifolia 200
Pulmonaria officinalis 273, **275**
Punta Alta 67, 74, 165
Purple Loosestrife (*Lythrum salicaria*)
Pyramidal Orchid (*Orchis pyramidalis*)

Quillaja saponaria 89

Ramsay, Marmaduke 46, 48–9
Ranunculus biternatus 70, 85
Ray, John 165
Reversion 122, 228–30
Rhamnus lanceolatus 273
Rhododendron azaloides 130
Rio Colorada 73
—— de Janeiro 62, 85, 87
—— Macae 62
—— Nigro 66, 73, 76
—— Plata 67, 71, 74, 77
Rivers, Thomas 224–5, 228
Rosa spinosissima 226–7; *R. villosa* 139
Rosas, General Juan Manuel 73, 75
Ross, James Clark 57, 134, 137
Royal College of Surgeons 15, 109. *See* Down House
Royal Horticultural Society 234, 260

Royal Society 233, 239, 295;
Copley Medal 233
Ruck, Amy see Darwin family
Rucker, Sigismund 205
Rudimentary organs see Natural
Selection

Sabine, Joseph 226–7
St Helena 103–4, 107; flora 103–4,
135, 138, 184; geology 122
—— Jago 59–60, 87, 108; plants 66
—— Julien: flora 85
—— Paul's Rock 61, 137
Salinas 73
Salvia grahamii 260, 272
Sanderson, Professor John
Burdon 238–9
San Fernando 87
Santa Antonia 60
—— Cruz 59, 81, 86, 89
—— Fé 75
'Sassafras' 83
Scalesia darwinii 95–6, **106, 110**
Schinus dependens 67
Schomburck, Sir Robert 202, 205
Scott, John 268, 274
——, Sir Walter 36
Sea-water germination 151–3, **152**
Sedgwick, Professor Adam 30,
41, 47–8, 58, 104, 107–8, 192
——, Sara see Darwin family
Senecio candicans 70; *S. darwinii*
81
Shelburne, Lord 25–6
Shrewsbury 24–5, 27–8, 36, 47,
51–2, 57, 76, 104, 107, 113, 120, 128,
148, 234; School 37–9, 86
Silene alba 128; *S. dioica* 128
Silliman's Journal 208
Sismondi, Jessie de 37, 124–5
Sleep of plants 278, 284–7
Smith, Dr Sydney 15–16, 121
——, William ('Stratum') 165
Snapdragon (*Antirrhinum majus*)
Socego 63
Solanum 87, 91, 94, 111, 113, 224;
S. maglia (Darwin Potato) 224;
S. tuberosum 224
Somerset, Duke of 118
Spectator, The 192
Spencer, Herbert 16, 19, 172
Spiranthes spiralis 200, **201**
Spotted Orchid (*Orchis maculata*)
Sprengel, C. K. 20; *Das Entdeckte
Geheimnis der Natur* 130, 199
Spring Rice, Thomas 118
Stachys coccinea 260
Stephen, Sir Leslie 292
Stephens, James Francis:
*Illustrators of British
Entomology* 40
Stokes, John Lort 57
Strait of Magellan 68–9, 79
Strawberry 273, 284
Sulivan, Lt. James 67
Sundew (*Drosera*)

Survival 124, 154–5, 159, 170–3, 220,
237, 248, 257, 274, 276, 285
Sweet-scented Orchid
(*Gymnadenia conopsea*)

Tahiti 99–100, 103
Tait, Alfred 242
——, William Chaster 242
Tasmania 101; flora 178
Telegraph Plant (*Desmodium
gyrans*)
Theophrastus 224
Thompson, Harry Stephen
Meysey 40
Thorley, Miss 149, 154
Thrift 78, 177
Thunbergia 215
Thwaites, George Henry
Kendrick 264, 273
Thyme 130, 273–4
Tierra del Fuego 68–71, 77–80,
82–3, 88, 101, 113, 116; flora 69,
71, 79, 177; plants 70, 85–6, 134–8,
177, **69**
Times, The 192–3
Toothache Tree 92
Tournefortia argentea 102
Transmutation Notebooks 117,
119, 122–4, 128, 130, 160–1, 172,
250, **161**
Travel 57, 115–16, 124, 136. *See*
Darwin, Charles
Treasury, The 118
Tree Cotton 281–2
Trees, variation of 226
Trimen, Roland 264
Trimorphism 263, 271–3, 276, **271**
Tropaeolum azureum 217;
T. tricolorum 216–17, **216**;
T. grandiflorum 216; *T. penta-
phyllum* 217; *T. tuberosum*
217; nasturtiums: 286, 'Dwarf
Crimson' 217–18, 'Tom Thumb'
281
Tulip, Darwin 298
Twayblade (*Listera ovata*)
Tweedie, John 66, 286

Urtica darwinii 88
Utricularia spp 242, 247; *U.
neglecta* 246–7; *U. nelumbifolia*
248; *U. vulgaris* 243, **244**

Valdivia 88
Valparaiso 83–4, 86–9, 91, 108, 112;
flora 86, 89
Vandellia 253
Vaucher, Rev. M. 271
Vegetable Mould and Earthworms
120, 292; experiment for 131
Veitch, James jnr. 202, 206
Venus's Fly-trap (*Dionaea
muscipula*)
Verbascum 253
Verbena melindres 66, 283–4
Veronica peregrina 78

Vicia sativa 261, 274
Viola nana 274; *V. odorata* 274,
275; *V. tricolor* 171, 227, 274
Volcanic Islands 60, 98, 131–2,
141; theory on 61
Von Mohl, Hugo 211–12, 215–16;
*Uber den Bau und das Winder
der Ranken und Schlingpflanzen*
212
Von Tschermak-Seysenegg,
Erich 231
Vulpia tenella 67

Wallace, Alfred Russel 155–7, 160,
162–3, 188, 291, 294–5, **159**; joint
paper at Linnean Society 163,
publication 171; paper in
Anthropological Review 233
Waterhouse, George Robert 143
Waterton, Charles 36
Watson, Hewett Cottrell 169, 228
Way, Albert 40
Wedgwood, Charlotte 107, 124
——, Elizabeth 107, 120, 164, 194
——, Emma see Darwin
——, Fanny 37
——, Hensleigh 118, 124, 131, 142
——, John 234
——, Josiah (1730–1795) 22, 27
——, Josiah (1769–1843) 26–7, 37–8,
51–2, 104–5, 107, 121, 124, **52**
——, Josiah (1795–1880) 120
——, Susannah see Darwin
Went, Fritz W. 289
Werner, Abraham Gottlob 34
Wernerian Society 35–6
Westwood, John Obadiah 109
Whewell, William 118
Whitley, Charles 39
Wickham, John Clements 59
Wilberforce, Dr Samuel 190
Wild Cucumber (*Echinocystis
lobata*)
Winslow, John 148
Winter's Bark 83
Wirtgen 271
Wistaria 215
Woodruff, Professor A. W. 148
Wray, Leonard 273

Xanthopon morgani praedicta
202
Xanthorrhoea preissii 101

Yarrell, William 109
York Minster 68, 79–80
Yucca 257, **256**

Zacharius, Otto 119
Zoological Society 109
Zoology 36, 39, 58, 144; *Beagle*
specimens 15, 59–60, 67, 106,
109–11; *Zoology* of the *Beagle*
109, 111, 118, 122

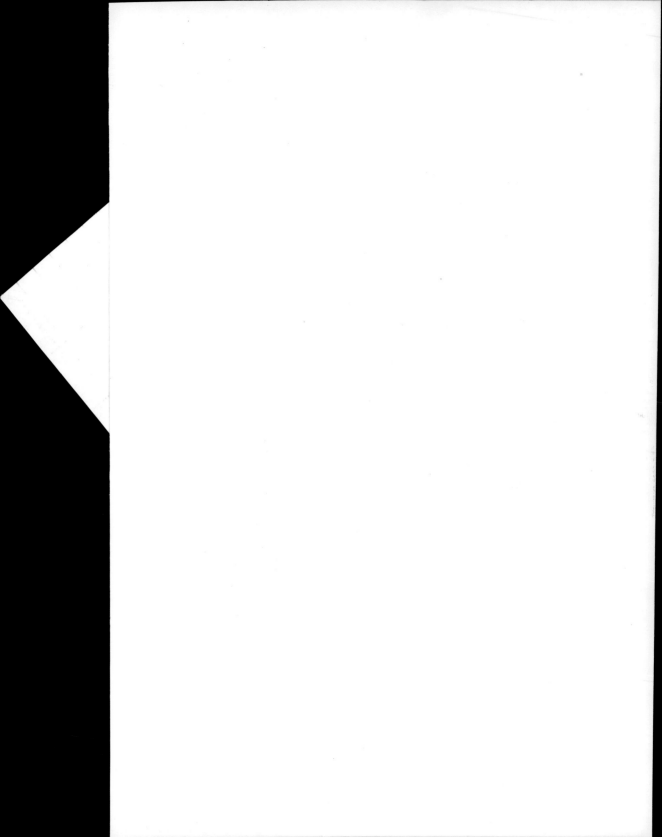

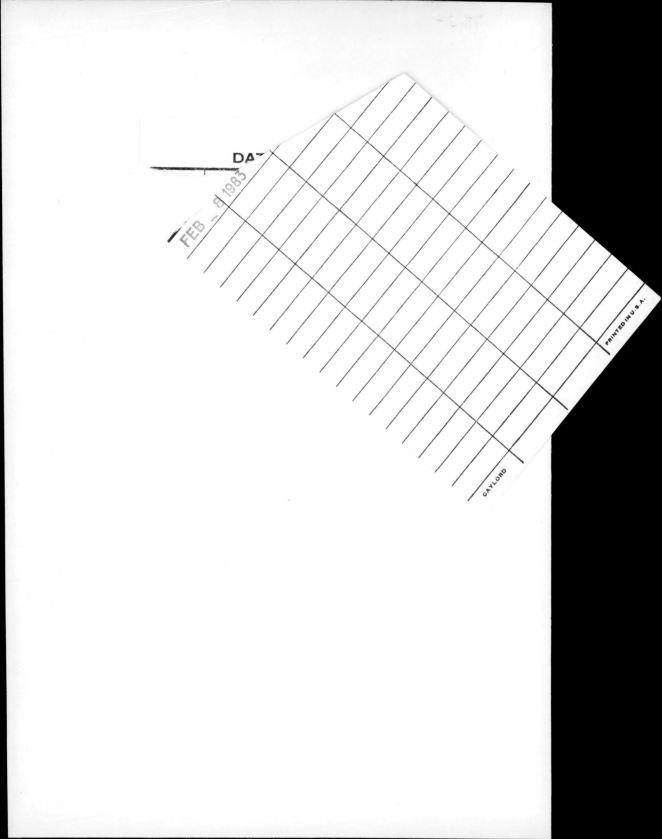